THE
ALIENS
ARE COMING!

ALSO BY BEN MILLER

It's Not Rocket Science

The Extraordinary Science Behind Our Search for Life in the Universe

BEN MILLER

NEW YORK

THE ALIENS ARE COMING!: *The Extraordinary Science Behind Our Search for Life in the Universe*

First published in the United Kingdom by Sphere, an imprint of Little, Brown Book Group
First published in North America by The Experiment, LLC, in 2016

The Experiment, LLC
220 East 23rd Street, Suite 301, New York, NY 10010-4674
www.theexperimentpublishing.com

Library of Congress Cataloging-in-Publication Data

Names: Miller, Ben, 1966-
Title: The aliens are coming! : the extraordinary science behind our search for life in the universe / Ben Miller.
Description: New York : The Experiment, 2016.
Identifiers: LCCN 2016028331 (print) | LCCN 2016028816 (ebook) | ISBN 9781615193653 (pbk.) | ISBN 9781615193660 (ebook)
Subjects: LCSH: Life on other planets. | Extraterrestrial beings. | Interstellar communication.
Classification: LCC QB54 .M5388 2016 (print) | LCC QB54 (ebook) | DDC 576.8/39—dc23
LC record available at https://lccn.loc.gov/2016028331

ISBN 978-1-61519-365-3
Ebook ISBN 978-1-61519-366-0

Cover and text design by Sarah Smith
Author photo by Colin Thomas

Manufactured in the United States of America
Distributed by Workman Publishing Company, Inc.
Distributed simultaneously in Canada by Thomas Allen and Son Ltd.
First printing November 2016
10 9 8 7 6 5 4 3 2 1

For Sonny, Harrison, and Lana

CONTENTS

CHAPTER ONE

EXTREMOPHILES

In which the author gets his starships in a row, and discovers that the solar system is an oasis, not a desert.

WE COME IN PEACE

On August 25, 2012, the first of our ships reached interstellar space. It was unmanned. Launched three and a half decades earlier, it had skirted Jupiter and Saturn, and was now heading out of the solar system toward Camelopardalis, a little-known constellation close to the Big Dipper. Although clear of the solar wind, it was not quite out of reach of the sun's gravity, nor would it be for a further thirty millennia. By then it would finally have traversed what is known as the Oort Cloud, a thick outer shell of icy rubble that encases our home star and its eight planets like the flesh around a peach stone. At that point it would be nearly a light-year out. Forget the galaxy; even the solar system is unimaginably large.

The ship's name was *Voyager 1*, and on board was a message from the people of Earth, encoded on what became known as the "Golden Record." This gold-plated phonographic disc, curated by the distinguished American cosmologist Carl Sagan, spoke on behalf of all humanity. It began with a recorded message from the Secretary-General of the United Nations, Kurt Waldheim. Reading haltingly, with a strong Austrian accent, he made the following statement:

> I send greetings on behalf of the people of our planet. We step out of our solar system into the universe seeking only peace and friendship: to teach if we are called upon; to be taught if we are fortunate. We know full well that our planet and all its inhabitants are but a small part of this immense universe that surrounds us and it is with humility and hope that we take this step.

After this greeting came a choir of voices speaking in fifty-five languages:* everything from Akkadian, the language of ancient Sumer, to Wu, the contemporary Chinese dialect spoken around Shanghai. Some, such as the Japanese, appeared shy: "Hello, how are you?" Others were more forthcoming, such as the Amoy of southeastern China, who offered: "Friends of space, how are you all? Have you eaten yet? Come visit us if you have time." The speaker of Ancient Greek, on the other hand, issued a barely concealed threat: "Greetings to you all, whoever you are. We come in friendship . . . to those who are friends."

These Greetings of Earth were accompanied by twenty-odd Sounds of Earth, among them echoing footsteps, hard rain,

* For this, as in all matters Voyager, I refer to the peerless *Haynes Manual NASA Voyager 1 & 2* by Christopher Riley, Richard Corfield, and Philip Dolling.

and a handsaw cutting fresh wood. Over one hundred Scenes of Earth showed images such as a hand being x-rayed, the chemical structure of DNA, and a man and a pregnant woman in silhouette. And, finally, there was the Music of Earth, with over twenty of humanity's finest recordings, including the first movement of Beethoven's Fifth, Bach's *Well-Tempered Clavier*, and "Johnny B. Goode" performed by Chuck Berry.

EARTH LEFT YOU A MESSAGE

That might have the flavor of science fiction—at least, I hope it does as I was trying my hardest—but it is all true. So far as we know, the Golden Record has not yet been intercepted by spacefaring aliens. If it is, what they are to make of it is anyone's guess. For a start, we have to hope that they aren't too big. An alien the size of a blue whale might have a hard time getting a needle in the groove, let alone building a hi-fi system for it to play on at the required speed of 16⅔ revolutions per minute. Equally, too small an alien—one the size of a microbe, say—might never realize the Golden Record, or *Voyager 1* itself, was even there in the first place.

Next, of course, we have to hope that they share our perception of time. As we shall see in a later chapter, not all animals on Earth do, let alone all aliens. To crows, for example, whose brains have a faster clock, human communication appears slow and deliberate. If "alien time" passes much faster than "human time," the aliens might not realize that human speech contains information; it might simply sound like long, unintelligible groans. To understand human speech, it helps to have a brain that chugs along at human speed.

And while we are on the subject of human speech, we had better hope that any aliens that find the Golden Record have ears, that the frequency range of those ears matches that of our own, and that they themselves communicate using vocalizations. On a deeper level, we had better hope that the concepts expressed in our messages—things like "peace" and "space" and "time"—have equivalents within their own language, or languages. And on an even deeper level, we hope they share the concept of a "message" within their culture, and don't just fire it straight back again.

It doesn't end there. To be able to see the instructions on the case of the Golden Record, detailing how the information inside is to be decoded, the aliens had better be able to see, and their vision had better be attuned to the same range of the electromagnetic spectrum as our own eyes. Again, we can see from life-forms on Earth that this is not a given. A race of superintelligent bats, for example, might see the Golden Record as nothing more than a metallic Frisbee. A community of superintelligent bacteria might just see it as a snack.

And, most importantly of all, we had better hope that the aliens have a good understanding of human culture. If they don't, they are going to have a hard time figuring out what we were up to. When storage capacity must have been so precious, why include so many greetings? Why are genitals shown in some of the drawings of humans, but not in others? What's the music for? Who are these people, and what the hell are they trying to tell us?

In short, we had better hope that the aliens are just like us.

LOVING THE ALIEN

Is there any question more fascinating than whether or not we are alone in the universe? The faint, ghostly light of the Milky Way is the glow of billions of stars. Is it really possible that Earth is the only habitable planet among them, and that we are the only intelligent species? And if there is intelligent life out there, might we be able to communicate with it?

The Ancient Greeks certainly thought we might. Epicurus, for example, one of the founding fathers of modern science, stated around 300 BC that "other worlds, with plants and other living things, some of them similar and some of them different from ours, must exist." Newton was also onside, as is plain from an appendix he added to his famous treatise on mechanics and gravitation, the *Principia*:

> This most beautiful System of the Sun, Planets, and Comets, could only proceed from the counsel and dominion of an intelligent and powerful being. And if the fixed Stars are the centers of other like systems, these, being form'd by the like wise counsel, must be all subject to the dominion of One.

Aliens are everywhere. They can be angels come to warn us of the follies of nuclear war or they can be demons that abduct us to carry out bizarre sexual experiments. Their shape changes, from the angry Little Green Men of the first half of the twentieth century to the placid Greys of the present day. They visit us in flying saucers, speak to us telepathically, or appear as strange lights in the sky. Yet so far as we can determine, all this is a product of our imaginations. Much as we might wish it were otherwise, there is no compelling evidence that intelligent, technologically advanced aliens have ever visited Earth.

But before you throw this book down in a pique of anti-scientific disgust and head for the Mind, Body, and Spirit section, stop. Because as is so often the case, the real science is so much more interesting than the non-scientific stuff. While alien autopsies grab the headlines, thousands of scientists—real, hardworking, peer-reviewed, genuinely qualified scientists—are slowly inching closer to the real thing. And trust me: If we do manage to make contact with an alien intelligence, those stories about flying saucers and pervy Little Green Men are going to seem very man-made indeed.

The plain truth is that the last few years have seen something of a sea change in the way we view life in the cosmos. Thanks to NASA's recent Kepler mission, we have discovered that planets like ours are common throughout the galaxy. We also know that life got started on Earth very early in its history, and that it thrives in some incredibly extreme environments. As our probes and manned missions venture out into the solar system, and we image Earthlike planets with ever-increasing accuracy, our first encounter with alien life is rapidly approaching.

Most scientists expect that encounter will take place via a telescope, and that the life in question will be in the form of single-celled organisms so small that they would be invisible to the naked eye. A second, slightly more remote possibility is that microscopic organisms will be found on an icy moon within our own solar system, or even living cheek by jowl with us right here on Earth. And if single-celled life is as widespread as we currently believe, complex intelligent life won't be far behind.* Just how far behind is the subject of this book.

Thrillingly, it turns out that life on Earth can teach us a surprising amount about life on other planets. Complex life, as we shall see, is rarer than single-celled life; exactly how much rarer is a subject of an intense but increasingly well-informed debate. Intelligence, as we shall shortly discover, is not unique to humans; in fact we share it with at least half a dozen other species, and maybe more. Some of those other intelligent species even have language, and decoding it may be an important first step toward communicating with extraterrestrials.

Forget science fiction. You are living through one of the most extraordinary revolutions in the history of science, the emergent belief of a generation of physicists, biologists, and chemists that we are not alone. Our journey to understand how this revolution has come about will lead us through some ravishingly beautiful science, and hint at answers to some truly deep existential questions. All of what follows is accessible if you have an open mind; in fact, a creative bent will be as valuable as a scientific one, because this subject goes right to the heart of what it means to be human.

Before we get started, here's the briefest of guides to the journey ahead. These three opening chapters will give us an overview of the hunt for extraterrestrials to date, UFO crazes included, and try to answer the question of why the Search for Extraterrestrial Intelligence, or SETI as it is known, has gone from pariah to pontiff in less than a decade. In the meat of the book, we'll get a handle on what our latest studies of life on Earth can tell us about the possibilities for "life-as-we-know-it"

* Complex in the sense of being made up of connected parts. In complex life, cells are grouped into tissues, which are in turn grouped into organs.

and "life-as-we-don't"—in other words, the chances of finding intelligent extraterrestrial organisms that are based on carbon, and those made of something else entirely. Finally, we'll look at how we might decode an alien message, should we be lucky enough to receive one, and what kind of messages—if any—we should be sending in return.

If our science is right, within the next decade we will have hard evidence that there are other living things out there in the universe. As we shall see, it's an outside bet, but if we are very lucky, some of those living things will have been just at the right stage of development at just the right time to have sent us a message that we are capable of understanding. Some of those messages might be traveling through you right now, as you read this book. If you are at all interested in how we might intercept them, and what they might say, read on . . .

CALLING OCCUPANTS OF INTERPLANETARY CRAFT

As a child of the Space Age, I have always been fascinated by the idea of life beyond Earth. Born in 1966, I was three and a half when *Apollo 11* landed on the Moon, and although I was too small to stay up and watch the live broadcast, I clearly remember the bulletins that swamped the news the following day. Even now, as I watch the footage of Neil Armstrong stepping down from the lunar module, I feel the same exquisite mix of elation and disappointment. Elation, of course, because such an extraordinary thing is possible. And disappointment that an enormous multicolored tentacle didn't reach out from behind a rock and give him a high five.

There was enough novelty in the Moon landings for my contemporaries and me to overlook the absence of aliens; the bizarre effects of low gravity and no atmosphere were more than enough to hold our attention. Looking back, it's almost comical how little science is going on in those first few Apollo missions. If you ever doubt that humans are descended from chimps, just watch a few weightless astronauts turning somersaults and attempting to have a tea party as they while away the hours on their three-day journey. To me now, all that hyperactivity seems like an attempt to distract the watching billions from one disquieting central fact: The Moon is about as dead as it is possible to be.

It didn't help that kids of my age had high expectations that we might meet aliens within our lifetime. For a start, we had inherited a vault of alien-invasion Golden Age science fiction, such as Ray Bradbury's *The Martian Chronicles* and John Wyndham's *The Kraken Wakes*. Most often in these stories, aliens were out there in the darkness, watching. Mankind had ascended the throne of Technology, and it was high time we were usurped. The conventional wisdom is that these chilling stories were a manifestation of the Cold War and the threat of Soviet attack, but if you ask me there was another equally important inspiration: the birth of broadcasting.

Radio had come first, with Marconi claiming the first transatlantic radio communication in 1901, and international broadcasts playing an important part in Germany's propaganda machine in the run-up to the Second World War. Television followed soon after, dominating by the late 1940s. Both media were blasted out by giant transmitters that sent just as much signal out into the cosmos as they did to the horizon. By 1950, when Ray Bradbury published *The Martian Chronicles*, somewhere lodged within the collective unconscious was the idea that if there were technologically advanced aliens on our neighboring planets, they knew exactly where we were and what we were up to.

Both radio and television signals, of course, are carried by electromagnetic waves, which travel at the speed of light.* That leaves us with a disquieting thought. Those signals, and all of those broadcast since, have been billowing away from Earth for the best part of seventy years. There are now hundreds of star systems within range of our TV and radio signals, not just the dozen or so that would have been in range in 1950. Maybe aliens are on their way right now, enraged by the unsatisfactory ending of *Twin Peaks*.†

For me, this story was told best by Carl Sagan. At the beginning of *Contact*, the 1997 movie based on his book of the same name, the camera surfs the spreading wave of talk shows, news bulletins, and popular music as it leaves Earth and makes its way out into the galaxy. As we gather pace, we catch up with earlier and earlier broadcasts. To begin with, we hear thrash metal and the Spice Girls. Further out, we pass Madonna, then the theme from the first *Star Wars* movie, then, further on still, we overtake Neil Armstrong's "giant leap for mankind." Finally, with the Milky Way Galaxy receding into the distance, we hear the announcer of *The Maxwell House Good News of 1939*, then Morse code, then silence.

* A quick reminder of the electromagnetic spectrum, from long wavelength to short wavelength: radio, TV, microwave, infrared, visible, ultraviolet, x-ray, gamma ray.

† On second thought, aliens may be exactly who David Lynch was writing for.

FROM RUSSIA WITH LOVE

But I'm getting ahead of myself. The point is that even as recently as the 1960s, many distinguished scientists believed there might be technologically advanced alien societies within our very own solar system, let alone the galaxy. We have sent surprisingly few radio messages specifically with the intention of making contact with aliens, and the very first, the so-called Mir Message, targeted Venus. Composed in Morse code, and transmitted on November 19, 1962 from a radar dish in Ukraine, it said simply "MIR, LENIN, SSSR." In case you are wondering, Mir is Russian for "peace" and SSSR was the Russian acronym for the Soviet Union. Again, full marks to the Venutian who worked that out.

But I'm here to tell you that as the 1970s wore on, and our knowledge of the solar system increased, this optimism waned. Apollo was canceled, as was its follow-up, Orion, which aimed to put men on Mars. Instead we switched our attention to unmanned missions, launching a series of robotic probes. By 1972, the Russians had managed to land *Venera 7* on Venus, and we knew for sure that not only was the surface temperature a face-melting 500°C, but its carbon dioxide atmosphere was so thick the air pressure was more than ninety times that of Earth. Three years later, *Venera 9* sent the first black and white photos of the Venusian skyline. It looked like an abandoned slate quarry.

It got worse. NASA's *Mariner 10* made a flyby of Mercury, the closest planet to the Sun, in 1973. Whereas Venus is practically a twin of the Earth, Mercury is just a little larger than the Moon. As you might expect, it turned out to have no atmosphere, and was pock-marked with craters, indicating that, like the Moon, its core was cold.* Less expected was that, unlike the Moon, it had a weak magnetic field, partially shielding it from the solar wind, but with surface temperatures that regularly shot up to 400°C it was not the kind of place you'd want to call home.† By the time *Viking* landed on Mars, I was ten. That really was a blow. Mercury, Venus, and the Moon looked inert, even from Earth, but Mars was different. It was red, the color of iron, earth, and life. When the camera powered up, would a herd of bouffant-haired crabs scuttle for cover? Sadly not. The Red Planet did have a thin atmosphere, so there was a sort of pinkish daylight, but as far as habitability went, that was about it. Mars was a desert.

VOYAGER'S REST

For me, the final nail in the coffin came with the Voyager missions. Throughout the 1980s, wondrously depressing photos emerged as these twin probes flew past Jupiter and Saturn, and *Voyager 2* then continued farther, past Uranus and Neptune. Beautiful as each of these four giant balls of gas was, with no solid rock and no liquid water, how could life ever take hold?

We had higher hopes for their rocky moons, but they too were cruelly dashed. The Galilean moons of Jupiter—Io, Europa, Ganymede, and Callisto—ranged in size between the Moon and

* Craters tend to indicate that there is no plate tectonics, and therefore no molten core within a planet. One of the shocks of the recent *New Horizons* flyby was how few craters Pluto has on parts of its surface. It's too small to have retained much heat from when it was formed, or to have enough radioactive material in its core to drive plate tectonics, and it isn't heated by tidal forces like the moons of the four gas giants.

† I know; Mercury is closer to the Sun than Venus, but colder. The reason is that it has no atmosphere and therefore no greenhouse effect.

Mercury, and were every bit as barren. There were a couple of surprises. Io had active volcanoes, busy spewing sulfurous gases, and Europa was as smooth as a billiard ball, but that was about it. With no atmosphere, out in the freezing boondocks of the solar system they were a biological nonstarter.*

With the Jovian moons out of the running, attention turned to Saturn. One of the main objectives of the Voyager mission was to investigate Titan, thought at the time to be the largest moon in the solar system.† *Voyager 1* plotted a course a mere four miles from its surface, but saw only an impenetrable haze. That meant Titan, unique among moons, had an atmosphere, but it also meant we had no idea what was going on underneath. It seemed pointless for *Voyager 2* to follow up, so instead it was diverted to take a look at Uranus and Neptune. On its way, it managed to grab a photo of another Saturnian moon, Enceladus, which appeared to be a lump of solid water ice.‡

If the moons of Jupiter and Saturn are chilly, those of Uranus and Neptune are bone-numbing. In January 1986, *Voyager 2* made it to Miranda, a tiny world less than a seventh of the size of our own Moon, with an average surface temperature of −210°C. It had a truly bizarre surface, made up of a patchwork of cratered and smooth sections, leading some to dub it "Frankenstein's Moon." Either Miranda had been smashed and hastily reassembled following an impact, or somehow the

* The four Galilean moons are Jupiter's largest—hence Galileo being able to see them in the first place. Thirteen moons were known at the time of the Voyager mission, which discovered three more. Today the tally is sixty-seven moons.

† It turned out to be slightly smaller than Ganymede, the moon of Jupiter.

‡ Saturn has sixty-two moons, of which seven are big enough to be spherical under their own gravity. They are, in order of increasing orbit, Mimas, Enceladus, Tethys, Dione, Rhea, and Iapetus. Titan, the giant, sits farthest out.

gravitational pull from Uranus was warming its interior, giving it the icy equivalent of plate tectonics.*

Finally, in the summer of 1989, *Voyager 2* made it to Triton, by far the largest of Neptune's fourteen moons and three-quarters the size of our own satellite. Like Miranda, Triton was truly alien, and not in a good way. Even its orbit was peculiar. As you may know, the planets and their moons all tend to spin and orbit in the same direction; counterclockwise if you are looking down on the solar system from above. Not so Triton, which orbits Neptune clockwise, betraying the fact that it isn't a homegrown moon, and was most likely kidnapped from the band of icy rubble outside Neptune's orbit known as the Kuiper Belt. Although small in comparison to Earth, Triton was geologically active with very few craters, and had a surface made of solid nitrogen. Unsurprisingly, it was also one of the coldest places in the solar system, with a temperature of –240°C.

And that was about it. In the 1950s, we had dreamed of battling angry expat Martians and being seduced by blonde-haired Venusians in a tropical paradise. By the end of the 1980s, it was painfully clear that we were going home from the party on our own. Summing it all up was the image *Voyager 1* took on February 14, 1990, as it looked back at the solar system from an orbit halfway across the Kuiper Belt. In a vast expanse of lifeless black, Earth appeared as a single fragile pixel, what Carl Sagan famously called the "pale blue dot." His words are so well turned they are worth repeating.

> The Earth is the only world known, so far, to harbor life. There is nowhere else, at least in the near future, to which

* Uranus has twenty-seven moons, named after characters from Shakespeare, the largest being (usual drill) Puck, Miranda, Ariel, Umbriel, Titania, and Oberon.

our species could migrate. Visit, yes. Settle, not yet. Like it or not, for the moment, the Earth is where we make our stand. It has been said that astronomy is a humbling and character-building experience. There is perhaps no better demonstration of the folly of human conceits than this distant image of our tiny world. To me, it underscores our responsibility to deal more kindly with one another and to preserve and cherish the pale blue dot, the only home we've ever known.

TRAINING THE BEAGLE

The Space Age, which had begun with such optimism, had ended on a cosmic downer. We were alone. Commerce, not exploration, became the driving force behind space science, and the satellite industry boomed. Public interest in space waned, and at one or two dinner parties in north London which I had the misfortune to attend during the early noughties, intelligent and educated people expressed doubt that we had landed on the Moon at all. Internet rumors suggested that mankind's greatest achievement had been a US government hoax, staged in a movie studio by Stanley Kubrick and shot with TV cameras in a bid to demoralize the USSR and to win the Cold War. The astronauts hadn't risked their lives; they had all been fakers.

If I had to pick a low point, for me it would be the launch of the *Beagle*, the life-seeking robot lander that formed part of the European Space Agency's 2003 Mars Express mission. Just as Darwin's voyage on HMS *Beagle* had inspired his theory of evolution, so it was hoped that this plucky little sniffer dog would root out signs of Martian life and rewrite the rules of biology. With a call sign composed by Blur, and a Damien Hirst spot painting as a test card for its on-board video camera, the *Beagle* was basically Britpop on steroids, and about as long lived.

To be fair, it had an ingenious design. Its mother ship, *Mars Express*, had only ever been intended to be an orbiter rather than a lander, but thanks to the charisma and media savvy of the UK's Colin Pillinger, the ESA higher-ups were outmaneuvered and passage secured for a stowaway roughly the size of a bin lid. After jettisoning from the *Mars Express*, two consecutive parachutes would slow the *Beagle*'s descent, and three airbags would cushion its landing. Once on the ground, the airbags would detach, the lid of the main housing would flip open, and out would flop four petal-shaped solar panels. A mechanical arm would then emerge, like the stigma from a giant flower, bristling with sampling tools.

Their variety was impressive. As well as the aforementioned video camera there was a microscope, a rock grinder and corer, a wind sensor, a wide-angle mirror and a telescopic drill called the Mole, which was capable of digging up to 1.5m into the Martian soil. Once a sample had been collected, the mechanical arm would then maneuver it into an inlet port in the central housing, ready for a well-equipped on-board lab to identify exactly what types of molecules were present. If there was—or ever had been—life on Mars, there was a good chance that the *Beagle* would be able to find it.

Touchdown was planned for Christmas Day 2003, and at the allotted hour patriotic Britons waited patiently by their radios and television sets listening out for the otherworldly strains of Blur's call sign. Instead there was silence. The *Beagle* had vanished without trace. The smug mediarati of north London were right, it seemed. The Moon landings were fake, and the ineptitude of the *Beagle* was but so much grist to their infuriatingly self-satisfied mill.

Yet help was at hand. As our interplanetary odyssey languished in the doldrums, a sudden gust of enthusiasm blew in from the most unexpected quarter, driving all before it. While space scientists had focused all their efforts on nearby planets and drawn a series of depressing blanks, their colleagues in the altogether less glamorous world of microbiology had been quietly coming up trumps. Because as it turned out, something akin to aliens had been found, and in the most unlikely of places. They were right here on Earth.

LIFE, BUT NOT AS WE KNOW IT

Tom Brock loved the outdoors. Canoeing and backpacking were favorites, and in July 1964 he paid a visit to Yellowstone National Park in Wyoming. This, of course, is the home of the famous geyser Old Faithful, which every hour and a half spouts scalding hot water over 150ft in the air. It wasn't this attraction that caught Tom Brock's eye, however; it was the multicolored scum in the hot springs nearby. Luckily for us, Tom Brock wasn't just an outdoorsman, he was a microbiologist. He knew a microbial mat when he saw one, and he also knew that they shouldn't be growing in near-boiling water.

A microbe is the technical name for a single-celled organism such as a bacterium. As the name suggests, individual microbes are too small to be seen with the naked eye, but, given a suitable environment, they are more than happy to club together to form what is known as a mat. The ones at Yellowstone often contain pigments such as chlorophyll, which you'll know is green, and carotenoids, which can be anything from yellow to red.

Chlorophylls and carotenoids are key players in photosynthesis, the process whereby microbes, plants, and algae use the energy of light to build long-chain carbon molecules from carbon dioxide, known in the trade as carbon fixing.* The effect at Yellowstone can be spectacular, particularly at the Grand Prismatic Spring, where the deep blue of the central basin is surrounded by concentric circles of green, yellow, orange, and red microbial mats as the water shallows out.

Microbes, of course, are living things, and the conventional wisdom at the time was that they should only exist within a narrow range of temperature. After all, all living things are made of proteins and contain water. Freeze them and they'll go solid. Heat them and their proteins will start to break apart, or denature, a process that in the everyday world we call cooking. Warm your meat or your microbes to anything above 60°C, and you can expect even the most resilient proteins to become gelatin in your hands.

That, at least, was what we believed back in the 1960s. Yet to Tom Brock's astonishment, in the broiling pools of Yellowstone, microbes were positively thriving. Soon he had shifted his research to what became known as "extremophiles"—organisms that love extremes. There seemed to be no limit to their audacity; during successive visits to Yellowstone throughout the mid- to late sixties, Brock and his research

* Fixing is basically the action of taking any gas from the air and converting it to a solid or liquid form. Nitrogen-fixing bacteria, for example, take atmospheric nitrogen gas and convert it to nitrates, which are then hungrily consumed by plants. Soon we shall meet a manganese-fixing bacteria, for which I make no apology.

team found strains of bacteria in the Yellowstone pools that thrived at temperatures as high as 90°C.

Extraordinary as this news was, it was bewilderingly slow to catch on. Truth be known, when it comes to reality, we humans are not the most reliable of creatures. Not only are we capable of seeing things that aren't there, but we are also capable of not seeing things that are. There were two million visitors* to Yellowstone Park in 1964, all of them gazing with wonder at the multicolored hot springs. Many of them must have been scientists, and one or two may even have been of a distinctly microbiological persuasion. Yet no one other than Tom Brock spotted what seems now to be glaringly obvious: The scalding waters were festooned with living creatures that really shouldn't have been there.

Brock wasn't slow to publish his findings, but, even so, it wasn't until the late seventies that news of his discovery started to reach mainstream scientific journals. And at that point our story takes another twist. It's one thing when a diligent microbiologist finds some unusual bacteria photosynthesizing in a hot spring in Yellowstone Park; it's quite another when geologists stumble across a whole zoo of unfamiliar creatures more than a mile deep in the Pacific Ocean.

THREE MEN IN A DEEP SUBMERGENCE VEHICLE

We should be grateful to *Alvin*, a three-man deep-sea submersible owned by the US Navy, for two reasons. The first is because in 1966 it recovered an unexploded hydrogen

* 1,929,300 according to the National Park Service.

bomb from the bottom of the Mediterranean Sea, after a B-52 bomber collided with a tanker plane while refueling in mid-air. In that case, it may well have averted nuclear Armageddon and the end of the human race. In the second, some would say it found the very place that the ancestor of all life—the human race included—got its start.

The theory of plate tectonics was put forward by the German geologist Alfred Wegener in 1922, and, as you probably know, proposes that the Earth's crust isn't static, but is made up of a patchwork of plates, each of which is moving. The joints between plates tend to be where all the geological action happens, in the form of volcanoes, islands, mountains, and trenches. Exactly what you get depends on what's on top of the plates—ocean, for example, or a continent—and whether they're being pushed together, pulled apart, or are slipping side by side.

At least, that's how it usually works. Sometimes, however, you get volcanoes in the middle of a plate, well away from the edge. In these cases, there seems to be something deep beneath the crust, a "hot spot" if you like, which the plate is riding over. As the plate moves, the hot spot punches a series of volcanoes up through it. The Hawaiian islands are the classic example, sitting as they do right in the middle of the Pacific plate, which is currently moving northwest toward Eurasia. As fresh Pacific plate moves over the hot spot, plume after plume of hot magma shoots up through it, creating a chain of volcanoes on the ocean floor. The tips of these volcanoes form the Hawaiian islands.*

* If you look at a map of the northern Pacific, you'll see that the Hawaiian islands form one long diagonal chain. The island of Hawaii is the most recent addition, and has already moved off the hot spot. A new island, the Loihi Seamount, is forming underwater about 22 miles off Hawaii's coast.

Another example, funnily enough, is Yellowstone Park, though in that case we are presently between eruptions, with the last one having taken place around 640,000 years ago. When a volcano erupts, the encircling area can often collapse, leaving a depression known as a caldera, after the Spanish for "cooking pot." It's in the caldera from the last eruption at Yellowstone that we now find Old Faithful and the Prismatic Springs. A third example is the islands of the Galápagos, and that's where the plucky little submarine known as *Alvin* comes in.

THE GARDEN OF EDEN

On February 8, 1977, *Alvin* set sail from Panama aboard a purpose-built catamaran named *Lulu*, heading for a deep-ocean volcanic ridge just north of the Galápagos Islands known as the Galápagos Rift. Once there, the plan was to try and find hot springs. The saltiness of the world's oceans suggested they were getting a supply of salt water from somewhere, and the smart money said that somewhere down on the sea floor there had to be the equivalent of Yellowstone's pools and geysers, pumping out salts and minerals. Nevertheless, at the time of *Alvin*'s dive, no one had yet found a real hydrothermal vent that would settle the issue one way or another.

The previous summer, a survey of the Galápagos Rift had used an unmanned deep-sea camera to hunt for hot springs, but without success. At one point, however, the umpteen photos of barren sea floor were interspersed with a few brief shots of a pile of dead white clam shells, along with a beer can. The team assumed it was just rubbish thrown overboard by a ship having a party, and named the site "Clambake." After all, nothing could possibly be living at that depth, because there was no light.* Without light, there would be no plants, algae, or bacteria. And without them there was nothing for anything else to eat.

How wrong they were. On February 17, 1977, *Alvin* took a dive, piloted by one Jack Donnelly and carrying two geologists, Jack Corliss and Tjeerd van Andel. As they neared the ocean floor, the water began to shimmer. Sure enough, hot water was pumping out of the dark volcanic rock, and forming black sulfurous clouds as it cooled, earning such vents the nickname of "black smokers." But that was far from all. As *Alvin*'s searchlights scoured the surrounding rocks, they revealed a ghostly menagerie of extraordinary creatures. There were giant white clams, white crabs, and even a purple octopus, all very much alive. Confused, Corliss picked up the acoustic telephone and called his graduate student Debra Stakes, above them on board *Lulu*. "Isn't the deep ocean supposed to be like a desert?" Corliss asked. When Stakes confirmed that it was, a puzzled Corliss replied: "Well, there's all these animals down here."

It got weirder. Subsequent dives revealed more hot springs, and even more strange creatures. At another spring they found an orange animal that resembled a dandelion; at another that they breathlessly christened the Garden of Eden, they found a forest of giant tubeworms with bright red tops, swaying in the water like a field of flowers. They did their best to collect specimens, but being a geology expedition they had little in the way of formaldehyde to preserve them in. Instead, they used the next best thing: some bottles of Russian vodka they had

* As you dive, the pressure increases by roughly 1 bar every 10m. Atmospheric pressure is 1 bar.

bought in Panama. Tjeerd van Andel began to lose interest in his original goal of finding hot springs, and lay awake at night, his mind buzzing with questions. Where had these creatures come from? What could they possibly be eating?

Two years later, in 1979, a team of biologists returned to find out. *Alvin* was modified with a new collecting basket and a second mechanical arm, and fitted with a movie camera. On each dive, the team returned with a zoo of creatures that had never been seen before: new species of mussels, anemones, whelks, limpets, featherduster worms, snails, lobsters, brittle stars, and blind white crabs. The delicate orange dandelion-like creature seen on the 1977 dives turned out to be a relative of the Portuguese man-of-war, though it quickly disintegrated after being brought to the surface. Finally, the mystery of what all these creatures were eating was solved by a biologist called Holger Jannasch. At the base of this baroque food chain was a microbe. Rather than getting its energy from sunlight, this bacterium was feeding on a chemical in the vent fluid; specifically, hydrogen sulfide.

Suddenly, all bets were off. If life didn't need light, or moderate temperature, where else on Earth might it be lurking? Suddenly, extremophiles seemed to pop up everywhere we looked. We found microbes in nuclear reactors lapping up radiation ten times stronger than that which would kill the hardiest cockroach. We found both fish and microbes thriving under extraordinarily high pressure 11,000m underwater in the Challenger Deep and the Marianas Trench. We found microbes and fungi bathed in acid so strong it has a pH of zero. We even found bacteria that live inside rocks.

All of which raised an interesting question: Who was really living at the extremes, them or us? To a bacterium which lives in the sweltering heat of a Yellowstone spring, aren't we the extremophiles, able to endure dessicatingly dry surroundings so cold that the little water that is available regularly freezes solid? What was the natural environment of the first life? Had the first cells incubated in the shimmering heat of a black smoker, and only later migrated to cooler, sunlit shallows? If microbes could live in rocks, could they travel between planets on meteorites? Had life begun elsewhere—on Mars, maybe—and taken a joyride to Earth on a space rock?

Life, in short, simply wasn't what we thought it was. It wasn't delicate, or precious, or in any way predictable. Far from it: It was tenacious, commonplace, and infinitely adaptable. Here on Earth, all it seemed to require was water, carbon, and a source of energy. Maybe Mars, Venus, and Mercury weren't quite the inhospitable deserts we had once feared them to be. If they contained even trace amounts of moisture, they might be home to some sort of bacteria. Could some of those icy moons, orbiting giant gas planets in the outer reaches of the solar system, be habitable after all?

THE ORIGIN OF THE SPECIES

There's something wonderfully poetic about *Alvin* finding a whole new raft of life near the Galápagos, of course, because it was on these volcanic islands that the great Charles Darwin collected the specimens that were to inspire his theory of evolution. The story is worth retelling, not only because Darwin was the astronaut of his day, boldly going where no naturalist had gone before, but also because evolution is such a linchpin in our search for intelligent extraterrestrial life. What follows may seem like a diversion, but by taking it we will have an easier approach to the summit, so here goes . . .

Galápago is a Spanish word meaning "tortoise," and, according to Darwin's journal, on September 18, 1835, the crew of the *Beagle* brought fifteen giant tortoises on board from Chatham Island,* ready to supply a feast. With few natural predators, the animals of the Galápagos were curiously trusting; in fact, hunting and collecting were pretty much the same thing. Darwin himself reports knocking a hawk off a branch with the tip of his rifle, and rather surreally recalls midshipman King killing a bird with a hat.

Understandably, these giant tortoises made a strong impact on our young hero, and he was intrigued by the observation of the vice-governor of the Galápagos, Nicholas Lawson, that "he could, with certainty tell from which island any one was brought."† In other words, each island had its own species of tortoise. Sadly, Darwin wasn't that successful in finding specimens that proved the point. He collected three giant tortoise shells, each from a different island, but they were from young animals and there was little to tell them apart.

Back in London, Darwin presented all his specimens to the Geological Society of London. The birds he gave to John Gould of the Royal Zoological Society for examination. Among them were those from the Galápagos, which Darwin had identified as blackbirds, wrens, and finches. When Gould returned the surprise result that they were all finches, "so peculiar as to form an entirely new group, containing twelve species," Darwin began to formulate an audacious idea. What if there

* Now known as San Cristobal.

† From *The Voyage of the Beagle*.

had originally been no finches on the Galápagos, which were, after all, relatively new volcanic islands. Could it be that a mating pair of finches had flown there from the South American coast, and somehow their descendants on each of the various islands had metamorphosed into new species?

To prove the point, Darwin needed to be able to show that, as with the vice-governor's giant tortoises, each island was home to a different species of finch. Unusually for the meticulous Darwin, he had failed to label his own birds accurately, but fortunately his servant Syms Covington had not been so sloppy. By combining Covington's specimens with those of the *Beagle*'s captain, Robert Fitzroy, Darwin was able to reconstruct the locations where he had found his own finches. It was true: Each island had begat its own species. Gould had returned his result on January 10, 1837. That March, Darwin wrote in a notebook the words that would change the course of biology forever: "One species does change into another."

Species could change, but how? For Darwin, the argument went something like this. The mating pair had prospered, and their offspring had populated the various islands. Since reproduction is never exact, within each island population there were a variety of traits. Some finches, for example, had thick beaks, while others had thin beaks. If the seeds on a given island were hard to crack, finches with thick beaks would have a survival advantage, and would therefore have more offspring. Eventually, given sufficient generations, the entire population of finches on that particular island would have thick beaks. Nature, in other words, did not favor all creatures the same. Some she selected, and some she did not. As Darwin put it, species evolved through a process of natural selection.

In July that same year, barely eight months after he had

returned on the *Beagle*, Darwin picked up his notebook and wrote the words "I think," and below them sketched the first tree of life. Starting from a single trunk—the first living creature—he drew branch after bifurcating branch, with each new outgrowth representing a new species. It was a simple drawing, but its implications were profound. Starting with a single organism, life on Earth had evolved into an ever-increasing number of species. Take any two living things, the figure said, and you could trace back their lineage to find a common ancestor. All life on Earth was one.

IT'S ALL IN THE GENES

That's such a piquant thought it's worth taking a moment to digest it. Every single living thing on the planet is related to every other living thing. Not only are you a descendant of your great-aunt Ada, but you are a distant cousin of a flatfish, and a kinsman of an amoeba. The creatures that we find at black smokers, strange as they are, perch on the same tree of life that we do, as do the most bizarre fossils we have ever found, those of the Ediacarans.*

Completely central to Darwin's theory of evolution by natural selection, of course, is the concept of inheritance: the passing of traits from parent to offspring. In Darwin's day, the mechanism of reproduction was unknown; today we understand that every organism on Earth carries its own blueprint in every cell of its body, coded into the long-chain carbon molecule known as deoxyribonucleic acid, known as DNA.

* The Ediacarans are the first known complex life forms, ruling the planet some 575 million years ago, and lacked eyes, mouths, and limbs. More on them later.

In short, you resemble your parents because you inherited their DNA.

Or to be more accurate, you inherited almost all of it. The system by which DNA is copied isn't perfect, and that's vital; it's this less than perfect copying that gives rise to what Darwin called "variation": the appearance of a new trait in the offspring that wasn't inherited from its parents. Most of the time the new trait makes no difference. Sometimes it is harmful, and the offspring will be less likely to reproduce as a result, meaning the new trait dies out. In some rare cases, however, it bestows a survival advantage, and increases the chance that the offspring will, as Daft Punk might put it, "get lucky."

Traits are coded by small sections of DNA known as "genes."* Take any trait—the thickness of a finch's beak, say—and it is possible to identify the genes which control it. Indeed, as the British biologist W. D. Hamilton showed, it is at the level of genes that the struggle for survival is best understood, rather than that of an organism or species. Put simply, genes are doing it for themselves.† All that matters to your genes is that they make as many copies of themselves as possible; we host organisms are simply a means to an end.

I'M A MAC, ZARG IS A PC

So there we have it. Evolution is the key that unlocks the mystery of life-as-we-know-it, and is able to explain two extraordinary and seemingly unrelated facts. Firstly, the older the

* Just so you know, the word *gene* is sometimes used in a different sense, to mean "bit of DNA that codes for a given protein."

† Hamilton's work has been rather brilliantly popularized by Richard Dawkins, who summed the whole thing up with one pithy phrase: "the selfish gene."

fossils we dig up the more primitive the life-forms we find. No one has yet found a mastodon in the same layer of rock as a trilobite, nor do we ever expect them to. Speciation—the process by which natural selection creates two species where previously there was merely one—is irreversible, and the total number of species,* both living and extinct, can only increase with time.

And, secondly, it explains the extraordinary similarity between the different branches of life-as-we-know-it. To use a computer analogy, everything we find is a Mac; nothing is a PC. Every creature on Earth is made up of one or more cells, relies on water as a solvent, stores its blueprint in DNA, and burns carbohydrate to release energy. Every molecule of living protein is made from the same twenty amino acids,† and every molecule of living DNA is coded using the same four nucleobases.‡ On an even deeper level, you might say that all life-as-we-know-it is carbon-based, because almost every single molecule that you can think of that has a biological or biochemical function is a compound of carbon.

So what does this mean for our search for intelligent alien life? Well, as the Galápagos giveth, so the Galápagos taketh away. On the one hand, its black smokers show us that there's no limit to the kind of environments where life might thrive. On the other, its finches tell us there's only one kind of life on Earth. So are we alone or not? What if life-as-we-know-it is a

* Estimated at 8.7 million by the United Nations Environment Programme in 2011, give or take around 1.3 million.

† As the name suggests, amino acids are made up of an amine (NH_2) group attached to an acid.

‡ The common nucleobases are made up of one or two rings of carbon atoms with a couple of nitrogen atoms inserted into the ring. The ones we find in DNA are guanine, adenine, thymine, and cytosine.

colossal fluke, a one-in-a-gazillion random event, never to be repeated? One thing is for sure: If we had just one other example of a second tree of life here on Earth, we'd be a lot more confident that life is common in the galaxy. And yet there's nothing. Or is there? It's time to talk about desert varnish.

LIFE IN THE SHADOWS

The godfather of desert varnish research was the pioneering German naturalist and explorer Alexander von Humboldt. In 1799, during his groundbreaking expedition to South America, he noticed a strange metallic coating on the granite boulders in the rapids near the mouth of the Orinoco River in northeastern Venezuela that made them appear "smooth, black, and as if coated with plumbago."* Intrigued, he had the coating analyzed by one of the leading chemists of the day, Jöns Jacob Berzelius,† who informed him that it was made up of manganese and iron oxides. That was puzzling, because granite contains only small amounts of manganese and iron.‡ Where were these metals coming from? Presumably from the waters of the Orinoco, but what was attaching them to the rocks?

In *The Voyage of the Beagle*, Charles Darwin also describes an encounter with mysterious rock coatings. His were found below the tide line on a beach on the coast of Brazil and were a "rich brown" in color. Darwin had a bit of a thing about Humboldt, and one of his prized possessions on the *Beagle* was

* "Plumbago" is Humboldt for "graphite."

† Most famous for coming up with the letter symbols for the chemical elements.

‡ Most granite contains iron oxide at 1.68 percent by weight, and manganese oxide at 0.05 percent by weight.

a seven-volume translation of his hero's *Personal Narrative* of his South American voyage.* He knew that Humboldt's coatings had been much darker, and wondered if the redder color of those formed on the beach in Brazil was because they contained less manganese and more iron. Struck by how they "glitter[ed] in the sun's rays," he too was puzzled by what could be causing them, remarking that "the origin . . . of these coatings of metallic oxides, which seem as if cemented to the rocks, is not understood."

Both Humboldt and Darwin found their coatings in the tropical climate of South America, but—perhaps unsurprisingly given the name—it turns out that what we now call desert varnish is found just as often in arid surroundings. The sandstone deserts of the Colorado Plateau in the southwestern United States are a classic example, where the shiny black patina on the rocks has often provided a handy surface into which Native Americans could scratch their art. Within the canyons, the swathes of varnish can be particularly striking, sometimes covering entire walls, or forming vertical stripes alternating between black and red and tan.

We've learned a bit more about desert varnish since the time of Humboldt, but not much. As Darwin suspected, its color does indeed vary from red to black depending on the relative quantities of iron and manganese oxide, with varnishes with equal amounts of both appearing tan in color. We know that the varnish also contains silica in the form of clay, and that it grows more readily on rocks and walls that are intermittently wet and get good sun, as if the rapid drying of water somehow helps the varnish grow. We know that the varnish

* Full title: *Personal Narrative of Travels to the Equinoctial Regions of America During the Years 1799–1804.*

accumulates slowly, growing by less than the thickness of a human hair every thousand years. And we also know—and this is where it gets controversial—that it contains microbes.

Barry DiGregorio, for example, an Honorary Research Fellow at the Buckingham Centre for Astrobiology in the UK, thinks that these microbes are photosynthesizing manganese-fixing bacteria, and that the varnish is, like the blooms that Brock found in the hot springs of Yellowstone Park, a type of microbial mat.* On the other hand, Randall Perry, a researcher in the Earth Science Department of Imperial College London, believes that it's the clays in the varnish that are doing all the heavy lifting. He thinks they react with moisture to form a gel, which then traps all sorts of other stuff including stray microbes, as well as catalyzing some pretty funky chemistry which concentrates the metal oxides. When this gel dries in the sun, it hardens into a varnish.†

Who's right? Well, maybe neither, says Carol Cleland, a philosopher at the University of Colorado Boulder in the US, and an affiliate of the NASA Institute of Astrobiology. She is fascinated by the fact that we don't know whether desert varnish is chemical or biological. As she points out, it's a stretch to see how rocks could become coated with metal simply as a result of chemistry, but on the other hand, when we excavate desert varnish—if you can call scratching away at something a hundredth of a millimeter thick an excavation—we don't find lots of cells, just a few fragments. What, she asks, if desert varnish is another kind of life entirely?

* It is thought that manganese-fixing bacteria were a crucial stopping-off point in the evolution of photosynthesis; indeed, all plants today use manganese as a building block of chlorophyll.

† The gel is silicic acid, and the reaction can be written as clay + water → silicic acid, or if you are that way inclined, $SiO_2 + 2\ H_2O \rightarrow H_4SiO_4$.

At first glance, that might seem slightly unhinged, but she has a point. After all, extremophiles have been around for billions of years, but it took Brock to notice them; once he had, we started to find them everywhere. In 2005, Cleland proposed the existence of a "shadow biosphere"; a microbial ecosystem living in parallel with our own, but that we have yet to identify. As she rightly points out, all of our tests assume that there is only one kind of life: our own. What if there's something out there that doesn't have the same DNA code, or uses different amino acids to build its proteins? What if it doesn't have DNA or proteins at all? Or cells? Or is based on something other than carbon? How would we recognize it?

Cleland feels that we should be actively searching for life-as-we-don't-know-it, and that desert varnish is a good place to start. Another is manganese nodules, the strange metallic boulders that populate the seabed in many of our oceans, and which always remind me of the egg hatchery in the movie *Alien*. These, again, are assumed to be the result of chemistry rather than biology, but how can we be sure? Once we can rid ourselves of our preconceptions of what life should look like, maybe we will start to find it everywhere, even in the hot springs of Yellowstone National Park.

Of course, all of this begs a question: If we do find a second genesis of life here on Earth, what do we call it? Seeing as it's from Earth, the word "alien" doesn't seem right. The cosmologist Paul Davies has suggested ditching the word "alien" altogether, and using the term "weird life" to describe anything that doesn't share a common origin with life-as-we-know-it; he has even proposed a "mission to Earth" to seek it out. And what do we call the kinds of life that are emerging from the ever-growing field of synthetic biology? Is that weird too, or just weird-ish?

IN OUR BACKYARD

However you slice it, one thing is certain: The discovery of extremophiles has opened our minds to the idea that our own solar system might be teeming with life after all. There may not be herds of wildebeest sweeping majestically across the plains of Jupiter's moon Europa, but there may well be blooms of microorganisms in the giant oceans we now know to be hidden beneath its icy crust. Likewise Ganymede, the largest of Jupiter's Galilean moons, is now known to be hiding a salt-water ocean sandwiched between layers of ice. Knowing what we now know about life's appetite for weird and wonderful environments, what might be lurking in its depths?

Remember Enceladus, the tiny moon of Saturn which appeared to *Voyager* to be a solid lump of ice? Well, in July 2005, NASA's *Cassini* spacecraft arrived to finish the job. To everyone's surprise, Enceladus wasn't a cold, dead world after all, but violently active. At its south pole was a giant volcanic hot spot, from which plumes of ice particles and water vapor were erupting hundreds of miles into space; in fact, it's this fire hose of material that supplies Saturn's vast outer ring. By 2014, further measurements by *Cassini* had confirmed what many suspected: Beneath the pack ice at Enceladus' south pole was a giant superheated ocean.*

Although Mars appeared lifeless to the *Viking* lander in the late seventies, circumstantial evidence has grown that it too was home to microbial life in the past, and may still be today. Thanks to recent missions like the Mars Reconnaissance Orbiter (MRO) we know that the Red Planet had copious

* By the way, in 2013 a water plume was also reported on Europa. It now seems as if that was caused by some sort of freak event, such as a meteorite impact.

amounts of water until as recently as two billion years ago. That means conditions on Mars were right for life-as-we-know-it round about the same time as they were on Earth; some have even suggested that our kind of life seeded on Mars first, and then came to Earth on a meteorite. As recently as 2015, NASA's *Curiosity* rover found nitrates, a compound essential to many forms of life, and the MRO even found evidence of running water.

Last, but most definitely not least, is the glorious Titan. Again, it may not have been love at first sight, but he's growing on us. *Voyager 1*, as you will remember, saw nothing but an orange methane smog, and *Voyager 2* decided it had better things to do than follow up. Yet many wouldn't let it go. Just as the temperature of the Earth sits near the triple point of water—that is, the temperature at which water at atmospheric pressure can exist as a solid, a liquid, and a gas—*Voyager 1* had confirmed what we had suspected since the middle of the twentieth century: Titan was at the triple point of methane, with a methane-rich atmosphere.* When NASA's *Cassini* spacecraft made its flight to Saturn, hitching a ride was a probe named *Huygens*, built and paid for by the European Space Agency.†

If it's weird life we're looking for, the images beamed back by *Huygens* showed us that Titan is just the place to find it. Martian meteorites reach us so frequently, and Mars's early climate was so similar to ours, that any life we find there is quite likely to share a common genesis, or at least a similar biochemistry. And that's leaving aside the issue that the

* For the physicists, Titan's surface temperature is 94K, and the partial pressure of its methane is 117 millibars; the triple point of methane at that pressure is 90.7K.

† The Dutch astronomer Christiaan Huygens discovered Titan in 1655.

Viking lander wasn't sterilized properly, and may have contaminated Martian soil with earthly microbes. But Titan is a whole other deal. This is a really big moon with a proper atmosphere, with methane clouds, methane lakes, methane ice, and maybe even methane snow. What kind of abominable methane snowman might be lurking there?

Abominable microbial methane snowman, I should say. While our solar system no longer appears to be the graveyard we once thought, no one is expecting to find anything within it other than simple, single-celled life. That's extraordinarily exciting just in itself—imagine what we could learn from just one weird microbe—but its implications would be even more exciting still. It would mean that biology is as universal as chemistry. And where there's biology, there's evolution. And where there's evolution, there's complex, intelligent life.

CHAPTER TWO

SETI

In which our author introduces us to the scientific search for alien radio signals, known as SETI, acquaints us with the famous Drake Equation, and investigates the strange phenomenon of UFOs.

It was the summer of '67, and Jocelyn Bell Burnell's* wildest dreams were about to come true. As a research student under the eminent astronomer Antony Hewish, she had spent the previous two years helping build a brand new radio telescope at the Mullard Radio Astronomy Observatory just outside Cambridge. Now that same machine was ready to explore the virgin heavens, and she alone would be the confidante of its secrets.

A radio telescope, of course, makes an image of distant stars and galaxies using the radio waves they give off. Not that the average radio telescope looks particularly like the sort of thing Admiral Nelson defiantly held up to his blind eye; in fact, the

* Then Jocelyn Bell.

one that Bell Burnell was in sole charge of looked more like two rugby fields laid end to end and covered in TV antennae.

This being the mid-sixties, the output of Bell Burnell's telescope was not the hard drive of some freon-cooled supercomputer but four three-pen chart recorders, whose inkwells and chart paper needed replenishing every morning and which produced 96ft of recorded chart paper every day. And after a few weeks spent getting her eye in, Bell Burnell noticed something very strange indeed.

The telescope she was using had been purposely designed to investigate a newly discovered kind of radio source called a quasar. A quasar is a galaxy at the very beginning of its life, blasting out radio waves as the supermassive black hole at its center feasts on extremely high-temperature gas and dust. Bell Burnell was soon able to pick out good candidates for quasars, and to discard unhelpful noise from earthbound radio sources such as dodgy spark plugs on passing mopeds on the nearby A603 highway. But there was another type of signal she could not account for: a rapid juddering of the chart pens that produced a quarter-inch or so of what she called "scruff," which cropped up roughly once every 3000ft of chart paper.

It didn't take Bell Burnell long to work out that the "scruff" must be coming from the same patch of sky; in fact, it was in step with the distant stars, implying it was well outside the solar system. To see the "scruff" in more detail, Bell Burnell began to set the chart paper to run faster every time the telescope scanned that particular corner of the cosmos. The results were extraordinary. The "scruff" resolved into a signal. There on the chart paper was a regular pulse, with each pulse precisely 1⅓ seconds apart.

Bell Burnell was stumped. What on earth, or rather what not-on-Earth, could it be? Stars and galaxies glow, they don't pulse. Pulses mean life. And then an extraordinary thought struck her: Could this be a message from an alien civilization?

FLYING SAUCERS

We'll return to the story of Jocelyn Bell Burnell and her mysterious radio pulses at the end of the chapter; suffice it to say that her exemplary detective work and scientific nous will provide an exquisite counterpoint to some of the undeniably entertaining but rather bonkers UFO stories which follow.

I can imagine that some of you already feel offended. You picked this book up in the hope that it would be about UFOs, and you feel like you've been short-changed. Maybe you've seen a UFO—which simply stands for Unidentified Flying Object—or you know someone else who has. I think it's worth getting something clear from the start: I have an open mind. To me, that's what science is all about. It also means not accepting a theory as correct unless it is supported by high-quality evidence, no matter how much you want it to be true. UFO stories are great fun, and I enjoy them as much as anyone; I just don't believe that they have much to do with real-life extraterrestrials.

That said, I think it's worth giving a potted history of the UFO phenomenon so that we can put the real science of alien life in context. So many people see UFOs, report contact from UFOs, and recount abduction by UFOs that something must be going on. What is that something and when did it start?

The term UFO was coined by the US Air Force in the 1940s to describe anything seen in the sky that cannot easily be explained in terms of known craft or natural phenomena. Toward the end of the nineteenth century, witnesses began to report sightings of alien airships; these were then followed by sightings of alien rockets in the first part of the twentieth century. But the phenomenon as we know it really took off with the appearance of that alien design classic, the flying saucer.

So when did the first flying saucers appear? Strangely enough, there's a precise answer to this question: Tuesday, June 24, 1947. Because that was the year amateur pilot Kenneth Arnold made an extremely memorable business trip.

UP UP AND AWAY

At two o'clock that afternoon, the thirty-two-year-old Arnold took off from Chehalis in Washington State in a three-seater, single-engine Callair, heading for Yakima some 120 miles due east, on a course that would take him past Mount Rainier in the Cascade Mountains. A military plane had crashed on the mountain the previous winter, killing thirty-two marines, but because of the snow the wreck had never been found. It was a beautiful clear day, and with the snow receding—and the incentive of a five thousand dollar reward—Arnold decided to go check it out.

As his plane emerged from searching one of the canyons at the foot of Mount Rainier, Arnold saw a bright blue flash, and thought for a moment that he must have caught a reflection from a plane very close by. Alarmed that he might be on a collision course, he scanned the sky around him, but could see no craft in the immediate vicinity. A second bright blue flash then lit up his cockpit with the brilliance of a "welder's arclight," and in the distance he saw "to the left of me a chain of objects which looked to me like the tail of a Chinese kite, kind of weaving and going at terrific speed across the face of Mt. Rainier."

At the center of each craft there was a bright blue light, pulsing in a way that he later said reminded him of a human heartbeat. From their diagonal formation and high speed he thought at first they must be some kind of military planes. For Arnold, the really strange thing was that the craft didn't have any tails; silver in color, they looked "something like a pie plate that was cut in half with a sort of convex triangle in the rear." Never having seen anything similar before, Arnold decided that the tails must be concealed with camouflage paint, and "didn't think too much of it."

Being a good pilot, Arnold glanced down at the second hand on his watch and timed the fleet of ships as they made the journey from Mount Rainier to Mount Adams, clocking the trip at one minute and forty-two seconds. That got his attention. The two peaks were separated by some fifty miles, so the speed of the fleet must have been something like twenty-five miles per minute, or 1,500 mph. That was truly extraordinary. At the time the air speed record was less than half of that, at around 620 mph. He counted nine craft in total, judging their closest approach to be twenty-three miles, and their wingspan to be at least 100ft across.

After landing at Yakima, Arnold went to the office of his friend Al Baxter, the general manager of a crop dusting company called Central Aircraft. Bemused by the story, Baxter called in two of his flight pilots and a helicopter pilot for a second opinion. The best explanation they could offer was that Arnold had spotted some test missiles from the nearby army air base at Moses Lake.

One of the local papers, however, saw something else in the story. Were the strange craft really military planes, or something else altogether? The day following Arnold's encounter, a very brief column appeared in the *East Oregonian* inaccurately stating that Arnold had observed a "saucer-like aircraft." The paper's editor, Bill Bequette, decided to put the story on the Associated Press wire to see if the US military would respond and clear the matter up. The wire stated that Arnold had seen "nine bright saucer-like objects flying at "incredible" speed . . . "It seems impossible," Arnold said, "but there it is."

When Bequette got back to the office after lunch, the phone was ringing off the hook.

SURFING THE WAVE

Arnold's sighting caused something of a media frenzy, and launched what is known as "The 1947 Wave" of UFO sightings. Many of these sightings were also of flying saucers, metallic, disc-like craft that traveled at great speed. Others were of rockets, balloon-like craft, and balls of light. Three days after the Arnold incident, a rancher named William Brazel found some strange-looking debris on his ranch outside Roswell, New Mexico, and called the local Army Air Force base saying he had found a flying saucer. The base's PR officer passed the report on to the press, and news spread that the US government had recovered a crash-landed flying saucer. The phenomenon spread, and by the end of July 1947 there had been a total of forty-five UFO sightings across America, seventeen of which were of flying saucers.*

After the flying saucer sightings of the 1940s, the 1950s saw the UFO phenomenon ramp up a notch with the emergence of the so-called "contactees." These were people who claimed to have communicated with aliens. One notable case was that of George Adamski, who said that an alien spaceship made of translucent metal had landed next to him in the Colorado Desert and a blond Venusian named Orthon had warned him of the dangers of nuclear war via telepathy.

The 1960s then saw the first abduction stories, beginning with Betty and Barney Hill's encounter with a UFO on a nighttime drive through White Mountain National Park, New Hampshire, in 1961. Both of them remained troubled by the event, and two years later they recalled under hypnosis that they had been kidnapped by little grey aliens with large black eyes who had, among other things, examined their genitals and showed a keen interest in Barney's dentures.

The crop circle phenomenon then briefly took over in the 1970s, reaching a peak in the late 1980s. In 1991 two Englishmen, Doug Bower and Dave Chorley, admitted responsibility for the hoax. They had been inspired by the 1966 case of the Tully "saucer nest," when a farmer from Tully, Queensland, Australia, reported seeing a saucer-like craft rise

* I have to say, checking data in the world of UFOs has to be one of the most mind-bendingly fruitless activities that has ever befallen me. Wikipedian beware is all I can say. One report I read on the internet quoted 850 sightings in July 1945 alone, which I suspect is closer to the number of media articles than to the number of actual eyewitness accounts, if it is based on anything at all and not simply plucked out of thin air. The figures I've quoted here are taken from the US Air Force's investigation Project Sign, which is available—along with declassified pages from the later Project Grudge and even later Project Blue Book—at http://www.bluebookarchive.org

up from a swamp and fly away, leaving behind a flattened circular area of grass.

And that, in a nutshell, is the UFO phenomenon. Kenneth Arnold is reported as seeing flying saucers in 1947, followed by a wave of sightings across the US. The first contactees of extraterrestrials appear in the 1950s, followed by the first abductees in the 1960s. Crop circles come and go in the 1970s and 1980s and are shown to be a hoax. UFO reports still occur today, and are now mainly of the abductee type. Clearly something is going on here, but what exactly is it? Could it really be true that aliens are visiting our planet?

THE ALIENS ARE COMING

The science on this is pretty clear: Yes, it could. It's true, the distances between the stars in the galaxy are prohibitively large, making any journey between alien worlds a considerable challenge. Our nearest star is a red dwarf called Proxima Centauri, which is 4.24 light-years away, and which may or may not have planets—we're not sure. As we learned at the beginning of the last chapter, the farthest mankind has managed to send a spacecraft is the *Voyager 1* probe, which thirty-five years after its launch is only just reaching the edge of our solar system at a measly distance of 0.002 light-years.* But who's to say what superior technology a long-lived intelligent civilization might create? Why shouldn't a souped-up alien spaceship be able to fly at an appreciable fraction of the

* In case you're interested, *Voyager 1* is heading in the direction of the snappily named star AC +79 3888 in the northern constellation of Camelopardalis, and is expected to arrive in around 40,000 years' time. That's obviously a lot of time to spend in an economy class seat.

speed of light? Or shortcut between distant regions of space by harnessing the energy of a star to create an Einstein-Rosen bridge, better known as a wormhole?

OK, the wormhole thing is pushing it. But basic calculations show that even aliens with reasonably fast spaceships could rapidly colonize the galaxy if they so choose. In fact, the astronomer Paul Davies has calculated that a single alien civilization with ships traveling at only (only!) a tenth of the speed of light could cross the entire Milky Way Galaxy within four million years.* Unless there's something extremely special about us or the Earth, the Milky Way should be awash with civilizations older and more technologically advanced than ours. Surely one of them would have visited us by now?

THE FERMI PARADOX

The most famous statement of this line of reasoning was made by the eminent Italian physicist Enrico Fermi and is known as the Fermi Paradox. Fermi is something of a hero figure among physicists, a brilliant theoretician who was also a gifted experimental scientist. Awarded the Nobel Prize in 1938, Fermi used the award ceremony in Stockholm as an opportunity to escape Mussolini's Fascist Italy and emigrate to the US with his Jewish wife Laura.†

* His calculation assumes that each colony takes 1,000 years to establish itself and that suitable planets are on average ten light-years apart.

† And Mussolini must have been very cheesed off that he did, because it's not too much of an exaggeration to say that Enrico Fermi helped alter the course of the Second World War. As a linchpin of the Manhattan Project, together with Robert Oppenheimer and Edward Teller, Fermi helped develop the atomic bombs "Little Boy" and "Fat Man" that were dropped on Hiroshima and Nagasaki, effectively ending the war.

In fact, among scientists Fermi is as famous for his powers of estimation as he is for his contributions to the Manhattan Project and his Nobel Prize. Should you be wondering, the relevance to the search for extraterrestrials is as follows. In 1950, Fermi paid a visit to Los Alamos, where Teller was working on the successor to the atomic bomb, known as the hydrogen bomb. That summer had seen a strange phenomenon in New York: the mass disappearance of public trash cans. It had been a good summer for UFO sightings, too, and as they walked to lunch one of Fermi's colleagues told him of a cartoon that he had seen in *The New Yorker*, where a flying saucer was pictured unloading New York trash cans on its home planet.

Fermi joked that the cartoon presented a reasonable scientific hypothesis, because it explained two separate phenomena. Teller recalls that a serious discussion then followed about whether flying saucers were real, with no one in the group feeling particularly convinced. Fermi then inquired as to the probability of faster-than-light travel. What were the chances that they would see material evidence of a solid body traveling faster than the speed of light within the next decade? Teller put the odds at a million to one; Fermi was more optimistic, putting them at one in ten.

The group entered the Fuller Lodge canteen and settled down to lunch, making small talk as only physicists working on the most powerful weapon in the history of humanity can. Then, halfway through the meal, out of the blue, Fermi suddenly asked, "Where is everybody?"

His companions immediately grasped that Fermi was talking about extraterrestrials, and burst out laughing. One of the scientists present, Herbert York, recalls that Fermi followed up with a series of calculations on the probability of earthlike planets, the probability of life given an earth, the probability of humans given life, the likely rise and duration of high technology, and so on. He concluded on the basis of such calculations that we ought to have been visited long ago and many times over. As I recall, he went on to conclude that the reason we hadn't been visited might be that interstellar flight is impossible, or, if it is possible, always judged to be not worth the effort, or technological civilization doesn't last long enough for it to happen.

In other words, after running the numbers Fermi concluded that intelligent civilizations must exist, but for one reason or another they are staying put. But if they aren't going to come calling on us, how do we get to meet them? The answer: radio waves.

OPENING THE WINDOW

At least that was the conclusion of a paper entitled "Searching for Interstellar Communications" by Giuseppe Cocconi and Philip Morrison of Cornell University, published on September 19, 1959, in the scientific journal *Nature*.* They pointed out that there happens to be very little noise on Earth over a certain range of the radio spectrum, known in the trade as the microwave window. For frequencies below this window there's lots of noise because of absorption and emission by interstellar

* Giuseppe Cocconi would later become one of the leading lights of CERN, while Philip Morrison famously narrated the outstanding science short "Powers of Ten." If you've never seen it, you have a rare treat in store at http://www.powersof10.com/film. Enjoy!

gas and dust, and above it there's lots of noise because of absorption and emission by the Earth's atmosphere. If aliens wanted to contact us, they reasoned, they would most likely send radio waves tuned to sit in this particular window.

THE DRAKE EQUATION

Frank Drake was nervous. It was November 1961, and in a few days' time he would be hosting the first conference of SETI, the Search for Extraterrestrial Intelligence. Among the attendees would be one of his heroes, the Russian émigré astrophysicist Otto Struve. During his navy scholarship in electronics at Cornell in the early fifties, Drake had attended one of Struve's lectures, and had been gripped by his claim that at least half of the stars were orbited by planets. Just think: If half of the stars had planets, that was a lot of potential alien real estate.

In April that same year, Drake, unaware of Cocconi and Morrison's paper, had come to much the same conclusion concerning the microwave window and its suitability for alien communication. Audaciously, he had pointed the newly built 85ft radio telescope at the National Radio Observatory in Green Bank, West Virginia, at our two closest Sun-like stars, Tau Ceti and Epsilon Eridani, and listened. If those two stars had planets, and those planets were home to intelligent life, then maybe he might be able to pick up a signal.

Throughout the spring and summer of 1960, Drake had listened to the two stars for a total of 150 hours but found nothing. Nevertheless, a vital first step in humanity's communications with extraterrestrials had been taken. A fan of L. Frank Baum, he had named the project Ozma, after the Fairy Queen of the far-off land of Oz, and now other eminent scientists wanted to join him on the Yellow Brick Road. Not only would his hero Otto Struve be among the ten at Green Bank, but so too would neuroscientist John C. Lilly, famed for his work on dolphin communication;* Melvin Calvin, a chemist who would win a Nobel Prize for his work on photosynthesis on the first night of the conference; Philip Morrison, one of the two authors of the Cornell paper; and the gifted astronomer and adviser to NASA Carl Sagan.

Drake needed to set the pace for the conference, and—much in the mould of Fermi a decade earlier—decided the best way to do that would be to estimate the number of civilizations in the galaxy, N, that we might be able to receive signals from. The equation he came up with has become a landmark in the quest to communicate with alien intelligence, and here it is in all its glory:

$$N = R^* \times f_p \times n_e \times f_l \times f_i \times f_c \times L$$

Trust me, it looks a lot more ferocious than it is. And as a way of breaking it down, I want to take you to a Dire Straits gig.

BROTHERS IN ARMS

Specifically, I want to take you to Wembley Arena in the UK for the British leg of Dire Straits' famous 1985–6 world tour. Their latest album, *Brothers in Arms*, has gone multi-platinum, and the 12,500-seat venue is packed to the rafters. The band close the show with their monster hit "Money for Nothing." Tired but happy, the army of fans make for the exit. But wait. The band

* And about whom we'll be hearing more in our final chapter. Much more.

come back onstage for an encore. Mark Knopfler strikes up the opening chords of the album's title track. Out in the corridors and turnstiles, the crowd hear that there is more to come, and about-turn.

Back in the auditorium, the seats start to fill. "These mist-covered mountains . . . " sings Mark. And one by one, they appear; cigarette lighters, held silently aloft. Human souls share the tender beauty of the song. Mark adjusts his sweat-band and looks out into the darkness. It's a deeply affecting sight, but let's put our practical hats on for a moment. How many lit lighters does he see?

You might think it depends on how much of the song Mark plays before he looks up. Surely, you say, the number of lit lighters will slowly build up during the song as more fans enter the auditorium. If Mr. Knopfler looks up after a minute, he'll see fewer lighters than if he looks up during the final few bars. But that's not necessarily the case.

It all depends on how long the lighters shine for. To see what I mean, let's imagine only one fan has a lighter, and that he can only keep it alight for thirty seconds before he burns his fingers. On average, will Mark see the light? Instinct will tell you he probably won't. For a start, he might look up before the fan has even entered. Or he might look up after the fan has entered, but after the light has gone out.

OK. So how many fans with lighters do we need to enter the auditorium during the four-minute song in order for Mark to see at least one light? This is where math comes in handy. Because to get the average number of lit lighters that Mark sees, we just need to multiply the rate at which fans with lighters enter the stadium by the length of time that they can keep them alight.* For example, let's say that eight fans with lighters enter the stadium during the song. In that case, the rate is 8/(4x60) fans per second, and the average number of lit lighters at any one time is

$$\text{Rate} \times \text{length of time alight} = 8/(4\times60) \times 30 = 1$$

Interesting, eh? In other words, even if eight fans enter with lighters during the encore, on average when Mark looks up he will see only a single light.

MONEY FOR NOTHING

Right, now we know the principle of how this works, let's get down to some gritty detail. Obviously with a capacity crowd of 12,500 fans, our current figure of eight fans with lighters entering during the course of the song is way out. How can we more accurately estimate the true number?

There's going to have to be a bit of guesswork here. Could all of the fans make it back into the arena during the four-minute encore? Based on the recent Bill Bailey gig I attended, I'm going to say "yes." There will inevitably be a bit of a bottleneck at the entrances, but Wembley Arena is well served by generously proportioned walkways with exemplary signage. In fact, I am going to be so bold as to say that 10,000 of the fans make

* In the case that there is only one obliging fan, the rate is of course 1 fan per 4 minutes, or 1/(4×60) fans per second. We already know that he can keep his lighter burning for a total of thirty seconds so we can calculate the average number of lit lighters Mark sees as (rate) x (length of time alight) = 1/(4×60) × 30 = 1/8. As this is less than 1, it means that on average Mark won't see a single lighter. Of course, if the scenario is repeated over the course of eight shows, we would expect that on one of those nights Mark would see a single light. I'm not sure how he would feel about that.

it back into the auditorium during the course of the four-minute song. That's a rate of 10,000/4×60 = 42 fans per second.

But wait. Not all of those fans will be carrying lighters. For a start, not all Dire Straits fans are smokers; in fact I would say quite the reverse. Given the band's demographic, everyone attending is liable to be quite clean-living. Nevertheless, this is 1985, when menthol-flavored cigarettes are considered to be a healthy option. Let's put the fraction that are smokers at 50 percent. And, of course, not all of the smokers will be carrying a lighter. For the sake of illustration, let's say that 50 percent of them are carrying two lighters apiece; their favorite Zippo, and a backup.

OK. So to estimate how many lighters Mark Knopfler sees, we need to multiply the rate at which fan-operated lighters enter the stadium by the length of time they are alight. In other words, if R is the rate at which the audience reenters the arena, f_s is the fraction that are smokers, f_l is the fraction of smokers that have lighters, n_l is the number of lighters that each lighter-owning smoker carries, and L is the length of time in seconds that it is possible to hold a lighter without burning your fingers, then—big breath—the number of lit lighters Mark Knopfler sees during the encore is:

$$N = R \times f_s \times f_l \times n_l \times L$$

Let's plug in the numbers. In which case we get

$$N = 42 \times 50/100 \times 50/100 \times 2 \times 30 = 630$$

So on average, at any one time, Mark sees roughly 630 flames twinkling all around him in the darkened arena. He knows he is loved and no longer feels alone.

SO FAR AWAY FROM ME

As with pretty lights at an eighties stadium rock gig, of course, so with detectable alien civilizations. More or less, anyway. To figure out how many alien civilizations a radio astronomer might be able to see with his telescope, there are two things we need to get a handle on: the rate at which detectable alien civilizations emerge, and how long they are detectable for.

OK, so at what rate do detectable alien civilizations emerge? Let's assume that the kind of intelligent civilization we are going to be able to communicate with is a lot like us; inhabiting a rocky planet in orbit around a Sun-like star. Then we can start with the rate of formation of Sun-like stars, R*, and whittle it down just like we did in the case of the Dire Straits gig.

First, let's work on the rate at which Earthlike planets form. If R* is the rate of formation of Sun-like stars, and f_p is the fraction of those stars with Earthlike planets, then their rate of formation is simply

$$R^* \times f_p$$

That's about as hard as it's going to get. Of course some solar systems will have more than one Earthlike planet, such as our own, where arguably the Earth, Moon, and Mars—and maybe even Venus—have been habitable to life at some point. This is a bit like our example of smoking Dire Straits fans who have more than one lighter, so if n_e is the number of Earthlike planets per solar system, then our running tally for the rate of formation of Earthlike planets is

$$R^* \times f_p \times n_e$$

Right. I'm sure you're getting the hang of this, so if f_l is the fraction of Earthlike planets that support life, the rate of formation of life-supporting planets is

$$R^* \times f_p \times n_e \times f_l$$

And if f_i is the fraction of life-supporting planets that have intelligence, the rate of formation of intelligent life is

$$R^* \times f_p \times n_e \times f_l \times f_i$$

We're so nearly there. We have found the rate at which intelligent life appears in the galaxy. All we need to do now is multiply by the fraction of alien intelligences f_c that have radio communication, and we will—at last—have the rate of formation of detectable alien civilizations. Here we go:

$$R^* \times f_p \times n_e \times f_l \times f_i \times f_c$$

Good. Now, just as in the case of cigarette lighters at a Dire Straits gig, the number of detectable alien civilizations will be equal to the rate at which they appear, multiplied by the length of time they last for. So let's put the cherry on top by multiplying by the length of time that an alien civilization is detectable, L. And, lo and behold, we have derived the Drake Equation:

$$N = R^* \times f_p \times n_e \times f_l \times f_i \times f_c \times L$$

Satisfying, eh?*

NARROWING THE ODDS

Back in the day, Drake and his colleagues at the first SETI conference judged the various factors of the Drake Equation to be as follows:

$R^* = 1$ (One Sun-like star forms per year)

$f_p = 0.2–0.5$ (Between a fifth and half of all Sun-like stars have planets)

$n_e = 1–5$ (Such stars will have between one and five planets in their habitable zone)

$f_l = 1$ (All such planets will develop life)

$f_i = 1$ (All planets that develop life will also develop intelligence)

* Readers of my previous tome, *It's Not Rocket Science*, may spot that for the sake of keeping our relationship fresh I have used a slightly different version of the Drake Equation. Instead of R* and L, that version uses N*, the number of stars in the galaxy, and f_l, the fraction of a star's lifetime during which a civilization is detectable. It all comes out in the wash, because R*=N*/(lifetime of a star) and f_l = L/(lifetime of a star). As the Americans say, "You do the math."

f_c = 0.1–0.2 (Between 10 and 20 percent will develop radio communication)

L = 1,000–100,000,000 years (The length of time for which signals are transmitted is between one thousand and one hundred million years)

Let's take all the lower limits:

$$N = 1 \times 0.2 \times 1 \times 1 \times 1 \times 0.1 \times 1000 = 20$$

In other words, on any given day, when we point our radio telescopes up into the sky there are twenty stars, spread throughout the galaxy, from which we might detect alien signals.*

Now let's take the upper limits:

$$N = 1 \times 0.5 \times 5 \times 1 \times 1 \times 0.2 \times 100,000,000 = 50,000,000$$

Meaning that there are fifty million stars from which we might pick up a signal. Simplifying things even further, we can see that, very roughly speaking, most of the factors are approximately 1 apart from L, the number of years that a civilization is detectable.† We can then write the Drake Equation in a stunningly simple form, like so:

$$N \approx L$$

* Or, rather, ten, since we can only see one hemisphere at a time.

† In fact, Frank Drake himself has gone so far as to buy a personalized license plate with the letters "NEQSL," which is Geek for N = L.

IS THERE ANYBODY OUT THERE?

Equations are like poems. There's what they seem to be about, and what they are really about. On the face of it, the Drake Equation simply tells us how to crunch the numbers to find out how many detectable alien civilizations might be out there in the galaxy, but of course there's something much more important going on. The real power of the equation is in the assumptions it forces us to make.

The deepest assumption is that the aliens will be just like us. We are presuming that the aliens will have technology like ours, societies like ours, and planets like ours. Now I don't think for a second that Frank Drake is being naive; rather, his equation says, "hey, we've got to start somewhere, so we may as well start here." If it provides anything, the Drake Equation gives us a lower limit on what we might expect to find out there in the galaxy. After all, who's to say that aliens don't inhabit dust clouds in deep space as well as rocky metal-rich planets like our own?

The Drake Equation shows us that in considering the problem—namely, are we alone?—we need to think deeply about the very nature of life, intelligence, civilization, and technology. What do we mean by these things? And once we have made our assumptions, what data do we have that can turn them into bona fide estimates? For example, if we assume that biology is as universal as chemistry, how can our knowledge of how life evolved on Earth help us to make a guess about its abundance throughout the galaxy?

As you can imagine, these are some of the most fascinating questions a human mind can ponder. To find answers, we will be foraging on the very fringes of scientific knowledge. What do the latest telescopes tell us about the abundance of Earthlike planets? What do the latest advances in biology tell us about the nature of life, and the chance of it being commonplace in the cosmos? What is intelligence, and how might we communicate with an alien intelligence that is vastly different from our own? What do we want to say? And why do we want to say it?

LONG-DISTANCE RELATIONSHIP

So why do scientists believe in radio contact with alien civilizations, but not in flying saucers? On a simple level, you might say it's because of a lack of evidence. Given the fact that everyone now carries a mobile phone with a camera on it, you might think that there would be some really good footage of alien contact, but there isn't. What's more, no alien artifacts are on display in any of our museums, and no alien spaceship has landed on the White House lawn. What we do have is eyewitness reports.

One of the fun things about this book is that, in examining aliens, we really get to think about what it means to be human. And it is time to face one of the more unpalatable truths about our species: When it comes to the world around us, we apes are not the most reliable of witnesses.

As young children, we are bathed in imagination. We truly believe that we can fly, that we can see monsters in the wardrobe, and that a fat, bearded Latvian man delivers all the world's Christmas presents on a fifteen-foot sleigh pulled by magic reindeer. We see a mixture of the world as it is, and the world as we imagine it to be. For children, wishing makes it so.

As adults, on the other hand, we pride ourselves on our impartiality. We are certain that the wild dreams that we have at night never intrude into our waking hours. We believe we see the world as it truly is, and consider ourselves the masters of our own imagination. By this logic, when an otherwise upstanding member of the community—a policeman, say, or a magistrate—sees a ghost in the middle of the night, his anecdote is all the proof that we need. Chris is a company director, Chris saw a ghost, therefore ghosts exist. But are our minds really as reliable as we think they are?

Your average scientist would say not. In fact you could say that one of the aims of science is to remove the so-called "human factor" from our observations of the world; to try and describe the universe in an objective, logical, self-consistent way that can be tested by experiment. Thus evolution is taken to be true not because it's a great story and Darwin was a really steady guy, but because it predicted the existence of certain fossils before those fossils were ever found. Any given scientific theory stands only for so long as it is supported by experiment. It doesn't matter how many Nobel Prize–winning biologists believe in evolution; if a fossilized dolphin suddenly turns up among the trilobites in a piece of Cambrian sedimentary rock then we'll all be back to the drawing board.

Compelling as the stories about flying saucers are, and as much as I for one would like to believe that they are true, the evidence is of poor quality. Kenneth Arnold was, no doubt, a reliable man not given to exaggeration. He truly believed that he saw a fleet of strange craft that summer, flying across the snowline of Mount Rainier, and was as puzzled by what he saw as anyone else. As an amateur pilot, he had additional credibility; he had the skills to tell a fleet of spaceships from a flock of geese, for example, or from a formation of conventional aircraft.

Yet, harsh as it may seem, in scientific terms the reliability of Kenneth Arnold is neither here nor there. Science doesn't care who you are or what you think you saw, it simply demands evidence. You saw a flying saucer? Show me the footage on your smartphone. An alien spaceship crash-landed in New Mexico? Show me a piece of the ship. You were abducted by aliens and subjected to an internal examination? Show me . . . actually never mind.

"Ah," I hear you say. "But scientists are people, too. Why should I believe some hippy with a test tube over a model citizen like Kenneth Arnold?" And you'd be right. Trusting a scientist purely because of his or her name and reputation is a dangerous game. Scientists make all the mistakes that everyone else makes. Their imaginations play tricks on them, they cherry-pick data that supports their pet theories, and they have a biased view of their own talents and abilities. But experiment saves the day time and time again. For a scientific hypothesis to gain weight, it has to be testable by experiment, and that experiment has to be repeatable. Scientists do make mistakes, but every time they do experiment puts them back on the right track. Without hard evidence to back them up, despite seven decades of sightings, crash landings, and abductions, the scientific case for alien artifacts is always going to be hard to make.

LITTLE GREEN MEN

Which brings us neatly back to Jocelyn Bell Burnell. Bell Burnell, you will recall, has picked up a series of pulses 1⅓ seconds apart, coming from within the constellation Vulpecula ("Little Fox"), which is itself smack dab in the middle of the so-called "Summer Triangle" of bright stars Deneb, Vega, and Altair. She has a dilemma. No serious astronomy graduate wants to tell their supervisor they have intercepted an alien signal; on the other hand, no serious astronomy graduate doubts that as far as aliens are concerned, they are on the front line. If anyone's going to take the call, it's probably them.

Summoning all her courage, Bell Burnell telephoned her supervisor, Anthony Hewish, who was teaching in one of the undergraduate laboratories, and told him what she had seen. "Must be man-made," said Hewish, and came out to the telescope the next day to see the string of pulses for himself. Sure enough, there they were. He decided that there must be something wrong with the equipment, and for the next month he and Bell Burnell tried to eliminate as many sources of error as they could.

First, they confirmed that the source was keeping pace with the stars rather than with the Sun. Astronomers refer to this as keeping sidereal time.* That implied that the signal wasn't coming from Earth, ruling out man-made interference. Except, of course, that produced by other astronomers, who also keep sidereal time. Could it be that some neighboring observatory was transmitting the signal as part of one of their research projects?

A letter from Hewish to all the neighboring observatories drew a blank. What other straightforward explanations could there be? The team eliminated radar reflected off the Moon, signals from satellites, and effects due to a large corrugated

* A sidereal day is simply the time taken for the Earth to make a full rotation relative to the stars. If you were looking down on the Earth's circular orbit from above, you would see that by the time the Earth has made a full rotation relative to the stars, it has not quite made a full rotation relative to the Sun. To do that, it needs to turn for another 3 minutes and 56 seconds. What this meant for Bell Burnell was that if the source was located in the stars and appeared at 4.04 p.m. on one day, she would expect it to appear at 4.00 p.m. the next.

metal building just to the south of the telescope. They then checked all the wiring, which to Bell Burnell's considerable relief turned out to be sound. She had helped wire it, after all.

They made a thorough analysis of the pulses. The pulses were 1⅓ seconds apart, and each one lasted less than 0.016 seconds. That meant that whatever was producing them had to be small. Basic physics says that nothing travels faster than the speed of light, so the object producing them had to be, at most, about three thousand miles across.* That's less than the radius of the Earth (3,959 miles), which in astronomical terms is on the tidy side.

To try and get a handle on how far away the source was, the team measured the dispersion of the pulses. Dispersion, as you probably know, occurs when higher frequency waves travel faster through a medium than lower frequency ones. The classic example is of white light dispersing into all the colors of the rainbow as it passes through a glass prism. Interstellar space may not be full of glass, but it is far from empty. In fact, the spiral arms of our galaxy sit in a sort of gas of free electrons. Just as glass slows down low-frequency red light more than it does higher-frequency blue light, this "gas" of free electrons slows lower-frequency radio waves more than the higher-frequency ones. Send a pulse through interstellar space, and after a while, thanks to dispersion, the lower-frequency parts of the pulse will start to lag behind the higher-frequency parts. In the case of Bell Burnell's mystery object, by the time the pulses reached Earth there was a noticeable delay.

* You can read all this in the original paper, "Observation of a Rapidly Pulsating Radio Source" by A. Hewish, S. J. Bell, J. D. H. Pilkington, P. F. Scott, and R. A. Collins, *Nature* 217, 709–13 (1968). At the time of writing you could view it for free on the *Nature* portal, http://www.nature.com/physics/looking-back/hewish/index.html.

By measuring this time delay between the highest and lowest frequencies in the signal, and then using a simple model for the number of free electrons that the pulse had passed through, the team was able to work out how far away the source was. Their calculations placed it well outside the solar system but well within the galaxy, at a distance from Earth of roughly 200 light-years.*

Next, they pondered what the setup of this alien civilization might be. What if the signal was coming from an Earthlike planet which was in orbit around a Sun? What could they test? If the aliens were in orbit, surely their signal ought to be in orbit, too. If it was, sometimes it should be moving away from Bell Burnell's radio telescope, and sometimes it should be moving toward it. That movement should produce an effect; in fact it's a commonplace one called the Doppler Effect.

There's a classic example of the Doppler Effect in the change in pitch of a passing train's horn. The train horn is producing one note. As it approaches, the sound of that note is pitched up. As it passes, the sound is pitched down. Put in more general terms, when the source of a wave signal is moving relative to a detector, it changes the frequency of that signal.

Was there a Doppler Effect in the signal? To the team's surprise, there was. Yet it wasn't due to the motion of the alien signal around some alien Sun. You see, Bell Burnell's telescope was itself in orbit, around our own Sun. The Doppler Effect that the team measured in the signal turned out to be due to the relative motion of the telescope to the source. As Bell

* Once you start dealing with the enormous distances between stars in the galaxy, it gets a bit inconvenient to use puny Earth units like a meter. A much handier unit is the light-year, being the distance that a beam of light travels in one year.

Burnell herself wryly puts it, the team had simply managed to prove that the Earth revolves around the Sun. Reassuring, but of itself not much of a breakthrough.

They did make some progress. Once they had isolated this small Doppler Effect, they could see that the pulses in the signal were extraordinarily regular, with the gap between them varying by less than one part in ten million. That meant that whatever was producing them had a lot of mass, and therefore a lot of energy. If it was aliens, they really meant business. Someone had built themselves a very powerful transmitter indeed.

THE THIRD EYE

Feeling more and more certain that the signal they had found was not the result of some random man-made interference or faulty wiring, and increasingly sure that its source was something massive and extremely compact that was situated well within the galaxy but beyond the nearest stars, Jocelyn Bell Burnell and her team decided to bite the bullet. They would go and ask a rival telescope if they could see it, too.

Keeping it in the family, Hewish approached his colleague Paul Scott and his research student Robin Collins, who were operating a radio telescope at the same frequency. They calculated that the signal should show up in the second telescope just twenty minutes after it appeared in Bell Burnell's. As soon as the signal appeared in Bell Burnell's telescope, the team moved over to the second chart recorder. Twenty minutes went by with no signal. Hewish and Scott wandered off down the hall, with Bell Burnell tagging behind, discussing what could possibly cause the signal to appear in the one telescope and not the other. Suddenly, there was a shout from the lab. Robin Collins had hung back, waiting, and there the signal was, pulsing away. They had miscalculated the delay by five minutes. The source was real.

On December 21, 1967, Anthony Hewish and the head of the group, Martin Ryle, held a meeting at the Mullard to discuss what to do about the object they half-jokingly called LGM, as an abbreviation for Little Green Men. If it really was a pulse from an alien civilization, then those aliens were a contrary lot. For a start, the radio frequency of the pulse, 80MHz, seemed an unlikely choice; although perfect for quasars, it happens to be a very noisy frequency.

If Bell Burnell's Little Green Man was signaling to Earth, or other Earthlike planets, surely he would tune his signal to sit in the microwave window? Instead he had chosen a part of the spectrum where it would most likely get absorbed by gas and dust. Why send a signal at a frequency where it was less likely to be picked up?

Nevertheless, there the signal was, and if science teaches us anything it is to be humble in the face of the facts. If this really was an alien radio signal, who should they tell first? An astrophysical journal, or the Prime Minister?

Bell Burnell returned home for supper decidedly disgruntled. She had spent two precious years of her life wiring up a state-of-the-art radio telescope, ready to search for quasars. Instead, her experiment had been hijacked by a bunch of numbskull aliens. She only had six months of grant money left, and the window for her to finish her PhD and secure some sort of academic career was rapidly closing. As she put it herself, "I was furious. For some reason, some silly lot of green men had decided to use my frequency and my aerial to signal to Earth."

IN THE BLEAK MIDWINTER

That evening, Bell Burnell returned to the lab, determined to get back on track. A backlog of 2,500ft of chart paper had built up and was begging for analysis. Just before 10.00 p.m., when the lab was due to shut, she was looking at a section that belonged to the constellation of Cassiopeia when she thought she spotted some more "scruff."

Hurriedly, she laid out all the other bits of chart paper she could find that corresponded to Cassiopeia. There the "scruff" was again. The timing couldn't be more acute. The next day she was going back home to Ireland for Christmas to announce her engagement. Calculating that the patch of sky she wanted would be in the telescope at around two o'clock in the morning, she decided on no sleep till Belfast, and headed over to the observatory.

This being the dead of winter, the equipment was cold and temperamental, but Bell Burnell "breathed on it and swore at it, and I got it to work at full power for five minutes. It was the right five minutes and at the right setting. In came a stream of pulses, this time at intervals of one-and-a-quarter seconds, not one-and-a-third."

That settled it. There was no fault with the equipment, no man-made interference; there was something out there in the stars. And it couldn't be Little Green Men. After all, what were the chances that there would be two lots of Little Green Men on opposite sides of the universe, both signaling at an obscure frequency to our little blue planet? Unlikely as it seemed, somewhere out there in the galaxy were massive, compact objects that produced pulses of radio waves. Jocelyn Bell Burnell had discovered the pulsar.

JOURNEY TO THE STARS

That might seem an anticlimax—we are hunting for aliens after all—but to my mind it's glorious. Null results might be the bane of pseudoscience, but they are a boon to science. No matter how much we want aliens to be out there, we have to go by the evidence. If SETI ever does pick up an alien radio signal, we can guarantee it will be subjected to the same kind of scrutiny that Bell Burnell's pulsar was, and that is a very good thing indeed. As the late Oliver Sacks put it: "Every act of perception is to some degree an act of creation, and every act of memory is to some degree an act of imagination." When it comes to things we dearly want to believe, we have to be on our guard.

A pulsar is like an enormous lighthouse: a fast-spinning, highly magnetized ball mostly made up of densely packed material rich in neutrons, radiating a beam of electromagnetic radiation into space. They are the highly compressed corpses of large stars, formed after they have run out of fuel and exploded as supernovae. Each is unique, with its own distinctive type of radiation and pulse rate. Since Jocelyn Bell Burnell discovered them, we have found pulsars which spin so fast there are millisecond gaps between pulses. We have found still others whose "beam" is made of x-rays, and others where it is visible light.

As we know, when NASA launched the *Voyager 1* probe, etched on to the casing of the Golden Record was a map showing the position of the Earth relative to its fourteen closest

pulsars, with the pulse period of each pulsar coded in binary.*
If ever an alien intelligence intercepts it and comes to pay us a visit, Bell Burnell's pulsar will be one of the landmarks that guides them here.

* As I'm sure you'd want to know, the base unit is the hyperfine transition of hydrogen, the most common molecule in the universe. This is to do with something you may have heard of called "spin." Basically both the proton and electron in a hydrogen molecule have spin, and they like to line their spins up, either parallel or antiparallel. Sometimes an electron will emit a photon and flip from the higher energy state (spins parallel) to the lower energy state (spins antiparallel). The frequency of that photon—an extremely precise 1420.40575177MHz—can then be used to define a length of time. That's if the aliens don't tear the Golden Record off and eat it.

CHAPTER THREE

PLANETS

In which the author searches for Earthlike planets, learns about the Wow! Signal, and takes a stroll around Vienna with the UN Ambassador for the Human Race.

There is something oddly futuristic about the United Nations. Though the squat Arrivals building of its Vienna offices bears more than a passing resemblance to my low-rise 1970s primary school, the enormous courtyard I step out into is unfeasibly impressive. Everything about the place should seem dated: the mountains of grey concrete; the jet fountains that strafe an enormous shallow circular pool; the towering Cold War-style flagpoles; but instead the overall effect is of vertiginous progress. The trappings may all be mid-twentieth century, but the very existence of a super league of sovereign nations, united in the common interest of mankind, still seems like pure science fiction.

And it's this unique position in the world of human affairs that interests me today, because it's been widely reported in the British press that thanks to the recent discoveries of Earthlike planets by the Kepler Space Telescope—and the possibility that they might harbor intelligent life that we can make radio contact with—the UN is appointing a spokesperson for the human race. This "Ambassador for Earth" has been named by no less a newspaper than the UK's *Sunday Times* as one Dr. Mazlan Othman of the United Nations Office for Outer Space Affairs (UNOOSA), and I have an appointment to meet her for lunch.

Yet as I mount the stairs to Dr. Othman's office, the strong scientific imperative for my visit suddenly evaporates. This is the opposite of "l'esprit d'escalier," a phrase nonexistent in French but which we English take to mean the inspiration which strikes as soon as an encounter is over and we are heading down the stairs on our way home. The more floors I climb, the drier my mouth gets and the sweatier my brow becomes, until all confidence in my mission has completely drained away.

In the world of extraterrestrial intelligence, this discomfort is commonplace, and is known simply as "the giggle factor." For some reason, when talking about the very real, scientifically sound possibility of communicating with aliens, everyone gets the urge to laugh. And here, where national flags flutter at the tops of impossibly tall flagpoles, and where international diplomats negotiate the gravest of choices while pursuing the loftiest of ambitions, what on Earth do I think I'm doing asking the UN's head space executive about flying saucers?

It doesn't help that Dr. Othman has an extremely impressive CV. Malaysian by birth and an astrophysicist by training, in the early noughties she spearheaded the Malaysian space program, ANGKASA, and built a space observatory on the island of Langkawi, launched a remote-sensing satellite, *RazakSAT*, in the world's first near-equatorial Low Earth Orbit, and oversaw the launch of the first Malaysian astronaut to the International Space Station in 2007. Since then, she has served as the Director of UNOOSA, and was appointed Deputy Director-General of the United Nations Vienna Office in 2009.

I needn't have worried. Once I have sweated and spluttered my way past her secretary in a manner even Hugh Grant would think was exaggerating, Dr. Othman greets me warmly, blaming the layout of the UN rather than my terrible sense of direction, and is disarmingly relaxed and informal. She leads me through to her office, a bright and breezy affair with a spectacular view across the Danube toward the Old City. Her desk sits in the far corner, half obscured by a jungle of luscious potted plants and, to my right, a sideboard displays glittering scale models of satellites and space stations.

We sit, and I do my best to try and convince her that I am not a crazy person, that I know my stuff about science, and that, while I think the evidence for UFOs is feeble, I am very interested in the possibility that there is intelligent, communicable life on other planets. I state my belief that biology is as universal as chemistry and physics, and that the recent discoveries of the Kepler Space Telescope have shown us plenty of places where that biology might get a chance to do its thing. In short, I do everything I can to try and reassure her that I am an emotionally well-balanced, scientifically literate individual with a passion for astrobiology. And, in doing so, I am fairly sure that I come across as a crazy person.

When I finally pause for breath, I see that the Director of the United Nations Office for Outer Space Affairs has a twinkle in her eye. "Come on," she says. "Say it. You want to talk about aliens."

TWINKLE TWINKLE LITTLE PULSAR

Strange as it now seems, as little as twenty years ago we still had no hard evidence that planets existed outside our own solar system. As an undergraduate student in the late eighties, I remember feeling very excited by reports that a planet had been discovered orbiting Gamma Cephei, a binary star some forty-five light-years away in the northern constellation of Cepheus.* This seemed almost too good to be true. One of the most iconic moments of the first *Star Wars* movie had been Luke Skywalker looking out from his home planet, Tatooine, at two setting suns. Could it be that this new planet, like Tatooine, was a lawless desert world, blasted by the heat of two stars, where humanoid beings farmed moisture in underground dwellings? What else had George Lucas got right? Is it really possible to dodge a blast from a laser gun?

Sadly, though the first reports of this first planet were published in 1988, the year I began my PhD, they were then retracted in 1992, the year I turned professional as a comedian.† I only hope there was no connection. If there was, I needn't have worried, because that same year Aleksander Wolszczan and Dale Frail, working at the Arecibo Observatory in Puerto Rico, found the first bona fide planet in the

* The constellation of Cepheus is, of course, home to the famous Delta Cephei, a variable star whose brightness pulses with a repeating period of five days and nine hours.

† There's a happy ending, by the way; the planet Gamma Cephei Ab was finally confirmed in 2002.

constellation of Virgo. In fact, they found two, orbiting at roughly half the distance that the Earth orbits the Sun. Only they weren't orbiting a star. They were orbiting a pulsar.

That, obviously, was not what we were expecting at all, and I doubt that Frank Drake was straining at the leash to point a radio telescope at PSR B1257+12 to try and pick up a message. For a start, it's a thousand light-years away, so the conversation would be a little stilted. And, secondly, pulsars are awesome things, but for life as we know it, being blasted by x-rays is always going to have its drawbacks.

Then finally, in 1995, we discovered 51 Pegasi b, the first planet orbiting a Sun-like star. That wasn't quite what we were expecting either. Fifty light-years away in the constellation of Pegasus, it was half the mass of Jupiter, but extremely close to its home star, taking only 4.2 days to complete an orbit. How had it gotten so close? After all, in our own solar system it's small rocky planets like Mercury, Venus, Earth, and Mars that sit close to the Sun. Gas giants like Jupiter and Saturn roam much farther out, and medium-sized icy stuff like Uranus and Neptune sit out in the boondocks. Could it be possible that, in some solar systems, planets didn't stay put?

Once someone had found something that was definitely a planet, the floodgates opened. Radio astronomers everywhere tried to get in on the act, and all manner of planets turned up. Many of them, like 51 Pegasi b, were so-called "Hot Jupiters": large gas planets tucked up close to their home stars. Others were so-called "Hot Neptunes": medium-sized planets that had also migrated into tight orbits. Still others were truly

monstrous creations patrolling at unfeasibly large distances.*
What we didn't find was anything like the Earth.

Which is to say we found nothing Earth-sized that was an Earth-type distance from a Sun-like star. One of the things we think makes Earth such a good home for life is the fact that most of it is covered in water. In fact, astronomers define something called the habitable zone, which means the range of orbits where the temperature of an orbiting planet is going to be neither so hot that water simply vaporizes (such as on Mercury and Venus) nor so cold that it freezes (such as on Mars). In the official jargon, we couldn't find any Earth-sized planets in the habitable zone of their home star. Could it be that, far from being mediocre, the Earth was incredibly special?

Formally, this became known as the Rare Earth Hypothesis. Microbial life might be common, the argument went, but intelligent life is rare because planets like the Earth are rare. After all, many things conspire to make the Earth ideal for life. Firstly, it's a good size. Much smaller, and its gravity would be too weak to hold on to an atmosphere, and without a decent atmosphere there would be no greenhouse effect to keep the surface warm. This, of course, is the problem with Mars, which has only been able to hold on to the thinnest of atmospheres.

Secondly, it's volcanic. As we shall shortly see, one of the most promising hypotheses for how life got started depends on volcanic springs, and, in any case, the gases released and consumed by the rock cycle play a vital role in maintaining a life-friendly carbon dioxide–rich atmosphere. Not only that, but plate tectonics also ensures that heavier elements like

* 2M1207b, for example, was found to be over three times the mass of Jupiter, with an orbit eight times wider.

metals get recycled into the oceans and atmosphere, providing lots of esoteric chemistry that life can make use of.

Thirdly, the Earth has a strong magnetic field. Not only is that handy if you are trying to find your way around, but it means that we are protected from the hail of damaging radiation known as the solar wind. Fourthly, it has a large Moon, which not only slows its rotation, giving milder weather, but keeps the Earth's axis pointing in the same direction. Without a Moon, the Earth's spin axis might flip, wreaking climatic havoc. And, finally, it has Jupiter as a cosmic vacuum cleaner to protect it from comets.*

One look at the Chicxulub crater in the Yucatán Peninsula will tell you that comets are bad news. That particular impact wiped out the dinosaurs, and we have Jupiter to thank for the fact that such Armageddons are relatively rare events. Basically, the bully planet's strong gravity hoovers up anything dodgy that the outer solar system throws our way. Incredibly, in 1993 we actually saw this in action when the comet Shoemaker–Levy was spotted orbiting Jupiter, disappearing in a spectacular collision just over a year later.†

In truth, none of these arguments have really gone away. Intelligent life may be rare. After all, if our nearest neighbors are in the next galaxy, rather than in the next star system, that goes a long way toward explaining the Fermi Paradox. Nevertheless, we have some cause for hope. Not everyone believes that a flip-flopping north pole would be a disaster, for

* Loosely speaking, a comet is an icy rock from either the Kuiper Belt—the doughnut-shaped ring of icy rubble that is home to Pluto, and currently being charted by NASA's *New Horizons* probe—or the Oort Cloud.

† We believe Shoemaker-Levy started life in orbit around the Sun, but was captured by Jupiter some two or three decades before it was discovered.

example; and there are arguments that, while Jupiter protects us from comets, it wreaks havoc in the asteroid belt. But by far the most encouraging evidence has come from the Kepler Space Telescope; in short, the size and orbit of the Earth aren't nearly as rare as we might have feared. In fact, they are decidedly run-of-the-mill.

PLAYING WITH A LOADED DECK

It turns out that back in the nineties and noughties, the method we had for finding planets relied on them being big. Called the radial velocity method, it basically relied on detecting planets via gravitational effects. As a large planet orbits a star, it causes that star to wobble, and the wobble affects the frequency of the light that the star gives out. If you analyze the way that the star's light changes frequency you can work out how fast it is wobbling, and once you know that you can then work out the mass of the planet and the size of its orbit.

All very clever, but the only planets you can detect tend to be large ones. Small rocky planets like the Earth don't cause their home stars to wobble, or at least the wobble is so small that it hardly affects the star's light and so is very hard to measure. To find out whether there were any Earthlike planets out there, we needed a new telescope. And in 2009, that's where NASA's Kepler mission came in.

THE MAN IN THE MIRROR

It's truly fitting that the Kepler Space Telescope is named after Johannes Kepler, the seventeenth-century German astronomer. Not only did he invent the modern refracting telescope,* beloved by amateur astronomers the world over, but he also placed an important foundation stone upon which we built our understanding of planetary motion. Using observations made by the Danish astronomer Tycho Brahe, Kepler proposed that the planets' orbits aren't circular, as had been believed since the time of Plato, but elliptical. What's more, he claimed that the planets' speeds aren't constant, as had also been believed since the time of the Ancient Greeks, but vary according to where they are in their orbit. One of the crowning achievements of Newton's *Principia* was that he was able to show that his Law of Universal Gravitation, when applied to the Sun and planets, produced all the effects proposed by Kepler.

The optical telescope that bears Kepler's name is a spectacular beast. As the name suggests, it is far from earthbound; it was launched into space on a three-stage rocket into a heliocentric, or Sun-centered, orbit. The advantages of being up in space are several: For a start, the atmosphere blurs starlight, which is why the stars twinkle, and, secondly, you don't have to wait for it to get dark to be able to use it. The business end points north, so that it never catches the Sun, and looks at a small area of the sky on the edge of the constellation of Cygnus, the Swan. I say small; within that area it monitors the brightness of some 145,000 stars.

Kepler uses what's called a transit method to detect planets. Put simply, it looks at the light coming from a given star, and

* OK, I can't let that go without a few qualifications. Galileo is usually credited with the invention of the refracting telescope, but what he actually did was improve on the 1608 design of the Dutch eyeglass maker Hans Lippershey. Kepler's trick was to use two convex lenses, rather than one convex objective lens and one concave eyepiece lens as used by Lippershey. It's Kepler's 1611 modification that forms the basis of contemporary refracting telescopes, though other advances have also been made, such as use of achromatic lenses (lenses which treat all colors of light the same).

if it sees a dip in brightness, it knows a planet is crossing, or, to use the jargon, transiting. Needless to say, it's very sensitive, as it's looking for something like the drop in brightness of a fruit fly passing across a car headlight. By measuring exactly how much light is blocked, it can work out how big the planet is, and by measuring the time between transits it can work out how large the orbit is and how hot the planet must be.

The results have been astonishing. To date, Kepler has turned up over one thousand planets, with over three thousand prime suspects awaiting confirmation. And what have we discovered? Well, hindsight is a wonderful thing, but in many ways it's obvious: Big and small planets are less common than medium-sized ones. In other words, most planets appear to be between Earth and Neptune in size. Smaller planets like Earth, and big ones like Jupiter, are plentiful, but not quite so abundant.

As a result, we now believe our own solar system is something of an outlier. For a start, Jupiter-sized planets are unusual, and tend to be hot rather than cold like ours. Secondly, we don't have that many medium-sized planets. And, lastly, there's nothing inside Mercury's orbit, whereas many of the systems found by Kepler have planets of all kinds in orbits of ten days or fewer.* That said, the solar system's not exactly rare: We currently think the odds of a cold Jupiter are something like one in a hundred. With 200 billion stars in the Milky Way, that still leaves around two billion planetary systems similar to our own.

* The so-called Grand Tack model links these three facts: Jupiter and Saturn formed closer to the Sun than they are now, at roughly 3 AU (1 AU is the radius of Earth's orbit). The remaining gas in the disk slowed Jupiter down, and it migrated toward the Sun, ejecting the material that would have formed super-Earths. When Jupiter reached roughly 1.5 AU, the gravitational pull of Saturn forced it to change direction. Eventually Jupiter came to rest in its present orbit of 5 AU, with Saturn at 9 AU.

When it comes to Earthlike planets, the Kepler data is also encouraging. The latest estimate is that over one in five Sun-like stars have an Earth-sized planet in their habitable zones, implying the nearest wet rocky planet might be as little as twelve light-years away.* Although Kepler stopped making measurements in 2013, we are still sifting through the mountains of data it produced, and finding more and more small rocky planets in the habitable zones of stars like our very own Sun.

The problem with Kepler, of course, is that to maximize its chances of success it was pointed toward a distant but dense clump of stars. That means that all the Earthlike planets we have found so far are well out of the range of even our most advanced telescopes. Kepler 186f, for example,† is a rocky planet roughly the same size as the Earth orbiting its home star at the right distance to have water on its surface. All right, that home star happens to be an M-type rather than a G-type like our own Sun—a "red dwarf" to you and me—and such stars are prone to scorching solar flares.‡ But Kepler 186f happens to be at the outer edge of its star's habitable zone, just out of harm's way. And that means it may be suitable for life.

* Tantalizingly, there is some evidence that Tau Ceti, the thirty-fifth most distant star from the Sun and our nearest lone Sun-like star, may have an Earth-sized planet in its habitable zone. Tau Ceti just so happens to be twelve light-years away. It's also one of the two stars that Frank Drake targeted in Project Ozma.

† Discovered April 7, 2014. There's an article on NASA's website here: http://www.nasa.gov/ames/kepler/nasas-kepler-discovers-first-earth-size-planet-inthe-habitable-zone-of-another-star/

‡ Stars are classified by color, from blue to red, large to small, O B A F G K M, producing one of the most questionable mnemonics in science, "Oh Be A Fine Girl Kiss Me." The Sun is a G-type star. Stars become increasingly abundant as you move from O to M. Roughly 8 percent of stars are G-type, 12 percent are K-type and 77 percent are M-type. All the other classes make up just 3 percent of the total.

SETI, of course, wasted no time in pointing a radio telescope at Kepler 186f, and searching up and down the dial for anything that looked suspicious. They found nothing. That doesn't mean there were no radio signals being transmitted, because Kepler 186f is nearly 500 light-years away, and to be detectable any transmitter on Kepler 186f would have to be ten times the strength of the Arecibo Radio Telescope here on Earth. Or to put it another way, if a civilization like ours exists on Kepler 186f, the SETI search wouldn't have found it. There might be sentient beings on Kepler 186f right now, uploading the secrets of the universe on to the intergalactic internet; we'll never know.

Happily, NASA's next generation space telescope, the Transiting Exoplanet Survey Satellite (TESS), launches in 2017 and will pick up where Kepler left off, surveying half a million of our nearest stars and hopefully pinpointing thousands of Earthlike planets. No doubt SETI will be quick to target them, but so too will other next-generation space telescopes like the James Webb. Designed to work in the infrared, this fabulous piece of equipment will be perfect for analyzing planetary atmospheres. After all, molecules like oxygen, carbon dioxide, and nitrogen have a distinctive "bar code," emitting and absorbing light at well-defined frequencies across the spectrum. If we know what we're looking for, there's every chance we may be able to detect alien life remotely. Our next-door neighbors may have their lights off and their curtains closed, but telescopes like the James Webb can tell us whether or not they are home. Then Frank Drake will know exactly where to point his telescope.

THE RESTAURANT AT THE BEGINNING OF THE UNIVERSE

As we walk to lunch Mazlan tells me how the media came to refer to her as the Alien Ambassador. "I was due to give a talk at a Royal Society conference about extraterrestrial life. I was going to say that if we do receive signals, the United Nations is the best way to coordinate a response."

We pass UNOOSA's space display, and I am temporarily distracted. There's a beautiful model of the *Shenzhou* spacecraft, which will be the shuttle craft for the Chinese Space Station, together with its *Long March* launcher rocket. There is something vaguely familiar about both—the technology is essentially Russian, after all—but it is remarkable to see Chinese characters on the side of spaceships. What a different world it will be in the 2020s, with the International Space Station decommissioned and only Taikonauts in orbit.

Most of the display items are scale models of satellites; one of UNOOSA's tasks is to provide a registry of all items launched into space, and then help keep track of anything that ends up in orbit. There are many reasons to love satellites, and one of them has been early warnings of climate change: Their data have given us 90 percent confidence that the planet is warming due to carbon dioxide emissions. That said, what most people love about satellites is the money they make. With GPS, television, and the internet all relying on satellite transmissions, virtually every nation on the planet is trying to get a piece of the action.

Only that morning I had read an article in the *Daily Mail*, the gist of which was that while we were pouring aid into Nigeria, they were squandering it on a space race. After feeling suitably furious, it struck me that Nigeria's space race was probably little to do with planting the Nigerian flag on Mars and more to do with satellites. After all, launching satellites is probably the most sensible way of supplying infrastructure to a developing country that you can possibly imagine. I checked online, and surprise, surprise, that is indeed the purpose of Nigeria's space program, with telecommunications and Earth observation satellites bringing internet services, weather-mapping, and food security to one of the fastest-growing populations on the planet.

And, joy of joys, among the display items there's also a moon rock, found by the astronaut James Irwin of *Apollo 15* on the rim of the Spur Crater in the Mare Imbrium, better known as the right eye of the Man in the Moon.* The *mare*, or seas, on the Moon are basically enormous impact basins, formed by collisions with asteroids or comets. These basins then flooded with molten lava, which cooled to form huge flat plains of dark basalt, making them ideal landing sites for the early Apollo missions. It's a sobering thought that the Earth has been similarly disfigured throughout history, though of course thanks to weathering and recycling of its crust via plate tectonics its impact basins have been Botox-ed away.

"So there's nothing in it?" I ask. She smiles, and we continue our walk. "A journalist called after the story first broke. She asked me, 'Are you the alien ambassador?' I said, 'I have to deny it. But it sounds pretty cool.'"

When it comes to messaging aliens, of course, the UN has got form. As we heard in the opening chapter, it's

* To be clear, I am taking his right eye to be the large dark circle on the top left as we look at him, handsome chap. By the way, if you haven't already done so, I heartily recommend downloading the latest version of Google Earth and taking a trip to the Moon.

Secretary-General Kurt Waldheim's voice that opens *Voyager*'s Golden Record, and I think I know an audition speech when I hear one. The way things are going, very soon the peoples of Earth are going to need someone to speak on their behalf. Isn't this the role the UN was born to play? Someone must have thought this through. If the aliens call, surely somewhere among all those reports and resolutions there has to be a protocol? Dr. Othman laughs. "Here at the UN, we simply serve. We don't create protocols unless we are mandated to by our member states."

Suddenly it hits me. There's only one thing worse than the aliens talking to the UN, and that's them talking to just about anyone else. After all, we kind of know how this goes. In 1996, when American scientists in Antarctica thought they had found fossilized bacteria in a Mars meteorite, the first the rest of the world knew about it was when President Clinton announced it on TV.* We need to keep politicians out of it; they'll just hog the glory. The last thing any of us wants to see is a humanoid alien on the White House lawn, hand in hand with Donald Trump.

It's time to put Dr. Othman on the spot. What if an alien ship lands tomorrow? I wince, expecting her to tell me not to be so silly. To my great surprise, she hardly breaks stride.

"It depends where they land. If they land in Mali, they will be the provenance of Mali."

"Really? But what about the UN?"

"If the government of Mali requested that we became involved, we would get involved."

"And if they did make that request?"

* More on the so-called Allan Hills meteorite in Chapter Five.

"Then we would need to get it verified. We could help assemble a team of scientists, and assist in obtaining visas, but that could take a couple of months."

"But SETI have a protocol, don't they, for what to do if a ship lands?"

"There's a SETI protocol, sure. But it has never been adopted [by the UN].* There has never even been a debate."

WILD MEN OF THE MOUNTAINS

The Search for Extraterrestrial Intelligence has always struggled to get taken seriously. First of all there's that acronym, "SETI," which sounds a bit too close to "yeti" for comfort, and immediately puts the reader in mind of the Bigfoot hoaxes involving blurred camerawork and out-of-work actors blundering about in bad costumes made of 1970s shagpile carpet. And then there's the Steven Spielberg movie *E.T.*, in which an alien lands on Earth in a spaceship, forevermore linking the word "extraterrestrial" with UFOs. After all, it's hard to make a serious argument for SETI when your mental image of an alien contact is a wrinkly brown baby-faced midget with a glowing forefinger.

This mistrust is completely undeserved, if you ask me, and not a little unfair. An admittedly subjective sample of the public, based on taxi drivers, people I've sat next to at weddings, and fellow travelers on Network SouthEast rail tells me that people take UFOs very seriously indeed, but are—initially, at

* Sorry, I've always wanted to do that square brackets thing. The UN's position on all things extraterrestrial is to be found in the Outer Space Treaty, a copy of which you can find on the UNOOSA website, here: http://www.unoosa.org/oosa/SpaceLaw/outerspt.html

least—extremely dismissive of SETI. I can't understand it. SETI is conducted by professional astronomers on state-of-the-art radio telescopes, while the search for UFOs is conducted by drunk people on their way home from the pub. The science behind UFOs is nonexistent; the science behind SETI is sound. And SETI had its greatest champion in the gifted astronomer Carl Sagan; UFOs have their strongest advocates in the shape of the Church of Scientology. I rest my case.

As a result of the public's ambivalence, SETI has been a bit of a stop-start affair. The project began with two independent events. Let's have a quick refresher on what those were. Firstly, in 1959 two Cornell physicists, Giuseppe Cocconi and Philip Morrison, published a paper in the journal *Nature*, where they pointed out that the microwave radio band would be a good way for extraterrestrial civilizations to communicate with us, because shorter wavelengths tend to get absorbed by the Earth's atmosphere and longer ones by the gases in the interstellar medium. They put forward the idea that within this band there was one obvious marker frequency emitted by neutral hydrogen, the most common molecule in the universe.*

Frank Drake had independently come to the same conclusion, and in 1960 pointed the large telescope at Green Bank at our two closest Sun-like stars, Tau Ceti and Epsilon Eridani, and tuned his receiver to look for signals close to 1,420MHz. As his receiver had a bandwidth of 100Hz, in the technical jargon we might say that he searched one "channel" of 100Hz.

* The hyperfine transition of hydrogen, used as a measurement of time on the Golden Record. The corresponding "21cm hydrogen line" has a precise wavelength of 21.10611405413cm in free space, which corresponds to a precise frequency of 1420.40575177MHz.

He found nothing, and thereby expanded our knowledge of extraterrestrial civilizations: there wasn't one on either of those two stars. Or, to be more precise, he found no extraterrestrial civilizations that were sending us radio signals close to the hyperfine transition of neutral hydrogen for the 200 hours that he listened in April 1960.

After Frank Drake's promising start, the US initiative in SETI failed to attract US government funding and stalled. Drake and others continued to beg, borrow, and steal time on radio telescopes whenever they could, but NASA was slow to take up the cause. In the Soviet Union, however, where the scientific community had not been weaned on *War of the Worlds* and *John Carter of Mars*, SETI seemed like less of a joke, and more like a subject for serious scientific study. So after a promising start in the West, with the 1959 paper by Cocconi and Morrison, closely followed by Frank Drake's vigilante Project Ozma, most of the running on the theoretical side of things was made by the Soviet Union.

Interestingly, the Soviet take was very different from the American one. Whereas the Americans took it for granted that they would be able to recognize an alien signal, the Soviets weren't nearly so sure. After all, Ancient Egyptian had only been translated with the help of the Rosetta Stone. Even then it had taken twenty years of academic slog, and Ancient Egyptian was a human language. What hope did we have of recognizing an alien signal even if we found one?

RUSSIAN DOLLS

Instead, the Soviets focused on something much more fundamental: energy. Drake's counterpart in Soviet SETI is the

Russian astrophysicist Nikolai Kardashev. In his 1963 paper "Transmission of Information by Extraterrestrial Civilizations," Kardashev classified civilizations by their energy consumption. It was statistically likely, he argued, that most civilizations in the galaxy are much older than our own. In other words, the chances are we've just rocked up at a party that's been swinging for billions of years.

The older a technological civilization is, argued Kardashev, the more advanced it would be, and therefore the more energy it would require. He decided to classify them as one of three types, solely on the basis of their energy consumption. A Type I civilization was one which had harnessed all the energy of its home planet. By this measure, our own civilization isn't quite a Type I, but it's not far off. A Type II civilization had command of all the energy of its home star; a Type III civilization had harnessed the energy output of its home galaxy.

Kardashev's paper said little about what these civilizations would actually be like, but did give a nod to a contemporary paper by the British-born American physicist Freeman Dyson, who had proposed a kind of Type II civilization which has come to be called a Dyson sphere. The gist of Dyson's 1960 paper is this: An advanced extraterrestrial civilization might completely surround its home star with a swarm of artificial structures, to make use of every scrap of light. These structures would effectively enclose the star, completely blocking out most forms of electromagnetic radiation.

Objects like the Kepler Space Telescope, which orbits the Sun, are, of course, the first bricks in a Dyson-sphere-like wall. But if an advanced civilization completely surrounded its home star with light-blocking satellites, how would it show up in our telescopes? Dyson's answer was heat. Heat, of course, is just another part of the electromagnetic spectrum, known as the infrared. So in principle, these sorts of civilizations should show up in infrared space telescopes if they are sensitive enough. Look for a large object radiating a lot of heat but little visible light and you just might find ET.*

So while the Americans were searching nearby Sun-like stars for microwave radio signals near the hydrogen line, the Russians were scouring the skies looking for large objects radiating in the infrared. Of the two strategies, you'd have to say that the Russians seemed to have the edge. A Type II civilization might be difficult to spot with a ground-based infrared telescope, which was all that was available at the time, but what about a Type III civilization? Who could fail to spot something the size of a half-blotted-out galaxy, where the dark bits were radiating strongly in the infrared? The Americans, on the other hand, were searching for a needle in a haystack. And yet it was the Americans who found something first.

THAT'S, LIKE, WOW!

I hope that got your attention. Because if you think that SETI has so far drawn a complete blank, you need to know about the Wow! Signal. It's a bittersweet story, because unfortunately it's the alien version of a kiss on the cheek rather than a committed long-term relationship. But again, I urge you: What follows is not the dramatic story of a UFO "enthusiast," but a measurement that a very real SETI scientist made with a very real scientific instrument.

* Bloody hell, I'm doing it now. Needless to say, I'm not talking about a wrinkly brown alien with a glowing finger, I mean extraterrestrials in general.

That scientist was the astronomer Jerry Ehman, and the instrument was the Big Ear Radio Observatory, a now defunct radio telescope belonging to Ohio State University. In fact, it's more than defunct; it's an eighteen-hole golf course. The Big Ear was a Kraus-type radio telescope, meaning it wasn't the dish type you may be more familiar with, but instead consisted of two huge rectangular reflectors at either end of an enormous aluminum sheet the size of three US football fields. One reflector was flat, the other curved, and they were set up so that incoming radio waves bounced off the flat reflector, on to the curved reflector, and then into a detector. In fact, it's such an ingenious setup I think it's worth drawing, so here it is:

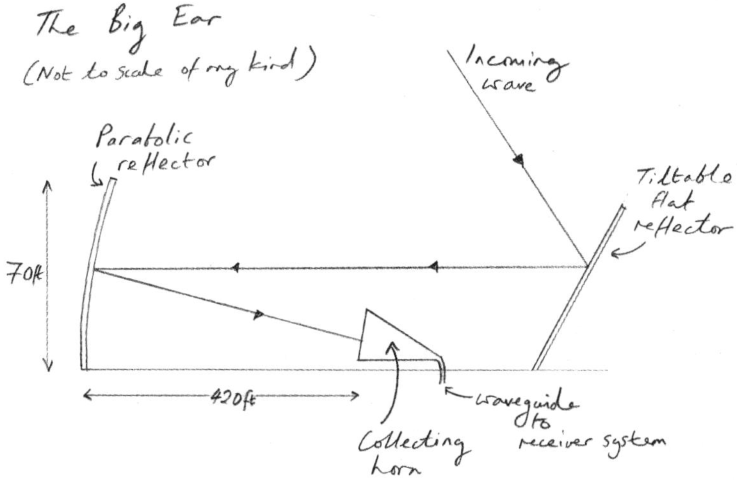

The Big Ear started its working life in 1965, and for twelve years it provided the Earth's most detailed mapping of cosmic radio sources, known as the Ohio Sky Survey. Jerry Ehman had been a big part of that project, joining in 1967 after completing a PhD in astronomy at the University of Michigan. Sadly, in 1972 the Sky Survey had its funding withdrawn and Jerry lost his job.

Unable to secure another research post at Ohio State, he started teaching business classes at nearby Franklin University while continuing to work on the Big Ear as a volunteer.

Rather than see their state-of-the-art telescope go to waste, Jerry and others continued working on it pro bono, rebooting it as a SETI project. Together, they converted the Big Ear from what's known as a wideband instrument to a narrowband one. Most natural radio sources are wideband, meaning that they give off electromagnetic waves with a wide band of frequencies, sometimes from high-frequency x-rays all the way out to low-frequency radio waves. A good example would be quasars. Narrowband sources—no prizes for this—are bunched into a narrow band of frequencies. They are nearly always artificial; the best example would be a radio station.

The flat reflector at the Big Ear would be set into position, then as the Earth rotated it would scan a strip of the entire sky. After twenty-four hours, the angle of the reflector would be minutely adjusted, and the next strip would be scanned. Jerry and his ex-colleagues set up the Big Ear's computer so that it would sample the intensity of the signal arriving in each of its fifty channels, convert it to a number between 0 and 35, then print it out.* Or, rather, a number between blank and Z, because space on the printout was limited and that way two-digit numbers could be printed as a single digit: 10 was A, 11 was B and so on.

On Friday, August 19, 1977, Jerry Ehman sat down at his kitchen table to go through the latest batch of printouts. There

* The observant among you will notice that fifty channels was a big improvement on Frank Drake's single-channel setup. In other words, the Big Ear was searching a much wider band of frequencies around the hydrogen line than Frank's telescope at Green Bank had been sixteen years earlier. What's more, it was searching them remotely.

on the right-hand side of the printout were the coordinates of the patch of sky that the Big Ear had been looking at, together with the time it had made each observation. And down the left-hand side were the signal intensity readings: the usual motley crew of blanks, 1s and 2s as the telescope soaked up the silence of the spheres. But wait. There, in channel 2, was something extraordinary: 6EQUJ5. Ehman took his red pen and circled the six digits, then next to them scrawled a single word: "Wow!"

So what did those digits mean? To you or me they might make an excellent personalized license plate for the presenter of a popular game show,[*] but to Jerry Ehman they described a signal of unusual intensity, some thirty times more intense than the cosmic background radiation, flashing through the detector. Let me convert it to asterisks for you, so you get a feel for it.

Usually, Jerry would see this:

*

*

**

*

But this time he saw this:

******

****

****

****

****

* I am more than ready to admit that's a minority interest gag. I was going for 6E QUJ5 = SEXY QUIZ. *Pointless*, maybe?

*****

And just to remind you, this compares to a maximum possible intensity of Z, or

****

After a bit of number-crunching, Jerry was able to determine that the rising-and-falling shape of the signal was due to the movement of the telescope rather than any change in the intensity of the broadcast, which had remained more or less constant. The detection had lasted only seventy-two seconds, but it was impossible to say with certainty whether the source had been broadcasting for days, months, or even years before that because, as luck would have it, this was the first time that the Big Ear had scanned that particular strip of sky looking for narrowband signals. It was also impossible to tell whether or not the signal was carrying information; it could have been AM, FM, a single frequency switching rapidly or slowly on and off, or not varying at all.*

The most logical explanation seemed to be that an earthbound artificial signal had been bounced back into the telescope somehow, yet all efforts to explain Wow! in this way foundered. There had been no planets in the right position to reflect Earth signals back into the telescope, nor had there been any large asteroids or satellites. Aircraft, spacecraft, and ground-based transmitters were also ruled out. There was no getting away from it: The detected signal was exactly what you would expect from a point source located in the furthermost stars. What's more, it was confined to a very narrow

* Quick refresher: For a signal to carry a substantial amount of information, it has to vary. For AM signals, the information is coded into changes in the signal's amplitude; for FM signals it's coded into changes in frequency.

band of frequencies near the hydrogen line, just as predicted by Cocconi and Morrison all those years before.

Working back from the position data on the printout, Jerry and his team were able to narrow down the source of the signal to two small patches of sky in the constellation of Sagittarius.* As you may know, from our point of view here on Earth, Sagittarius hangs like a net curtain over the center of the Milky Way Galaxy, and behind it lies a clot of distant stars. The same strip of sky was scanned again the following night, but the Big Ear heard nothing. In fact, the team kept the Big Ear in the same position for the next sixty days and nights, but the signal never returned.

So was the Wow! Signal the first message received by mankind from an extraterrestrial civilization? Possibly. For us to be certain, we'd have needed to be able to repeat the observation, but so far all attempts to pick up more signals from that particular corner of the constellation of Sagittarius have proved fruitless. But it's interesting, eh? Suddenly, the idea that we might pick up a signal in the very near future doesn't seem so crazy after all . . .

* Two patches, because the Big Ear had two receiving horns at the focus of its parabolic reflector. The signal had only entered one horn before it cut out, but it was impossible to tell which one. Two horns implied two slightly different patches of sky.

THE REAL JODIE FOSTER

Jill Tarter certainly doesn't think so. Jill is the astronomer that Jodie Foster's character in the movie *Contact* is based on.* From 1995 to 2004 she ran Project Phoenix, a SETI search of our nearest Sun-like stars. As ever with SETI, Phoenix relied on private funding, and had no dedicated radio telescope. Instead, it hopped around the globe, from the Parkes Observatory in New South Wales,† up to the National Radio Astronomy Observatory, Green Bank, West Virginia, and finally to the Arecibo Radio Telescope in Puerto Rico. Rather than scan the whole sky like Big Ear, Phoenix targeted 800 of our nearest Sun-like stars, listening in on billions of narrow channels across a wide range of radio frequencies.

Sadly, at the end of this impressive search the result was nil: not one solitary alien signal. Does this mean we are alone? Not at all. As Jill Tarter put it in a recent interview, "the haystack we are searching in is vast." In fact, let's put some numbers to this. Even if our galaxy contains 10,000 communicable civilizations among 200 billion stars, we'd need to search around ten million stars before we found anything. So far we have searched a paltry 10,000 stars, and we've only really searched for one kind of signal, the kind we would send if it was 1950 and all we had was radio telescopes. So the silence doesn't mean there's nothing out there, it just means we live in the sticks. Who knows? As far as the

* If you haven't seen it, I suggest you put that right immediately. Based on the novel by Carl Sagan, Jodie Foster plays a SETI researcher who picks up an alien signal and Matthew McConaughey. And the most believable of those two things is the aliens.

† Famous for being the main receiving antenna for the Apollo 11 moonwalk, as exaggerated to great comic effect in the movie *The Dish*.

Intergalactic Federation is concerned, maybe we live in a conservation area.

Clearly there is more work to be done, and thanks to a generous donation from the co-founder of Microsoft, Paul Allen, since the late noughties SETI has had its own purpose-built radio telescope. The Allen Telescope Array at Hat Creek Radio Observatory in Northern California represents something of a departure from the norm. Instead of building one large dish, the feeds from a large number of smaller twenty-foot dishes are combined using sophisticated electronics to create a much more versatile instrument, able to survey a wide area of the sky at a wide range of radio frequencies with extremely high resolution. In its current incarnation, the ATA has forty-two dishes; later stages will see it grow to 350 dishes.

The ATA has three very exciting goals. Firstly, it is going to search the nearest million stars for artificial signals over a large chunk of the microwave window, from 1 to 10GHz. So if there are any alien civilizations broadcasting radio waves within 1,000 light-years of the Earth, we are going to find them. Secondly, it is going to do the same thing for all the exoplanets that the Kepler Space Telescope has discovered. And thirdly—get this—it is going to survey the forty billion stars of the inner Galactic Plane for strong artificial signals near the hydrogen line. I don't know about you, but I think it's time the UN got itself a protocol.

FOLLOW THE METHANE

Back at the UN, Mazlan Othman and I have finished our lunch. In fact, we've been talking so long that the waitresses are wondering whether to offer us another meal. We've somehow got on to the topic of what a highly evolved species might look like, and whether they would even be that interested in the physical universe at all.* As Mazlan puts it, why go to the trouble of building a spaceship to visit new worlds when you can just build virtual ones? That's one answer to the Fermi Paradox: The aliens aren't here because they are playing nine-dimensional *Tetris*.

The dream of SETI is to receive an extraterrestrial message that we can understand and respond to. But would it ever really be possible to communicate with species much more evolved than our own? "You and I need this," says Mazlan, knocking on the tabletop. "We rely on the physical world for our existence. When I no longer have need of the material world, what am I? Maybe the purest form that a being can take is energy."

I start to worry that someone might overhear us and call security. Isn't she worried that this kind of loose talk is somehow, you know, unprofessional? "I think it's a missed opportunity when people don't want to talk about aliens. I want to hear what people believe. If it engages people, draws them in, then what have you got to lose? Because then we can ask ourselves, 'what's true?' And then the discussion becomes meaningful."

So what will it take for one of the member states to bring the SETI debate to the UN? What are the hard facts that would close the case for communicable extraterrestrial civilizations? A signal, obviously, would be pretty conclusive. But what about other evidence, like the number of potential habitats? After all, the Kepler Space Telescope has shown us that there are at least as many planets in the Milky Way Galaxy as there are stars. What's more, those planets aren't just found orbiting

* That's certainly a valid question in the case with my nine-year-old son, who spends more time building virtual worlds in Minecraft than he does constructing dens in the garden.

Sun-like stars, they are everywhere: orbiting red dwarf stars, binary stars, pulsars, and even wandering rogue, answering to no one. With so many places that life could take hold, surely the case for other intelligent, technologically advanced civilizations is easily made? What do we need to do to get people's attention?

"You need to find life. Bacterial life." Does she mean on Mars? After all, NASA's *Curiosity* rover has proved that Mars was once habitable. "Maybe. Though Mars is so close to Earth that life could easily have traveled between them." She's got a point. As we shall see in the next chapter, evidence is growing that Mars, not Earth, had the chemical catalysts necessary to kick-start life. Once microbial life got going on Mars, the theory goes, it was then carried to the Earth by a meteorite. If that does turn out to be the case—and that's a very big if—Martian life will turn out to be a branch on the same family tree as Earth life, and we'll all be back to the drawing board. So where else should we be looking?

"I find Titan exciting. There's a methane ocean, with lots of carbon, and that's what we're made of. If we found bacterial life on Titan, we'd really be making the case that life is widespread." Titan, you'll remember, is the largest moon of Saturn, with sandy beaches lapped by methane oceans, and methane clouds scudding across a sky pumped full of nitrogen. Could it be that a completely independent type of microbial life has made its home there?

And if Titan fails us, there's always the icy moons of Jupiter. Beneath a crust of ice, Europa has an ocean of liquid water, and an atmosphere rich in oxygen. What's more, unlike Titan, which currently has no scheduled mission, Europa is going to get a flyby when the European Space Agency's *JUICE* probe launches in 2022. Maybe there are microbes lurking deep down in Europa's hydrothermal vents? Curiosity may have proved that Mars was once habitable, but Europa is habitable right now.

THE HITCHHIKER'S GUIDE TO VIENNA

The art of being a good visitor is knowing when to leave, and every fiber of my being tells me that time was several hours ago. Mazlan asks me what my other business is in Vienna that day, and I give her the honest answer: to buy an ice cream and wander around the Old City. How about her? It is then that my genial host confesses that today is her day off. I am truly impressed. There are few people who would willingly spend time with me when paid, let alone when they could be watching daytime television with their feet up. As she is leaving the office, she says, how about she points me in the right direction, as the Old City is on her way home?

What follows is a tram ride around old Vienna, taking in the Opera House, the Austrian Parliament, and the Natural History Museum, though not necessarily in that order. With my bearings well and truly set, we then wander through the cool green of the Volksgarten, alone but for four silent American tourists riding on Segways. The other side of the Volksgarten gives out on to the Heldenplatz, and there looming in front of us is what looks like four Buckingham Palaces tongue-and-grooved together.

This, I learn, is the Hofburg Palace, one-time stronghold of the Habsburgs, and even today the official residence of the Austrian President. It is also home to the National Library, which is what Mazlan wants to show me. We cross the Heldenplatz, past two fierce statues of military heroes on rearing horses.* Usually it wouldn't strike me as at all odd to celebrate war and learning in the same monument, but in my current mindset I can't help but wonder how all this would come across to the crew of an alien star cruiser.

Mazlan leads me into the library, where we find a riot of marble pillars, detailed frescoes, and Ancient Greek statuary. And it's then that I have that day's second light bulb moment, only this time it's of the slightly less embarrassing, insightful kind. And that insight is this: Culture is key. In the Drake Equation, it's just another number, the length of time for which a civilization is detectable. But culture is more than just the glue that holds a civilization together. Once I can write something down in a book, I no longer have to remember it. Statues tell a story that survives long after the heroes they commemorate have passed on. Culture is our way of storing information outside ourselves; a leap from the physical to the virtual.

I look up at the vaulted ceiling of the Hofburg to find an enormous biblical fresco, and I remember something else Mazlan had said, back when we were discussing what highly evolved beings might be like: "Maybe God is the alien. He is in all of our minds. He is pure consciousness."

* Later I learn that these are Prince Eugene of Savoy and Archduke Charles of Austria, two of the most successful military commanders in Austrian history. It was Prince Eugene who defeated the Turks on behalf of the Habsburgs in the seventeenth century, while Archduke Charles—himself a Habsburg—defeated Napoleon in 1809.

CHAPTER FOUR

UNIVERSES

In which the author brushes up on his astrophysics, builds a pocket cosmos, and discovers that if the universe had been created almost any other way then none of us would be here to witness it.

So a picture is emerging. Since the middle of the last century, the search for aliens has been divided into two camps: the UFO enthusiasts, who definitely aren't scientists; and SETI, who definitely are. No prizes for guessing whose side I'm on. The irony, for me, is that so far the UFO lot have had the best press. Most people, I would hazard, have heard of Roswell while few have heard of the Wow! Signal. So far, SETI may have drawn a blank, but given the paltry number of stars we've searched, that's exactly what you would expect. And thanks to NASA's Kepler Space Telescope, in the future we'll have a much better idea of where to look.

As the discovery of Kepler 186f makes clear, when it comes to SETI it's all about timing. Kepler 186f is 500 light-years away, so for a signal to reach us today not only would it have to have been sent by a transmitter ten times more powerful than any we have on Earth, it would have to have been sent 500 years ago. We know that civilizations with the ability to communicate are possible, because we are here. But to detect our neighbors, they need to transmit a signal at just the right moment in the past for us to intercept it in the present. All this is expressed in the Drake Equation, of course, but there's nothing like a real-world example to bring it into focus.

It's stating the obvious, but the reverse is also true. To message our neighbors, we need to allow our signals time to cross the galaxy. And if we want those signals to reach millions of planets, they need to be strong. Back in the day, our TV was broadcast by powerful transmitters. *Fawlty Towers*, for example, which first aired in 1975, is only just reaching alien worlds thirty-five light-years away. *The Office*, on the other hand, may never arrive. By that point, we were watching on cable, satellite, or DVD, none of which leak much of a signal into space. If you think about it, as far as radio transmissions go, we were only noisy for a handful of decades.*

And even if we are lucky enough to catch some neighboring aliens during their equivalent of our twentieth century, there's no point striking up a conversation if they live a thousand light-years away. No signal—be it radio, laser, or modulated cosmic ray—can ever travel faster than the speed of light, so each time we sent a message, we'd have to wait 2,000 years before we heard anything back. Even if Frank Drake's guess is

* If we ever pick up a stray transmission from an alien world, my guess is that it will be multi-camera with a studio audience.

correct, and the average civilization lasts 10,000 years, we'd just about have time for "Hello, what's the secret of life?" . . . "Thank you very much, goodnight."

To be truly communicable, our neighbors need to be close by. Luckily, they may well be. As we've learned, the Kepler data has shown us that over a fifth of Sun-like stars have Earthlike planets, implying the nearest one might be as little as twelve light-years away. Spookily close, eh? We've been scouring the distant galaxy, and they were in our backyard all along. Maybe our nearest M-type star, Proxima Centauri, is home to an Earthlike planet? If that's the case, then they probably can pick up *The Office*. In fact, one or two of them may even be in it.*

Kepler has given us a good feel for the fraction of Sun-like stars that have Earthlike planets, all of which is grist to the mathematical mill that is the Drake Equation. Now we need to know the fraction of Earthlike planets where life blooms. Thankfully, the extraordinary story of how life on this planet got its start provides vital clues as to what we can expect to find out there in the galaxy. Once again, I have to say you are reading this book at just the right time. Our understanding of Earth-based life has been making extraordinary progress over the past few years; in fact, to such an extent that many prominent astrobiologists are beginning to wonder whether it really is Earth-based at all.

* I always did think the sudden appearance of Ricky Gervais in the UK series was suspicious.

CRAZY LITTLE THING CALLED LIFE

Like all great endeavors, the story of life on Earth is basically a game of two halves. In the first half, a microscopic universe emerges in a Big Bang of impossibly dense energy, then rapidly inflates, then settles down into a slow expansion, eventually cooling to form stars and galaxies.* In the second, the Earth forms, develops an atmosphere, and is bombarded by water-laden comets which create the oceans. Life appears, first as single-celled microbes, then becomes multicellular,† then evolves into complex organisms like you and me. And here's the rub. In the first half, the game is rigged in life's favor. In the second, everything is left to chance.

To see what I mean, let's look at the first half of that story. It's often said that the Earth is a "Goldilocks" planet, because, just like Mommy Bear's porridge, it is neither too hot nor too cold, but just right. "Just right" in the case of the Earth means at the right distance from the Sun to have oceans of liquid water, and in fact we talk about a "habitable zone" as being the band of orbits that a planet can sit in where it is neither too close to its home star so that water vaporizes nor so far away that it freezes.

But it's not just the Earth that is incredibly well suited to life. Kick the tires of creation and almost everything you find

* Actually, for the last billion years or so the expansion has been speeding up, due to a mysterious thing we call dark energy. More on this shortly.

† An organism is multicellular if it is made up of more than one cell stuck together; a chain of bacterial cells would fit the bill, but a colony wouldn't. Multicellularity has evolved independently nearly fifty times, and has been around for at least 1.5 billion years. Complex life, on the other hand—the special kind of multicellular life in which cells are grouped into tissues, and tissues into organs—has been around for roughly 575 million years. Arguably, complex life has evolved six times: in red algae, brown algae, green algae, fungi, plants, and animals.

seems rigged in life's favor. Consider this: The age of the universe, its mass, the rate at which it's expanding, its lumpiness (more on that later), and the relative strength of its four fundamental forces are all incredibly finely tuned. Change any one of them slightly and life would cease to exist. Forget about a Goldilocks planet, we live in a Goldilocks universe.

A LITTLE HISTORY OF THE UNIVERSE

Let's just remind ourselves of what an extraordinary place the universe is by brushing up on our basic cosmology. Firstly, no one knows its true size. It may be infinite, it may be one of an infinite number—we don't know. One thing is certain, however: It has a finite age. The latest data, from Europe's Planck satellite, clocks the cosmos at 13.8 billion years. For comparison, the age of our home galaxy, the Milky Way, is 13.2 billion years; the Earth is a mere stripling at 4.55 billion years.

The fact that the universe has a finite age means that there's a limit to how far out we can see with our telescopes, because this is also the age of the oldest light. At the time of writing, one of the most distant objects known is the galaxy UDFj-39546284, which we are seeing a mere 480 million years after the Big Bang.* The light from UDFj-39546284 has therefore taken over thirteen billion years to reach us, which is even longer than the average pharmacist takes to locate your medicine after you hand them your prescription.

A finite age, of course, also implies that the universe had a beginning. We call that beginning the "Big Bang," a wry

* Discovered in January 2011 by the Hubble Space Telescope. UDFj-39546284 is contained within a famous image called the Hubble Ultra Deep Field, a tiny window of space in the southern constellation of Fornax.

phrase coined by the twentieth-century British astronomer Fred Hoyle, who was himself a proponent of the rival "Steady State" theory. Despite his gentle mockery,* it is the Big Bang which has proved the test of time. The universe is expanding, as we can readily see from the red-shifted light of distant galaxies and supernovae; if we rewind the footage that means it must have originated from a single point. Poetically, cosmologists call such a point a *singularity*, because it operates outside the known laws of physics.

Just so you know, the universe has expanded in four distinct phases. Immediately after the Big Bang came an extremely short, slow, steady expansion. The second phase, called inflation, was extremely rapid, lasting only from 10^{-36} to 10^{-32} of a second.† The third, lasting roughly thirteen billion years, was also slow and steady. The fourth, which has lasted roughly a billion years, has seen the universe's expansion speed up, as if some repulsive force is finally winning out over gravity.‡ This is the epoch in which we have the great fortune to find ourselves.

Some people worry about the Sun running out of fuel, which is predicted to happen in about five billion years' time. Not me. I worry about the runaway expansion of the universe. The fact that the furthermost galaxies are moving faster and faster away from us means that, in a few billion years or so,

* I have to point out that Hoyle himself claimed no ill feeling toward the idea of an expanding universe. You be the judge.

† A word of warning: Some take the words "the Big Bang" and "inflation" to mean the same thing. Not so here. Within these pages, "the Big Bang" means the beginning of the universe, and inflation is what happens next. You've got to have a system.

‡ The evidence comes from distant supernovae, which we can see are accelerating away from us.

they will be moving away from us at the speed of light. At that point, they will disappear from our telescopes, leaving only darkness. Eventually, of course, all but the very closest galaxies will do the same, and one by one the stars will go out, too. Tell me that doesn't give you the most horrendous claustrophobia. If the Sun dies, we can just move to another star, but jumping ship to another universe is a whole other matter.

We have a name for whatever it is that's causing this runaway expansion of space: dark energy. Many cosmologists believe it's a property of space itself, called vacuum energy. In essence, the idea is that space is never really empty, but contains a swarm of so-called virtual particles. As the name suggests, they don't stick around for long. An electron and a positron, for example, will spontaneously appear out of nothingness, have their fun, then annihilate one another within a vanishingly small fraction of a second. Cheeky as this sounds, it's permitted by quantum theory, where you can borrow any amount of energy provided you borrow it for a short enough time. The snag is that the rate of expansion caused by all the virtual particles in the universe should be much, much greater than that we see. In fact, it should be about 10^{120} times greater, a discrepancy which is known fondly in the cosmology community as "the worst prediction in physics."

Secondly, we aren't entirely sure what most of the universe is made of. In a nutshell, we can see that nearby galaxies are spinning too fast to be held together just by normal matter in the form of stars, gas, and dust; we call the missing stuff "dark matter." Dark matter, whatever it may be, exerts a gravitational pull on ordinary matter, but little else. Our best guess is that it is some sort of as-yet-undiscovered particle with no charge and a large mass. The leading candidate is a theoretical particle called a neutralino, though, despite a worldwide effort by thousands of scientists running an impressive array of detectors, no one has yet managed to track one down.*

Whatever dark matter is, it's definitely the boss of you. Incredibly, the ordinary matter of the periodic table makes up only 15 percent of the total matter in the universe; the other 85 percent is dark. There are some outlying theorists who suggest that dark matter is made up not of one but a whole zoo of particles, and is therefore also capable of forming dark black holes, dark planets, and possibly even dark life, but given that we have yet to find even one dark matter particle, the jury's out.

And thirdly, despite an unpromising start in a chaotic molten fireball, both dark and ordinary matter is full of glorious structure. Like wheels within wheels, planets orbit stars, which are grouped into galaxies, which club together to form galaxy clusters, which in turn federate into galactic filaments. Lumpiness now implies lumpiness earlier, and in fact one of the great triumphs of experimental cosmology has been to trace the structure of stars and galaxies right back to the fireball, only 380,000 years after the Big Bang. That needs a word or two of explanation. You'll remember that the microwave background radiation is the remnant of the light from the Big Bang, released when neutral atoms formed around 380,000 years after inflation. At the time it was set free it was extremely high-energy—think of x-rays and gamma rays—but as the universe expanded, it cooled, becoming microwaves.

* If you've heard of supersymmetry, you may be interested to know that the neutralino is the supersymmetric partner of the weak gauge boson. Like other gauge bosons, such as the newly discovered Higgs particle, the neutralino is also its own antiparticle.

Since the microwave background radiation was last in thermal equilibrium with the rest of the universe when neutral atoms formed, it should bear the imprint of any lumpiness in the form of temperature variations. Yet back in October 1988, when I started my cosmology course, every effort to find any temperature variation in the microwave background had returned a negative result. If the lumpiness was there, it was too fine-grained for our best telescopes to spot.

It was NASA's COBE satellite, launched in November 1989, that finally found what we were looking for. The temperature variations in the cosmic microwave background radiation were tiny, but they were there, winning the COBE team the Nobel Prize in Physics in 2006. At last we knew: The fireball had not been smooth, it had been lumpy. What's more, the lumpiness had a specific value of 1 part in 100,000, as predicted by our theories of inflation.*

As always with cosmology, there's a catch. Inflation theory predicts that on a scale greater than that of filaments there's no more lumpiness, a phenomenon that cosmologists rather grandly call "The End of Greatness." Interestingly, no one seems to have told the universe. As we improve our telescopes, bigger and bigger stuff seems to turn up.

The record used to be held by something called the Great Sloan Wall, a filament of galaxies a billion light-years across. Recently, however, astronomers found the first stirrings of something roughly ten times larger, called the Hercules–Corona Borealis Great Wall, or HCB GW for short. And it's not an isolated result. The latest data from the Planck satellite seems to show that there is more matter in one half of the

* The basic idea is that quantum fluctuations in density in the microscopic universe immediately after the Big Bang were amplified by inflation.

universe than the other, which implies there are even bigger objects out there yet to be discovered. It's beginning to feel like every time we increase the volume of space we survey, we find an even larger structure.

Yet impressive as such objects undoubtedly are, it's at the scale of galaxies that the universe holds a truly numinous beauty. Our own discus-shaped Milky Way is a stunning example. A giant swirl of hydrogen gas, dust, and stars circulating an enormous central supermassive black hole, within its spiral arms stars of all sizes are constantly forming as thick clouds of gas and dust collapse under their own gravity.* And it's within stars that atoms are made.

The bigger the star, of course, the hotter and faster it burns, and the bigger the atoms it can make. Smaller stars like our own Sun are capable of making lighter elements, but the biggest stars can make every element in the periodic table, right the way up to uranium. At the end of their lives these giant thermonuclear pressure cookers then explode in what is called a supernova, showering the surrounding gas with freshly minted atoms.

If this aforementioned gas then collapses under gravity to form a new star, some of these atoms then find themselves clumping together under gravity to form asteroids, moons, and planets. Our own Sun is just such an example, burning hydrogen to make helium, surrounded by a belt of rocky planets, then a rocky asteroid belt, then giant gaseous planets, then giant ice planets. It's extraordinary to think it, but all stuff

* Actually it's only a spiral as far as ordinary matter goes. If we could only see dark matter, it would look more like a globe, with the supermassive black hole at its center, and the glorious bright spiral of the Milky Way stretched across the equator.

that surrounds the Sun is a remnant of ancient supernovae. You, this book, and this planet are literally made from stardust.

A POCKET UNIVERSE

All very well, you might think, but what has all this got to do with life? What difference would it make if the universe was an infinite number of years old rather than 13.8 billion? If all we experience from dark matter is the pull of its gravity, couldn't we do without it? Does it really matter how big galaxies are, or whether stars make atoms?

Fascinatingly, the answers are: a lot; not in the slightest; yes, enormously; and you'd better believe it. To see what I mean, we are going to create our very own model universe, and then start messing with it. And if you are willing to accept that the very fact that you are reading this book requires stable stars, an abundance of hydrogen, carbon, nitrogen, oxygen, and maybe a little phosphorus, and enough time for evolution to take place . . . well, then, you are going to be in for a bit of a shock.

GIVE IT A WHIRL

To get a taste of what I mean, think of that most mysterious of forces, gravity. In daily life it feels like a hindrance, pinning us to our beds in the morning, exhausting us at the gym, and flinging each and every delicious slice of toast on to the kitchen floor butter-side down. One of the great joys of the Moon landings was seeing NASA's astronauts roam free, leaping across the lunar surface in giant strides despite being mummified in water-cooled suits and carrying backpacks the size of wardrobes.

Yet compared to the other forces of nature, gravity is extraordinarily weak. A hydrogen atom, for example, is made up of a sole proton orbited by a single electron. There are two forces at work: the electromagnetic force, which attracts the negative charge on the electron to the positive charge on the proton; and the gravitational force, which attracts the heavier proton to the lighter electron. Staggeringly, the gravitational force is roughly 10^{36} times weaker than the electromagnetic force. If gravity is so feeble, perhaps we could do without it altogether?

But consider this: Without gravity, there would be no life anywhere. Why? Because there would be no universe. Since the middle of the twentieth century we've known that all creation began in a Big Bang, inflating out from a minuscule fireball only a fraction of the size of an atom, and cooling into the colossal mix of ordinary matter, dark matter, and dark energy that we see today. In other words, though it started with nothing, today the universe contains a great deal of energy, present in several different forms.

So where did all this energy come from? How do you get something from nothing? Gravity is our best answer. In the broadest of terms, we think that the "positive energy" of all the stuff in the universe is balanced by the "negative energy" of its gravity. The total energy of the universe is therefore zero. For this reason, the originator of inflation theory, the American physicist Alan Guth, has described the universe as the "ultimate free lunch."

So without borrowing energy in the form of gravity, the universe would never have got its start. But that, of itself, doesn't dictate how strong gravity needs to be to produce a universe that is capable of bearing life. And here's the really interesting thing. If gravity were appreciably stronger, life would simply not exist. Intrigued? Then read on, because, quite frankly, that's not even the half of it.

WORKS STRAIGHT OUT OF THE BOX

OK, let's play this game. Let's imagine we are superbeings and we have been given our very own universe for Christmas. It's basically an empty glass box about half a meter across, sitting on a wooden base. It comes with a row of dials, a green button that resets to empty, and a red button that initiates the Big Bang. No need to look for batteries; it doesn't need any, because the universe comes for free.

To begin with, the dials are all pre-set. If we push the red start button, we see the universe emerge in a bright Big Bang, then immediately inflate, then expand and cool to form stars and galaxies. If we are really keen-eyed, we'll notice that the expansion speeds up over the last second of its 13.8 second lifetime.* The whole thing then freezes, preserving the universe as it is at the present moment. If we switch the lights out in our superdwelling, we can see the whole observable universe, with the Virgo supercluster at the center. And though it's much too small to see when viewed on this small scale, at the center of the Virgo supercluster is the Sun, orbited by the Earth.

Right, so let's find the gravity dial. It's there on the far left, next to the dials for the other three fundamental forces:

* In other words, as superbeings we experience a billion years as exactly one second.

electromagnetism, the weak force, and the strong force.* Straight away, we can see that something's up. The dial for the strong force is on a linear scale, and set to 1, but the dials for the other three forces work in powers of ten. The electromagnetic force is set to −2, the weak force to −6, and gravity at a whopping −38.† What happens if we reset with the green button, reduce gravity by a few clicks, then hit the red button to restart the universe?

Amazingly, we see something even more grand and life-friendly. We see the tiny flash of the Big Bang, followed by inflation, just as before. The first stars take a split second longer to form, but when they do they are truly gargantuan. Once again they go supernova, spreading the elements of the periodic table throughout their neighborhoods. Galaxies also form a little more slowly, grow larger, and the stars within them burn for longer. This is especially handy for stable hydrogen-burning stars like our Sun. With three clicks down on the dial, to −41, such stars last for much longer, giving evolution time to do its thing time and time again.

In other words, as far as life goes, lower gravity is good news. But what about an increase? Let's go up four clicks, to −34, hit the red button and see what happens. Fascinatingly, as far as life is concerned, it's a disaster. This time it takes much less matter to form stars, planets, galaxies and black holes. In comparison to the normal universe, galaxies are tiny and form much more quickly, with small stars so close-packed

* Of the four, gravity and electromagnetism are the only two we ever really witness in our everyday lives. The weak and strong forces are much more short-range, and hold domain over the world of the atom.

† Gravity is therefore 10^{36} times weaker than electromagnetism, as mentioned earlier.

that they are constantly stealing one another's planets. And the average lifetime of a star like the Sun decreases, lasting only hundredths of a second. Even if such a star manages to hold on to planets, there won't be nearly enough time for life to emerge. Strong gravity means no life.

KNOTS IN COTTON

OK, so we've toyed with gravity. What happens if we tweak the lumpiness of the early universe? Let's find out by locating another handy dial, Q, that controls the density variations in the fireball. Like the dials for the electromagnetic, weak, and gravitational forces, it's on a logarithmic scale, set to –5.* Let's dial up the magnification on our model, light up the dark matter and hit the slo-mo button, so we can see exactly how this process works.

Firstly, if we really zoom in, we can see that the density variations start life as quantum fluctuations, which then get amplified as the universe is inflated. Over the next 0.4 of a second—corresponding to some 400 million years in the real universe—pockets of dark matter form, growing denser and denser, which hydrogen and helium then fall into. Eventually the ordinary matter becomes so dense that it collapses under its own gravity, lighting up as the universe's very first generation of stars.

And some stars they are, too. Some are hundreds of times the mass of our own Sun, and when they explode as supernovae they shower their own particular corner of the universe with every chemical element imaginable. What's left of their

* In other words, such that the fireball density varies by 1 part in 100,000.

cores then collapses down to form a black hole, and continues to pull a swirl of matter and dark matter toward it. As the black hole starts to become supermassive, that swirl becomes a galaxy.* And galaxies are the nursery beds for generations of stars and planets.

AVENUE Q

OK, so that's the universe we've already won. That's safe. Now let's do what superbeings were put on this superplanet to do, and tweak the lumpiness of the primordial fireball to see how that affects the evolution of the universe. Let's start by dialing Q down, to 1 part in 1,000,000, and see what effect it has.

The difference is striking. After we press the red button, we see the flash of the Big Bang, then inflation, then darkness. The fireball is too fine-grained for any part of it to collapse under gravity. The expansion wins out, and once again we end up with a thin gruel of hydrogen, helium, and dark matter particles. There are no stars, no planets, and no galaxies. Once again, the universe is a no-go zone for life.

What about increasing Q? Surely that will make glorious giant galaxies, like when we reduced gravity? Sadly not. If we dial Q up to −4, reset, and push start, before a second is up, huge regions of dark matter begin to clump together, precipitating enormous clumps of gas, more massive than

* Cosmologists call these first stars the Population III stars. Obviously it would have made more sense to name them the Population I stars, but the stellar populations are named in the order that they were discovered rather than the order in which they arrived on the scene. I say discovered; we have found Population I stars (young and metal-rich, e.g., the Sun) and Population II stars (old and metal-poor, e.g., the stars in the bulge at the centre of the Milky Way) but for the time being Population III stars are theoretical only; everyone has their fingers crossed that one of them turns up in the James Webb.

entire galaxy clusters, that quickly collapse under gravity to form truly forbidding black holes. That's not a reality any of us wants to live in.

THE STRONG FORCE

By now, I'm sure you're getting the picture, but before we move on there's one more tire I want to kick: the strong force. Without it, of course, there would be no atoms, because its job is to bind protons together to make atomic nuclei. You'll remember that this happens in two places: firstly, in the Big Bang, where the nuclei of lighter elements like lithium and helium get made; and secondly, in big stars, which make all the rest.

That sounds pretty straightforward, and you might think that so long as the strong force is attractive, and can be felt when protons get within a short range of one another, all would be well. But there's a complication. Protons also carry a positive charge, which tends to push them apart. At the same time that the strong force pulls together all the protons in a nucleus, the electromagnetic force is pushing them apart. In fact, one of the puzzles of the early twentieth century was how atomic nuclei managed to exist at all.

"Very well," you might say, "let's just make the strong force between two protons in a nucleus stronger than the electromagnetic force between them." But that's not how the problem is solved either. As it turns out, the strong force isn't quite up to the job of holding two protons together in a nucleus; instead, the electromagnetic force wins and the two protons are forced apart. So how do atomic nuclei remain stable?

The solution, ingeniously, comes in the form of a neutral particle with roughly the same mass as a proton, which also feels the strong force, but doesn't have any charge. No doubt you'll remember that such particles are called neutrons. These gentle peacemakers are the diplomatic glue that holds atomic nuclei together. They also feel the strong force, but they don't have any charge, so aren't repelled by the positive charge on the proton. A deuterium nucleus, for example, which is made up of a proton and a neutron, is stable, as is a helium nucleus, which has two protons and two neutrons.*

FAITES VOS JEUX

Right. With all that in mind, let's twiddle the dial for the strong force on our model universe and see what happens. First of all, let's dial it up a bit. Surely, if we increase the attraction between protons and neutrons, making nuclei will get easier? And with more elements knocking around the periodic table, there will be even more chemistry and hence a greater abundance of life?

I'm afraid not. Let's increase the strong force by a tenth, to 1.1 on the dial, and press the red button. Once again, we see the flash of the Big Bang, and half a second later the Dark Ages end, as the first giant stars burst into life and go supernova. Galaxies form just as before, but we notice something odd about the stars within them. They have exceedingly short lifetimes; in the compressed time of our model universe, they

* You'll remember that we call the number of protons in a nucleus the atomic number, Z. The total number of protons and neutrons is called the mass number, A. Atoms with the same number of protons but different numbers of neutrons are called isotopes. Protium (P), deuterium (D), and tritium (T), for example, are all stable isotopes of hydrogen with zero, one, and two neutrons respectively. As a shorthand, we tend to let the name of the element stand in for its atomic number, and specify the isotope using the mass number. ^{3}He therefore has two protons and one neutron; ^{4}He has two protons and two neutrons.

last maybe a millisecond or so. Once again, there's not nearly long enough for evolution to produce life. And even worse, if we zoom in on the asteroids and planets, we see that none of them have any water. What's going on?

It was all there in the Big Bang, but we missed it. Crucially, increasing the strong force has meant that all the protons fused into pairs, forming a new helium isotope with no neutrons, ^{2}He. In other words, we have made a universe with no hydrogen. It's the hydrogen burning phase in stars that takes time, so they burn much quicker. And without hydrogen to react with oxygen, there's no water. Maybe that doesn't rule out all life-forms, but it certainly rules out everything that has ever lived on Earth.

So what happens if we decrease the strong force to, say, 0.9 on the dial? Again, once we push the red button, we get a shock. We see the flash of the Big Bang and the short, furious burn of the first stars, but they fade without going supernova. Galaxies form as before, but again their stars burn only briefly before fading into darkness. All stars have short lifetimes, big, medium, and small alike, and, even worse, without material from supernovae there are no planets. We have made another universe that is barren of life. What went wrong?

The answer, again, was in the intense heat and pressure of the primordial fireball. A decrease in the strength of the strong force has meant that protons no longer fused with neutrons to make the hydrogen isotope deuterium (^{2}D), a vital stepping stone in the synthesis of ^{4}He. Without deuterium to pave the way, we effectively turned off nuclear fusion like a tap. There are no other atoms except hydrogen, no chemistry, and therefore no life. Stars still light up, heated by gravitation, but without nuclear fusion reactions in their cores, they quickly lose energy as heat and cool into black holes.

Sobering, eh? We can't tweak the strong force by more than a tenth in either direction without nixing the periodic table, chemistry, and life. Since the strong force competes with the electromagnetic force, that implies we can't tweak the electromagnetic force by much either without having the same effect. Why? Reduce the electromagnetic force, and ^{2}He will be stable, removing all hydrogen from the Big Bang. Increase it, and the Big Bang makes no deuterium, and hydrogen is all we get.

A PUT-UP JOB

And if none of that convinces you, try this extraordinary fact, first pointed out by Fred Hoyle: Every element heavier than helium should be rare in the universe. The reason is that there's another roadblock early on in the fusion process that goes on inside stars. The easiest way to make larger and larger nuclei is to start with a helium nucleus, say, and keep adding a proton; however, there's a problem. If we add a proton to a helium nucleus, we get lithium-5, which rapidly decays. If we squash two helium nuclei together, we get beryllium-8, which is also unstable. That means there's no easy way to make boron-10, which has five protons, or carbon-12, which has six protons, or oxygen-16, which has eight. Yet carbon is the next most abundant element after hydrogen and helium, and oxygen is the next most common. In fact, apart from a scarcity in lithium, beryllium, and boron, elements all the way up to iron are commonplace. So what's going on?

One possible route to making carbon is for three helium nuclei to fuse together, in what is called the "triple alpha" process. But there's a problem. The first step is for two helium nuclei to come together to make beryllium-8, which is unstable. The chance of a third helium nucleus colliding with beryllium-8 in time to make carbon-12 is therefore vanishingly slim. Hoyle's genius was to suggest that there might be a fortuitous resonance, in the form of an excited state of carbon, which exactly matched the energy of the beryllium-8 nucleus when capturing a helium nucleus. The experimentalists went to work, and, sure enough, exactly just such a state was discovered.

Without Hoyle's resonance, next to no carbon would be made in stars, and—since to make an oxygen nucleus all you need is for a carbon nucleus to capture another helium nucleus—next to no oxygen. Give or take the odd rogue metal ion, all life on Earth is essentially made up of just a few elements: carbon, hydrogen, nitrogen, oxygen, phosphorus, and sulfur. Can it be coincidence that these exact elements happen to be the most abundant in the universe? A change of a fraction of a percent in the strong force, and just a few percent in the charge on a proton, would cause the Hoyle resonance to disappear. And then where would we be? Literally nowhere.

AN ACCIDENT OF HISTORY

So how do we explain this fine-tuning? Is it written into the laws of physics? So far as we know, it's not. The present laws of physics are a blank slate. They tell us the relationship between physical quantities, but not their absolute values. Newton's Law of Gravitation, for example, tells us how the gravity of an object varies with distance. Double the distance, it says, and you get a quarter of the gravity. But how much gravity does a mass of one kilogram have at a distance of one meter? Newton can't tell you. To work that out, someone somewhere has to do an experiment, make some measurements, figure out the strength of gravity per unit of mass and distance, and then plug it into Newton's equation.

And it's the same throughout physics. The Standard Model doesn't predict the masses of the fundamental particles, for example, which is why the Higgs gave us the runaround. We only know their masses from experiment. With the partial information we had, we were able to put an upper limit on the mass of the Higgs, but that was about it. Likewise, General Relativity in cosmology doesn't tell us the strength of gravity or the size of the cosmological constant. We have to build ourselves a telescope and have a look.

The truth is that over the past two centuries, experimental scientists have done an extraordinary job of keeping the theorists in business by making incredibly accurate measurements of all the constants that we find in nature. Don't think for a second that any of the theorists like this situation. The grail they chase is a theory that will, of its own sweet accord, predict the universe that we find ourselves in. This elusive Theory Of Everything (TOE), as it is known, would make all experimentalists redundant overnight. Every detail of the universe—the mass of dark matter particles, the strength of gravity, the charge on an electron—would pop out like a rabbit from a hat. Instead of using quantum physics here and relativity there, everything would be a special case of the TOE. Cosmology and particle physics would be a job well done.

Big deal, you might say. We're just not there yet. It's not surprising our theories don't describe the universe exactly as we find it; after all, we haven't got the full picture. Our "laws" are really all special cases, like Kepler's Laws of Planetary Motion turned out to be special cases of Newton's Law of Gravitation, which are, in turn, a special case of General Relativity. Once we have a truly fundamental theory, we'll have a truly detailed description. When we do, General Relativity will prove to be a special case of some as yet undiscovered Theory Of Everything, and we'll be able to predict the rate of expansion of the universe before we measure it, not fudge it after.

But there's an alternate argument. It says that the reason we can't predict the fundamental constants is that they are a result of the universe's history rather than a feature of any fundamental theory. The universe wound up the way it is by chance, not design. Why should we be surprised that gravity is weak, when if gravity were strong we wouldn't be here? Maybe there are other universes out there where gravity is strong; they just don't have people in them. The universe is fine-tuned for life because, if it weren't, we wouldn't be here.

THE MAN WHO BROKE THE BANK

The idea that the universe is fine-tuned for life is called the anthropic principle. The fundamental constants are what they are because otherwise we wouldn't be here to ponder them. By sheer chance the strong force can overcome the electromagnetic force to create atoms, and gravity is weak, so those atoms form stars and planets. By remote accident, the atoms on those planets happen to have interesting chemistry, and the environment on at least one of those planets kick-started the very special set of chemical reactions we call Lady Gaga. We lucked out, simple as that.

Some would leave it there, content to believe that the universe is a lottery, and our numbers just happened to come up. Others—and I am one of them—would say that's not a satisfying answer. To me, a freak fine-tuned universe begs further explanation. How many other tickets were there in the lottery? How many other universes failed to create life before ours succeeded?

The philosophers among you will no doubt protest that I am falling for one of the oldest logical faux pas in the book, a version of what is known as the Monte Carlo fallacy, after a famous Monaco gambling incident. Purportedly, on August 13, 1931, the roulette wheel at the Grand Casino in Monte Carlo landed on black twenty-six times in a row. Did anyone make any money? Yes, the casino did. Because after the wheel had landed black around fifteen times in a row, everyone started putting all their money on red. Time after time they piled high their chips, and time after time the wheel came up black.

So what was the flaw in the punters' logic? Simply that the roulette wheel somehow "knew" how many times it had come up black, so was bound to come up red. Of course, the wheel knew no such thing, which is why everyone took a very deep bath. The lesson, of course, is that truly random processes know nothing of their own history.

Take lotteries, for example. The version in the UK asks you to choose six numbers between 1 and 49. After a short burst of excruciating live entertainment, the numbers are drawn live on television. The chances of matching all six numbers and winning the jackpot are 1 in 13,983,816.* Are the odds better if you've played a great many times before and lost? Your cheating gambling heart might say "yes," but your cool, probabilistic head definitely says "no." Guinevere (one of the UK's number-picking lottery machines) doesn't know how many times

* Chance of picking the right number = (chance of picking right ball) x (number of ways you can arrange six numbers) = $(6 \times 5 \times 4 \times 3 \times 2 \times 1) \times 1/(49 \times 48 \times 47 \times 46 \times 45 \times 43) = 1/13{,}983{,}816$ or 13,983,816 to 1.

you've played, or what your lucky numbers are. You could just as well win at your first attempt as at your nine hundred and ninety-ninth.* Win the lottery one week, and you've got just as much—or as little—chance of winning it the next.

By analogy, then, am I committing the Monte Carlo fallacy when I say that a rare event like a fine-tuned universe implies lots of previous attempts which weren't fine-tuned? Isn't that just like saying, "Hank just won the lottery. He must have been playing for years." At the risk of answering my own rhetorical question, yes, I think it is the same. Just because Hank's numbers came up once doesn't mean he's played before and lost. But, then again, it doesn't rule it out either. If we raid Hank's house on a hunch, and find a cupboard full of scrunched-up lottery tickets, we wouldn't be that surprised.

And if you ask me, that's exactly what we have found with the inflation of the universe. Most, but not all, theories of inflation imply that the universe we find ourselves in is just a small part of a much bigger whole. Maybe after the Big Bang, some regions of space-time inflated more rapidly than others, becoming isolated bubbles. Maybe what we call the universe is just inside one of those bubbles, and out there in Never-Never-Land are countless others. Our universe happens to be fine-tuned for life; maybe lots of the others aren't. It's time to meet the multiverse.

* She doesn't know what your numbers are either. I'm not saying that having lucky numbers isn't fun—of course it is. But 1, 2, 3, 4, and 5 have just as much chance of winning as 5, 7, 24, 25, and 30 (our family birthdays). Come to think of it, I'm off to the corner shop.

THE UNIVERSE'S STRETCH MARKS

As Carl Sagan said, extraordinary claims require extraordinary evidence, and, as far as the multiverse goes, we don't quite have the quality of proof that one might hope for. Nevertheless, we are making progress. We already had circumstantial evidence for inflation—and, by proxy, the multiverse—in the exact size of the temperature variations in the cosmic microwave background (CMB), which the theory manages to predict with impressive accuracy. The real clincher, however, would be the detection of primordial gravitational waves.

In order to understand what gravitational waves are and why they might provide evidence for inflation, I am going to have the pleasure of introducing you to one of my favorite bits of physics. In essence, it's this: Acceleration makes waves. Perhaps the most obvious example is sound, where the acceleration of a solid object produces sound waves. But the same principle is at work in field theories like electromagnetism, where an accelerating electric charge radiates photons, also known as light.

Einstein's Theory of General Relativity describes how matter and energy relate to gravitation, so it fits the picture that an accelerating mass radiates gravitational energy in the form of a ripple in space-time called a gravitational wave. That supernova on the far side of the galaxy, that pair of neutron stars orbiting one another in deep space, the tennis ball in the men's final at Wimbledon—all of them create gravitational waves. Those waves spread out through the cosmos, stretching and compressing space and time like the ripples on a pond. Eventually, they will pass through you, and your space and time will wobble. Only a bit, of course. You will shrink and stretch in height, but imperceptibly. Your watch will run fast and slow, but by such a minute amount as to be unnoticeable. And then the wave will pass, and your space and your time will be still again.

That's the theory anyway. Relativity tells us that in most cases the wobbling of space and time caused by gravitational waves will be so small as to be undetectable. To see the effects, you need to accelerate something really big. In the case of inflation, that something is the universe. During inflation, the universe expanded faster than the speed of light.* And when you accelerate something as big as the universe as much as that, you are going to get some pretty big gravitational waves. It's the after-effects of these waves that we are currently searching for.

FLEXING OUR BICEP

Theory predicts that when the universe inflated, the gravitational waves that were produced should have left an imprint on the cosmic microwave background. To cut a long story short, they should have polarized it, leaving a distinctive pattern that a telescope in Antarctica named the Background Imaging of Cosmic Extragalactic Polarization (BICEP II) has been built to detect.†

* I know what you're thinking: Nothing can travel faster than the speed of light. Actually, that's not quite true. The more accurate statement is that no signal can travel faster than the speed of light. That leaves the expansion of space to do what it pleases.

† BICEP II is a refracting telescope, and is based at the Amundsen–Scott Station in Antarctica.

In 2013 everyone got very excited when the BICEP team thought they'd cracked it, but the result has since been shown to be a false positive caused by cosmic dust. After a bit of a rethink and a chunky upgrade, they are expecting to publish more results in 2016. If they are successful, everyone will have to start taking inflation and the multiverse a lot more seriously.

There's a beautiful circularity to that. At the time of the Ancient Egyptians, we believed that mankind was special, and that the Earth stood at the center of the cosmos. We call this the Ptolemaic Model, after the Greco-Egyptian astronomer Ptolemy,* who first formalized it. Copernicus then demoted us to a middle-ranking orbit around the Sun with his heliocentric Copernican Model. Einstein's General Relativity taught us that even the Sun wasn't that big a deal; no part of space-time is privileged over any other. And now the entire universe turns out to be just one ticket in a colossal lottery called the multiverse.

The kicker is that makes us special after all. Or, to be more precise, our universe is special because it's the one where the very fabric of existence is just right for life. When we go looking for our cousins in the cosmos, we should bear that in mind. Because even before we came along, the universe was trying for a baby.

* Although they all hailed from Alexandria, Ptolemy the astronomer lived after, and was no relation to, Ptolemys I–XV, the last pharaohs of Ancient Egypt. Ptolemy's dates are c. AD 100–170; those of the Ptolemaic dynasty are 305–30 BC. As I say, more about the Ancient Egyptians coming up.

CHAPTER FIVE

LIFE

In which the author's attempt to grasp exactly what makes life special causes him to fall through a mathematical rabbit hole, only to emerge in a wonderland where everything is valued not by its beauty, but by the rate at which it dissipates energy.

On August 7, 1996, President Clinton announced to the world's media that NASA had found evidence of life on Mars. A team led by David McKay, the geologist who had trained the Apollo astronauts how to search for moon rocks, had found fossilized bacteria in a Martian meteorite. "If this discovery is confirmed," Clinton intoned, "it will surely be one of the most stunning insights into our universe that science has ever uncovered."

Sadly, it wasn't confirmed. McKay's discovery was soon submerged in a blizzard of critical academic papers. The main objection was to the fossils' size. The largest among them were only 100nm in diameter, whereas the smallest bacteria known at the time were nearly ten times that size.* Some critics pointed out that's too small to hold enough DNA for them to replicate. You might ask why Martian bacteria would necessarily contain DNA, but that's a whole other question.

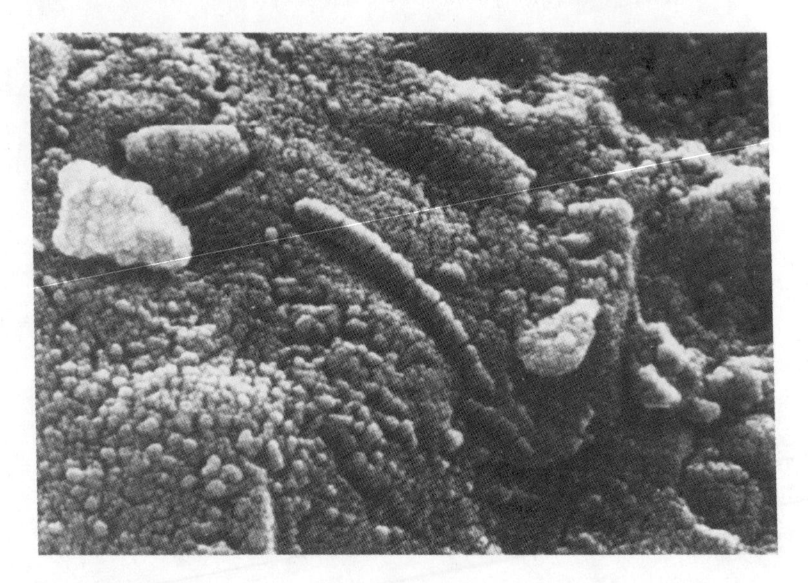

Take a look at the photograph above, reproduced from the McKay team's paper in the journal Nature. I hope to convince you that the sausage-shaped "bacteria" in meteorite ALH84001 may indeed be microscopic aliens. Even if I fail, the whole question of how we decide whether or not something is, or has ever been, alive is right at the heart of our quest to find life on other planets. If we hope to find life elsewhere, we have to know exactly what we're looking for.

* A nanometer is a billionth of a meter, or if you like, a millionth of a millimeter. The nano scale is really the doorstep to the world of atoms, which tend to be typically around a tenth of a nanometer in size, also known as an angstrom (Å). To put it mathematically, 1Å = 10-10m, 1nm = 10-9m.

BOLT FROM THE BLUE

A meteorite, of course, is simply a piece of rock which has come to Earth from space. All kinds of rubble from asteroids and comets are constantly hitting the upper atmosphere, but most of it burns up long before it reaches the ground. The result is a shooting star, or meteor. A classic example is the Perseid meteor shower, which happens every August in the northern hemisphere as the Earth trundles its way through the tail of the comet Swift–Tuttle, and very beautiful it is, too.

Occasionally, however, we get lucky—or unlucky—and a particularly large lump of space rock will touch down somewhere on our beautiful blue planet. And there's no limit, really, on how big such a rock might be. As we shall soon learn, the early solar system was buzzing with comets and asteroids, and the Earth took a severe battering from its birth right up until around 3.9 billion years ago.*

Things have calmed down a lot since, but now and then we still get a haircut. The last really big space rock to do the business hit us sixty-six million years ago,† wiping out the dinosaurs. Five thousand years ago an asteroid struck the Indian Ocean, causing a worldwide tsunami and quite possibly founding the myth of Noah's flood. And as recently as 1908, a comet exploded in the skies above Tunguska, Siberia,

* Some of those comets and asteroids were laden with water, which is where we believe the Earth's oceans came from. They also brought enormous quantities of carbon compounds, which is where we come from. The worst of the pummeling came at the end, and is known as the Late Heavy Bombardment. We don't know exactly what caused this final flurry, but one of the more convincing explanations is that of the Grand Tack model, which claims that the giant gas planets Jupiter and Saturn hit a resonance and changed their orbits, pulling the gravitational rug out from under the toddler solar system.

† The latest data has revised this figure; it used to be sixty-five million years ago.

with a thousand times the energy of the atomic bomb that was dropped on Hiroshima.

There's nothing that uncommon, in other words, about meteorites per se. What made ALH84001 so unusual is that it came from Mars. As you might have gathered, our atmosphere actually protects us from space rocks, as they tend to burn up in it due to friction. Other planets with thinner atmospheres get a rougher deal. When a stray asteroid or comet collides with Mars, the red planet takes it right on the chin. If the impact is big enough, lumps of Martian crust get billiarded out into space. If that debris escapes Mars's gravity, it can find itself on a collision course with the Earth. And in the case of our hero meteorite, the exact location of that collision was the Antarctic.

HOLES IN SNOW

The Antarctic is a meteorite hunter's heaven. One rather straightforward reason is that they show up really well against the snow. Another is that, generally speaking, ice piles up in the middle of Antarctica, and then flows down and out to its coasts, where it gets trapped at the feet of mountain ranges. Anything that falls out of the sky into the middle of the continent therefore ends up on a kind of slow-moving conveyor belt to the bottom of the nearest mountain, ready to be collected by some hardy geologist.

As a result, every summer in the southern hemisphere the USA's National Science Foundation supports the Antarctic Search for Meteorites, or ANSMET. Round about December, a small research team heads out from a base camp near the foot of the Transantarctic Mountains to hunt for meteorites. The rock that the McKay team studied had been found by one

Bobbie Score during a snowmobile ride on December 27, 1984, in the Allan Hills. As the first sample found that year, the rock had been labeled ALH84001.

To begin with, ALH84001 was assumed to be the remains of a common or garden asteroid, but by 1993 it was resident at the Johnson Space Center where it was identified as a piece of Mars's crust. It was at this point that it piqued David McKay's interest. Tests showed that it was an astonishing 4.5 billion years old, having formed on Mars just a squeak after the solar system itself. Sixteen million years ago some sort of impact cannoned it up and out of Mars's gravitational field,* where it wandered the solar system before rubbing up against the Earth's atmosphere and crash-landing in the Antarctic around 11,000 years ago.

The rock itself was riddled with cracks, having suffered some sort of impact around four billion years ago while it was still on the surface of Mars. And here was the really exciting bit. Lodged in the cracks were granules of calcium carbonate. To a geologist, calcium carbonate means water,† and water, as we all know, means life. Even more intriguingly, calcium carbonate granules of a similar size and shape to those in the meteorite were known to be produced by certain kinds of terrestrial bacteria.

"So what?" you may say, "the granules probably formed on Earth during the 11,000 years that the rocks had been lying in the Antarctic snow." But when the team dated the granules,

* To give you some evolutionary perspective, the great apes, or hominids, split from primates around fifteen million years ago, one million years after ALH84001 left the surface of Mars.

† Whenever water rich in minerals absorbs carbon dioxide, calcium carbonate is never far away. In fact you probably last saw some as limescale on the heating element of your kettle.

they found that they were billions of years old, not tens of thousands. There was no doubt about it; they had formed on Mars. But how? The McKay team decided to put their pet rock under an electron microscope to see if they could find out.

It was then that they saw the thousands of sausage-like shapes clinging to the granules of calcium carbonate. Have another look at that photograph, and tell me again whether you think they look like fossils of living things. If you think they do look like they were once alive, what is it about them that convinces you? And if you don't, why not?

WHAT IS LIFE?

We humans are expert at spotting life. NASA's recent Discovery mission to Mars was exhilarating, but there is something infinitely frustrating about a robotic probe scratching around for water in dusty soil when there could be purple alien bacteria crawling all over the rocks of a nearby mountain stream. Stick any microbiologist on the Red Planet for a day, and he'll be able to tell you whether there's anything living there or not. So what is it that we look for? How do we identify something as being alive?

One thing we look for is movement. Life gets around, be it a bacterium waggling its flagellum, or a flying squirrel spreading its paws and taking to the wind. As a physicist might say, living things do work, meaning they are capable of converting chemical energy into mechanical energy. The bacterium beats the surrounding liquid with its flagellum; the flying squirrel pushes down on the air to create uplift. Non-living things can do work, too, but unless they are man-made, they don't have an internal energy supply. Kick a football, and it starts to slow down the instant it leaves your foot. Kick a kangaroo and you may well end up on the other side of the nearest hedge.

And living things grow. If you're out for a Mars walk and you see a large black rock, and the next day it's twice the size, it's time to take some photos. And if the day after that the large black rock has a little black baby rock sitting next to it, you've really struck the motherlode, because the special type of growth that we call reproduction is another tell-tale sign of life. In fact, some would say it's a defining sign, because without it natural selection would have no way to work, and there would be no evolution of species.

Doing mechanical work, growing, and reproducing are all things that life does, but it's interesting to note that they don't define it. An avalanche is capable of doing work, even though that work might be flattening ski chalets. Likewise, rust can grow on a chain-link fence, but we wouldn't keep it in a jar as a pet. Simon Cowell was unable to reproduce without the assistance of Lauren Silverman; nevertheless he is most definitely alive. And it's not hard to imagine some highly evolved creature of the future that decides it doesn't want to die, thanks very much, uploads itself into some sort of virtual reality, and does away with natural selection altogether.* What is it that life does that non-life doesn't?

The thing about fossils, of course, is they don't move, grow, or reproduce. If David McKay's team had been really lucky, one of ALH84001's bacteria might have been preserved in the act of cell division, but no such luck. Neither were any of them spotted at different stages of growth, or flexing minuscule flagella. No, what makes the tiny sausage-shaped structures in

* For some reason I am still picturing Simon Cowell.

the Allan Hills meteorite look lifelike is a truly fundamental property of all living things: They are organized.

In organizing themselves, living things go against the grain of the entire universe. As we are about to see, the cosmos seeks one thing, and one thing only: equilibrium. The natural world doesn't want you to play backgammon, or learn to salsa, or even exist. It wants equal temperature, maximum disorder, and death. Eventually it will get its way, but hopefully not before you've finished reading this chapter, and come to grips with one of the most mind-expanding principles in the whole of science: the Second Law of Thermodynamics.

THE RULES OF THE GAME

The concept of energy occupies hallowed ground in physics. I don't think it's too much of an exaggeration to say that it underpins every other physical theory we have, and that goes for both Quantum Mechanics and General Relativity. It is enshrined in the four Laws of Thermodynamics, which can be roughly summarized as follows:*

(0) If two different objects are in thermal equilibrium with a third object, then they will also be in thermal equilibrium with each other.

(1) The total energy of the universe remains constant.

(2) The entropy of the universe always increases.†

(3) The entropy of an object approaches zero as its temperature approaches zero.

* I know, I know; although there are four laws, they are labeled zero to three. The problem was that the First, Second, and Third Laws were already well known before it was quite rightly pointed out that the Zeroth Law—which defines temperature—was fundamental to all of them.

† We'll get to entropy in a minute.

Everything that we can imagine an extraterrestrial life-form might do—be it thinking, communicating, growing, moving, or reproducing—requires energy. Although there's a slightly forbidding ring to their name, the Laws of Thermodynamics are really just a very simple set of rules for the way that energy works. They are all directly relevant to life, but it's the Second Law that has the most surprising implications for living things. As we are about to see, the reason you ate breakfast was not just because you needed a source of fuel. You ate it for its information content.

DOWN WITH EQUILIBRIUM

Balance, in psychological terms at least, sounds like a wonderful thing. In fact I have often experienced it myself, if only when swinging from high elation to severe depression. In people, it is generally and justifiably admired. Who doesn't seek to be stoic, good-humored, and harmonious of spirit? The Buddha, one imagines, didn't get into bar fights, and Buddhism would be a less impressive religion if he had. But while equanimity is an enviable quality in human beings, out there in the universe at large it is very bad news indeed.

The universe, I am sorry to tell you, is not a fan of yours for one very simple reason: It doesn't like hot things and cold things. Rather, it infinitely prefers something else: thermal equilibrium. You probably already have a rough idea of what that means, but, as a refresher, let's imagine putting something hot—a cup of hot tea, say—next to something cold, like a saucer. What happens next? No prizes for guessing that heat flows from the cup of hot tea to the saucer. Eventually, the cup and saucer will reach the temperature of their surroundings, and no more heat will flow.* We say then that they are in thermal equilibrium. This, of course, is what the Zeroth Law of Thermodynamics is telling us; thermal equilibrium is what results when two different objects have the same temperature.†

As you may know, this type of heat flow—from one body to another in direct contact—is called conduction. As far as a cup and saucer goes, the process is slow for the simple reason that the very thing they are both made out of, china, has been chosen because it is so bad at conducting heat. Our ancestors, in their wisdom, wanted their tea hot and chose their crockery accordingly. But that's not the end of the story. A hot cup of tea also cools by another method, known as convection. Basically the hot tea and teacup heat the cold air around them, and the hot air rises, only to be replaced by more cold air. Eventually the tea reaches the same temperature as the surrounding air, in which case—if you are my wife—at this point you decide to try and drink it.

Conducting and convecting heat is the kind of thing the verse loves to do, because both activities take it one step closer to its ultimate goal of universal thermal equilibrium. But there's another method by which hot things lose heat to cooler things, and as far as the universe at large is concerned, it's much more important. That method is called radiation, and it's really worth understanding in detail for two reasons. Firstly, because it supports the majority of life on our planet, and, secondly, because it will ultimately bring about a dull, dull Armageddon.

* If you really want to be precise, heat doesn't stop flowing, it's just that as much of it flows from the cup to the saucer as from the saucer to the cup.

† In case that still isn't clear, just substitute the word "thermometer" in place of "third object" and you'll see how the Zeroth Law is essentially a definition of temperature.

RISE AND SHINE

So here goes. Every object in the universe radiates light, centered around a peak wavelength. Light, as you'll remember,* can have a whole spectrum of wavelengths, starting with radio waves, then on to microwaves, through the infrared to visible light, and on up through the ultraviolet to x-rays, and finally gamma rays. The higher an object's surface temperature, the shorter the peak wavelength of the light it emits. In case that sounds a bit dry, let's talk about a concrete example. You.

The core body temperature of the average healthy human being is, as most people know, roughly 98.6°F. It won't surprise you to learn that your surface temperature when clothed is somewhat less than that. Assuming you are relaxing in a reasonably well-heated room at around 68°F, perhaps wearing a onesie, you might expect it to be something in the region of 82°F. Any object—and I mean *any* object, it could be a piece of granite, a plastic statue of Elvis, or a possum—with a surface temperature of 82°F will radiate light, and the peak wavelength of that light will be roughly ten millionths of a meter, in the region of the spectrum we call the infrared.

Maybe you've worn a pair of thermal imaging goggles, or, as in my case, seen Walter White doing so in an episode of *Breaking Bad*. Either way, you will know that people show up bright, and almost everything else shows up dark. That's because the goggles are capable of detecting infrared light and using it to form a visible image. The shorter the wavelength of the infrared, the brighter the image in the glasses. When Walter looks at Jesse, Jesse will show up brighter than his surroundings because he is at a higher temperature than

* If you've been diligent enough to read the footnotes.

they are. Should Jesse be holding a cup of hot tea, then that would appear brighter still.

HERE COMES THE SUN

So every object in the universe gives off light, and the hotter it is the shorter the wavelength of the light it gives off. The very hottest things in the universe, of course, are stars. Our own Sun has a core temperature of some 16 million °C, and a surface temperature of 5.5 thousand °C.* The light it emits spans the entire spectrum from radio waves to gamma-rays, but it peaks around the infrared to visible part of the spectrum.

Color, in other words, is closely linked to temperature. Your gas ring burns blue, and the embers in the fire burn red, because gas burns at a higher temperature than wood. Temperature, and temperature alone, dictates the color of the light an object radiates. Heat a banana to 5,500°C and it will give off the same characteristic blend of colors as the Sun.†

So what happens to the light radiated by the Sun, after it strikes the Earth? The short answer is that it bounces around like nobody's business until it finally gets absorbed. That Ferrari on your drive is red because the paint is absorbing

* Coincidentally, the surface temperature of the Sun is roughly the same as the core temperature of the Earth. The much-more-trustworthy-than-Wikipedia math and science website Wolfram Alpha quotes 5,777K for the Sun, and 5,650K for the Earth. 1 Kelvin = – 272.15°C.

† Viewed from space, the Sun is white, because it radiates across the entire visible spectrum, with no wavelength being dominant enough to affect the overall color. Here on the ground, however, direct sunlight is most definitely yellow. That's because whereas red light travels unhindered through the atmosphere, blue light is easily deflected by air molecules. Physicists call this scattering, and it means the atmosphere effectively "holds on to" the blue portion of sunlight, making the Sun yellow and the sky blue.

every light photon that hits it *except* those that are red. These reflected red photons will eventually land on something that will absorb them; your monogrammed black leather driving gloves, perhaps.

After they have absorbed the red photons, and any other photons that happen to strike them, the temperature of your driving gloves will increase. They will then radiate a spectrum of electromagnetic energy, peaking in the infrared, which will in turn be reflected or absorbed by any other object—the obsidian-grey leather steering wheel, the tan calf-leather seats, your pink jumpsuit—that they come into contact with.

I'm sure you can guess where this is heading. In this universe of ours, objects don't have to be in contact with one another for heat to spread between them until they reach the same temperature. It can happen just as easily in a vacuum, via the process of radiation. In fact, because most of the universe is a vacuum, radiation is by far the most common way that heat gets around. Seen this way, the reason the stars shine is indeed to guide your destiny. It's just that your destiny happens to be thermal equilibrium with the rest of the universe.

So what has this got to do with life? Everything, it turns out. Life is an extremely perverse cooling process. The flow of heat in the universe may be a one-way street, from teacups to saucers, planetary cores to atmospheres, and from stars out into the cosmos, but it can be harnessed by life-forms to do work, reproduce, and—most importantly—organize themselves. It's time to meet the First Law of Thermodynamics.

PINBALL WIZARD

The First Law is great, because it effectively provides an accountancy system for the entire universe. Let's remind ourselves of what it tells us:

(1) The total energy of the universe remains constant.

Put simply, this is nothing more than conservation of energy. Fire a pinball machine, and you may watch impotently as the ball careers around, missing every jackpot and reward channel before it finally rolls with unswerving accuracy directly between the flippers and into the drain. But the beauty of energy is this: If we know the energy of the fired ball, and the energy of the returning ball, by subtraction we can work out exactly how much energy has been absorbed by the machine, despite knowing nothing at all about what went on in the game.

How is this relevant to our search for life, you ask? Well, like our pinball machine, cells consume energy, and expend that energy as work and heat. To understand politics we follow the money; to understand cell biology we follow the energy. Take any kind of cell, and we can ask some simple but far-reaching questions. How does it get its energy? How does it store that energy? And what does it use it for? As we shall see in the next chapter, when we ask these questions of the very first bacteria we get fascinating clues as to how life on Earth got its start.

But back to the plot. The First Law tells us that energy may be converted from one form to another, but it can't be created or destroyed. The pinball machine is a classic example. When we pull back the pin, we convert chemical energy in our muscles into the mechanical energy of a compressed spring. Release the pin and the spring's mechanical energy is converted to kinetic energy as the ball flies away at speed. The tilt of the machine then converts some of the ball's kinetic energy into gravitational potential energy as it rolls up the slope of the machine, rising in the Earth's gravitational field. And that's before the game has even properly started.

In a perfect world, at every one of these conversion stages no energy would be lost. Each type of energy would be completely converted into the next, and, theoretically at least, a turn at pinball could go on forever. But real life, as we all know, is not like that. Compressing a spring, rolling a metal ball across a wooden surface, and just about everything else all generate heat. What's worse, that heat leaks out into the universe and gets absorbed. You can never get that energy to do anything useful again; it is lost to you forever. The quest to understand why—despite energy being conserved—you never get as much out of a system as you put in is what led to the extraordinary bit of physics that is the Second Law. Because it turns out that one of the reasons that the globules in the Martian meteorite look like they might have once been alive is that they appear to break the Second Law of Thermodynamics.

ONLY HERE FOR THE BEER

Even if you don't know what the Second Law is, you will almost certainly have heard of it. Thomas Pynchon, one of my favorite American authors, is obsessed with it, and if you are an enthusiast of abstruse fiction I thoroughly recommend his novel *The Crying of Lot 49* for its deep insights into what has to be the most fascinating of all physical laws. To really understand it, we first need a deeper understanding of something we have so far taken for granted: heat.

Our goal, as ever, is a deeper understanding of biological systems, but the work and character of a nineteenth-century brewer named James Prescott Joule are so especially noteworthy I can't resist a momentary diversion. The brewing game requires precise measurement of temperature, and Joule carried the art to virtuosic levels. He became fascinated with exactly what it was he was measuring, and in a landmark experiment demonstrated that what had previously been thought of as a fluid was in fact the random motion of atoms.

Put simply, heat is atoms jiggling about. The hotter something gets, the more its atoms jiggle. Put something cold—or, in other words, unjiggly—in contact with something hot, and the jiggly hot atoms will eventually jiggle the unjiggly cold atoms into some sort of intermediate jiggly-ness. Thermometers measure this jiggly-ness. This, of course, is why temperature has a zero, because in theory it's possible for something to have no heat and therefore not jiggle at all.*

Joule's trick was to drop a precisely known weight through a precisely known distance, making it do work on the way down by spinning a paddle in a bath of water. The spinning paddle pushes the water molecules around, increasing their speed and therefore their temperature. By accurately measuring the temperature increase in the water, Joule was able to demonstrate that the increase in its thermal energy exactly equaled the gravitational potential energy imparted by the falling weight. At a stroke, he showed that heat is basically a type of atomic motion. Even today, it still seems somehow

* Well, almost. Quantum Mechanics shows us that even at absolute zero, a particle has a residual jiggly-ness, known as zero point energy.

revolutionary. But there you have it. Sloshing water about increases its temperature.*

THE INFORMATION ENGINE

Right. We're ready for the Second Law. At first sight it appears to be one of the most unhelpful bits of science you could wish for, a non-event of the first water. Do not be fooled. You are about to go down a rabbit hole that will lead you deep into the innermost workings of the universe. Here she is:

(2) The entropy of the universe always increases.

So what exactly is entropy? Step by step, our understanding has deepened. It was first defined during the age of steam. A steam engine, as you will know, works by burning fuel to boil a body of water, and then using the steam created to drive its pistons. The steam then liquefies in a cold condenser before being reheated. The problem that quite rightly intrigued capitalists of the age was how to arrange this system of hot tank (boiling water) and cold tank (condenser) in order to generate the most mechanical work for a given amount of fuel.

The problem tested some of the finest scientific minds of the time. First out of the blocks was the French physicist Sadi Carnot, who in true Gallic fashion published one of the most whimsically entitled works in the history of the physical sciences with his 1824 memoir, *Reflections on the Motive Power of Fire*. Carnot rightly determined that what really

* The unit of energy called the joule (J) is named in his honor. The one used by physicists, anyway. Engineers use handier (i.e. much, much larger) units like the kWh, where 1 kWh = 3 600 000 J.

counted in a steam engine's efficiency was not the amount of fuel burned, but the temperature of the hot and cold tanks. But what was the formula?

If it was Carnot who framed the problem, it was the German physicist Rudolf Clausius who enabled its solution. He did so by recognizing that it is not the total amount of heat energy sloshing around in a steam engine that counts, but the portion of that energy that is available to do work. That led him to consider what had happened to the portion of an engine's heat that isn't available for work. In the case of a real steam engine, for example, there will be energy lost to its surroundings, waste energy generated by friction in its pistons, and losses due to the viscosities of both steam and water. None of that energy is ever coming back, or at least it is extremely unlikely to. How to characterize these unknown, unpredictable losses?

Clausius, ingeniously, had a way. He invented a measure of this unavailable energy, and called it entropy. Remember the pinball machine? In that case, the gain in entropy of the machine and its surroundings will be that fraction of the ball's initial energy that has been absorbed by friction and heating effects during the course of a turn. Entropy, as defined by Clausius, is simply a measure of how much of the input energy had become unavailable for work.

I know, I know; it's hard to see what any of this has to do with the origin of life. Your righteous indignation is understandable. No one, frankly, was expecting the Second Law of Thermodynamics to lead where it did. Clausius just wanted a working theory for steam engines. Instead, he ended up discovering nothing less than a brand new physical quantity, an entity that appears to be as fundamental to the way the universe works as energy or temperature. Because when the time came to try and understand entropy on the scale of atoms, an astonishing discovery was made. Entropy is directly related to information.

THE DEVIL IN THE DETAIL

The first man to put this discovery on a mathematical footing was the Austrian physicist Ludwig Boltzmann. His is a sad story. A brilliant man, he suffered vicious attacks from his German contemporaries, who were both unable to follow his deliciously nimble calculations and refused to subscribe to the atomic theory that underpinned them. Prone to depression, he committed suicide in 1906, aged just forty-two, but not before crafting the equation that appears on his tombstone:

$$S = k \log W$$

In this masterpiece, Boltzmann relates the entropy, S, of a gas* to the number of ways its particles can be configured, W, while still having the same overall pressure, volume, and temperature. The constant k, by the way, is respectfully known as Boltzmann's constant. All we're aiming for here is the gist, so the point to grasp is this: The greater the number of configurations a system has, the more uncertain we are about which one it is actually in.

How so? Let's take a concrete example. Say I take a party balloon and blow it up. I can easily measure the volume, pressure, and temperature of the air in the balloon. In how many ways

* To be precise, Boltzmann's formula is that for an ideal gas. This is a gas whose individual particles don't interact with one another, so troublesome things like viscosity can be safely ignored.

could the air molecules inside the balloon be configured* to create the precise values of volume, pressure, and temperature that I measure? In W ways, that's how many, and W is a big number. The entropy of the air inside my balloon is then k log W.

Now say I burst the balloon. The air molecules that were inside now start to mix with the air molecules in the room. Given enough time, they will escape the room and mix their way around the globe. In how many ways might they be configured then? I have no idea, but I can tell you that it's a lot more than W, the number of ways they could be configured within the balloon. My uncertainty about the configuration of the air molecules in the balloon has increased, and their entropy has therefore gone up.

INFORMATION IS THE RESOLUTION OF UNCERTAINTY

Following Boltzmann, the union between entropy and information was solemnized by the work of a brilliant American electrical engineer and mathematician named Claude Shannon. Although he is arguably the founder of the digital age, Shannon is criminally underacknowledged. At the tender age of twenty-one he used his master's thesis to invent digital circuit design, the foundation of all modern electronics, and following the Second World War he was employed at Bell Labs on a US military contract.

Radio communication had become essential to warfare, and Shannon was tasked both with improving the existing

* By configured, of course, I mean where are they, what direction are they going in, and how fast?

military systems and making them more secure. One of the great problems in electronic devices is how to reduce the effect of "noise"—random fluctuations in the signal. To help solve the problem, Shannon defined a new quantity, *H*, called the Shannon entropy, as a measure of the receiver's uncertainty as to the precise letters of a message. To be precise, he said that the Shannon entropy, measured in bits, was given by the expression

$$H = -\sum p_i \log_2 p_i$$

where H is the Shannon entropy, measured in bits,* of a message that is conveyed by *i* letters each with probability p_i and that funny squiggle is sigma, meaning "sum up the following for all values of *i*."

In case that's a little abstract, let's make these equations flesh. You and I arrange to go apple-scrumping† after lights out. At nine o'clock, when it's dark—it is autumn after all—I will creep into your garden and look up at your bedroom window. If your flashlight is on, you are going to shin down the drainpipe and join me in Farmer Benson's orchard. If it's off, well, we are going to have to go tomorrow night instead.

In Shannon's terms, the message—"Yes, I am scrumping" or "No, I am not"—is coded in two "letters," "flashlight on" and "flashlight off." At one minute to nine, while I watch my breath condense in the rhododendron bushes, I have no idea

* No doubt you recognize this unit, the bit, as fundamental to computing. Bytes and bits are the dollars and cents of information, where 1 byte = 8 bits. When we ask for a 15 terabyte hard drive, we politely request the capacity to store a message with a Shannon entropy of $15 \times 10^{12} \times 8$ bits.

† Or "apple-stealing," for those readers not familiar with this quaint British phrase.

whether you are coming or not. If both outcomes are equally likely, then Shannon's equation tells us that the Shannon entropy is as follows:

$$H = -\sum p_i \log_2 p_i$$

Two possible outcomes, each of which is equally likely, means $p_1 = p_2 = 1/2$, giving us

$$H = -\{ 1/2 \log_2 (1/2) + 1/2 \log_2 (1/2) \}$$

Realizing that ½ is a common factor, and remembering that $\log(A/B) = \log A - \log B$ we get

$$H = -1/2 \{\log_2 1 - \log_2 2 + \log_2 1 - \log_2 2\}$$

$$H = -1/2 \{2\log_2 1 - 2\log_2 2\}$$

Recalling that $\log_{anything} 1 = 0$ and $\log_2 2 = 1$, it all shakes down so that

$$H = -1/2 \{ -2 \} = 1 \text{ bit}$$

OK. This is heading somewhere, I promise. Let's now imagine that we decide to include a third "letter," where you waggle your flashlight about, meaning "wait because I am still deciding." Let's further imagine that each of the three possible letters is equally likely. What's the Shannon entropy then? Clearly it's more than in the previous case, but by precisely how much? Let's plug in the numbers to give us the answer.

$$H = -\sum p_i \log_2 p_i$$

$$H = - \{ 1/3 \log_2 (1/3) + 1/3\log_2 (1/3) + 1/3 \log_2 (1/3) \}$$

$$H = - 1/3 \{ - 3 \log_2 3 \}$$

$$H = \log_2 3 = 1.58 \text{ bits}$$

Maybe you're starting to see the pattern? For two letters, we had $H = \log_2 2$ bits. For three, $H = \log_2 3$ bits. And if we had W letters, each of which were equally likely, Shannon's formula tells us that the information entropy of the message in bits is given by

$$H = \log_2 W$$

Or changing the base of the logarithm to the natural number, *e*, we get*

$$H = K \log W$$

Where K is a constant. Wait a minute! That's uncannily reminiscent of Boltzmann's formula for the entropy of a group of particles with W possible configurations, each of which is equally likely:

* In case that means nothing to you, then e is a number that crops up so often in math it is called the "natural number." It's irrational, meaning that it can't be expressed precisely, and is roughly 2.718. Here's just one of the ways *e* can be written: $e = (1 + 1/n)^n$ as n tends to infinity.

$$S = k \log W$$

So what's going on?

SOD'S LAW

On the face of it, the progression seems unlikely. At one end, nineteenth-century physicists were looking for a way to improve the efficiencies of steam engines. At the other end, twentieth-century engineers were looking for a way to improve the efficiency of communication devices. Extraordinarily, they both turned out to be working on the same problem. Entropy, S, being our uncertainty about the microscopic configuration of a physical system, and entropy, H, being our uncertainty about the configuration of the letters in a message, are connected.

The link is information. When we calculate the entropy of a group of atoms, we can think of it either as our uncertainty as to which one of its possible W configurations it is actually in, or as how uncertain we are as to the content of a message that completely describes that configuration. Boltzmann entropy is, in fact, a special case of Shannon entropy.

So what does this mean? Well, for a start, it means that entropy is the enemy of information. The greater the entropy of a system, the less we know about the content of any message it might contain. But on a deeper level it means that information is more than just books, DVDs, and hard drives; it is a fundamental property of matter. A sugar molecule, a photon of light, an edition of *The Times*: All of them contain information. That information may have to do with the location of individual atoms, or the location of George Clooney's wedding; it's all the same to the universe. It simply doesn't care for it.

Now we start to understand the true nature of the Second Law. Energy dissipates. The entropy of the universe always increases, corrupting information, increasing disorder, and dispersing heat. Now we see why those fossilized grains in the Allan Hills meteorite are so striking, and, indeed, why all life-forms are so magically unusual; they contain an extremely high degree of order, far greater than could arrive by chance. From stromatolite colonies of cyanobacteria, to Venus fly-traps, ants' nests, and bridge clubs, life-forms constantly cheat the Second Law of Thermodynamics. From a world bent on chaos they concentrate energy, order, and information.

Life is shit, and then you die. Everyone is promoted to a position of incompetence. If something can go wrong, it will go wrong. These are all statements of the Second Law, and they are so ingrained in us that they feel like second nature. The joy that we feel when we tidy up the office, write a letter, or hold a newborn baby is a vaulting ecstasy at momentarily defying the Second Law of Thermodynamics. We know that it can't last, but somehow that makes it all the more sweet. We got one past the goalie. We passed the flaming torch of information, despite the best efforts of the universe. Life goes on.

We know, of course, that in the long run it's probably not good news. Life is swimming against the tide of creation, and it can't last forever. In a moment, I'm going to let the Laws of Thermodynamics run to their inevitable conclusion, but first we need to answer the all-important question: How does life do it? If the universe is painstakingly eradicating information wherever it can find it, how did life manage to get started, and how has it managed to become ever more complex?

INSANE IN THE MEMBRANE

For once I'm going to give a straight answer to a straight question: cells. The fundamental building block of all life on Earth is an ingenious way of piggybacking the Laws of Thermodynamics. All life on Earth is made up of cells, and they all have one thing in common. They have a means of keeping their insides separate from their outsides.

On the face of it, a cell membrane might not be the sexiest of features. There are other, far more photogenic things to get excited about like nuclei, Golgi bodies, and mitochondria, but arguably none of them would ever manage to scratch a living without something to protect them from the big bad world. As Claude Shannon might see it, a cell membrane is a great way of separating a low-entropy, high-information region—the cell's insides—from its high-entropy, low-information outsides, aka the universe. And it's this separation that enables the cell as a whole to perform a neat thermodynamic trick.

It works like this. It doesn't matter if there's a decrease of entropy inside the cell, so long as outside the cell it increases by an even greater amount. Overall, the entropy of the universe will have increased and the Second Law remains unbroken. All you need is a cell membrane to act as a gatekeeper, letting in low-entropy stuff and letting out high-entropy stuff. The low-entropy stuff we call food. The high-entropy stuff we call waste.

In other words, the cell membrane is crucial because it prevents equilibrium. Within it, entropy can be lowered, matter can be organized, and information can be stored. But it also has another crucial function that was essential to the first single-celled life-forms. It is a great way of storing energy.

MY NAME IS LUCA

In the next chapter we'll look in detail at what we know about the history of life on Earth. Our goal will be to try and understand how single-celled life got its start, and the series of innovations and coincidences that led to our own species of technologically accomplished and highly social apes. Once we have some kind of perspective on how likely our own intelligence is, hopefully we can get a feel for how commonplace our kind of intelligence might be in the galaxy, and how far away our nearest neighbors might be.

I'm not one for spoilers, but in the broadest of strokes we will find that the very first life on our planet was single-celled, and came in two types known as archaea and bacteria. Both originated in water. We can tell from analyzing the DNA and proteins within them that they are related, but so far we have no way of telling which came first. For the moment, we are just trying to understand the nuts and bolts of how life works, so we can dispense with the gory details. What's important for our present purposes is that they both store energy by pumping protons across their cell membranes.

A proton, you'll remember, is nothing more than a lone hydrogen nucleus. Whereas an electron carries a single unit of negative charge, a proton carries a single unit of positive charge. Archaea and bacteria are capable of getting their energy from a bewildering number of sources: from eating one another, from reactions with chemicals like hydrogen sulfide and ammonia, from rusting metals such as iron, and, of course, directly from sunlight. In every case, once acquired, they use that energy to drive protons across their cell membranes, storing it up for future use.*

Because protons carry charge, they essentially want to get away from one another. By creating an excess of protons in the water surrounding them, and a deficit within, these simple cells are effectively creating an energy source to be tapped at will. This is arguably a bit too much detail, but for some reason I can't resist telling you that, when the time is right, these energetic protons are used to make an energy-rich molecule called adenosine triphosphate, affectionately known as ATP.

This plucky molecule is essentially the currency of energy within all cells, able to donate energy wherever it is needed. All the processes that you can think of such as movement, making proteins, RNA, and DNA all extract their energy from ATP. Put bluntly, cells are miniature machines, capable of extracting energy and information from the environment, storing it, then making copies of themselves. Single-celled organisms such as bacteria reproduce by dividing; their human cousins by having dinner then progressing to a second date.† Eventually, however, entropy will have its way. It's time to glimpse the end of days.

* The chemist in me can't help but point out that there is no such thing, really, as a lone proton in water. Protons react with water molecules to form hydronium ions, H3O+. The net result is the same, though, as hydronium ions repel one another, and will readily give a proton up so that it can pass back through the cell membrane.

† Dying is an innovation of sexually reproducing complex organisms, with a division of labor between mortal "body" cells and immortal "reproductive" cells. Bacteria, for example, are entirely immortal, though it's not a life I feel particularly envious of. There's no escape though: Eventually the food will run out, and even bacteria will starve.

THE HEAT DEATH OF THE UNIVERSE

Let's begin by getting our bearings. The universe began from a microscopic, hot, dense state some 13.7 billion years ago, inflated in a so-called Big Bang for a fraction of a second, then settled down to a steady expansion which ended about seven billion years ago, when it was roughly half the age that it is now. At that point, some repulsive force that we call dark energy began to dominate over gravity, and its expansion began to speed up. Our best prediction is that this expansion will continue to accelerate, pushing more and more of the cosmos out of the reach of our most powerful telescopes. In fact, some two trillion years hence, the only galaxy we will be able to see will be our own.*

The present era is known as the stelliferous era, meaning quite simply the one where star formation takes place from the gravitational collapse of gas and dust. Before it came the primordial era, when the intense energy of the Big Bang cooled to form fundamental particles, then nuclei, then hydrogen and helium. The primordial era lasted about a million years; the stelliferous era will last a few trillion. So what happens next?

Well, for a kick-off, there'll be no new stars. All the hydrogen and helium gas will have been used up, and one by one the existing stars will start to burn out. The longest lived will be the smallest, the so-called red dwarves, but after a few trillion years even they will have exhausted their supply of nuclear fuel and will be beginning to cool. This is known as the degenerate era, after the extremely dense form of matter that remains when small stars cool.

* It will be a pretty big galaxy, though. The Milky Way is expected to collide with the Andromeda Galaxy in about four billion years, and in four hundred and fifty billion years' time, the fifty-odd galaxies of the Local Group will have merged.

There's no escaping this fate for any star, least of all our own Sun. As you probably know, our home star will run out of hydrogen in around five billion years' time, at which point it will bloat from a yellow dwarf into a red giant.* In roughly 7.9 billion years' time it will explode in a planetary supernova, and all that will be left is a small lump of hot degenerate carbon known as a white dwarf. In roughly a quadrillion years—that's a thousand trillion—it will have cooled to a temperature of just a few degrees above absolute zero. At that point it will no longer radiate any kind of light—radio, infrared, or otherwise—and will become what is called a "black dwarf."

It gets worse. As the degenerate era progresses, the swirling mass of dead stars that form the galaxy will slowly dissipate. One by one, near-misses and collisions will fling planets, black dwarves, and neutron stars out into intergalactic space. Any dead stars or rogue planets that remain will slowly be consumed by black holes. At the end of the degenerate era, estimated to arrive in some 10^{43} years' time, all that will be left of the cosmos will be a silent colony of black holes, gorged to the eyeballs on dead stars.

That marks the beginning of what is prosaically known as the black hole era. But it doesn't end there. Black holes, it turns out, aren't completely black. As Stephen Hawking was the first to point out, subatomic particles are able to "tunnel" across the event horizon, and return to the everyday universe. This, like any kind of radiation, has the effect of removing energy from the black hole, giving it a finite lifetime. In the case of a supermassive black hole that's something in the region of—here

* It will reach its maximum size in around 7.9 billion years' time, at which point its surface will extend as far as the present orbit of Mercury.

comes the biggest number in this book—10^{106} years.* All that will survive of them will be a thin gruel of subatomic particles.

And that's pretty much all that will be left of everything else, too. Even things like protons and neutrons are expected to decay back into subatomic particles—and all so-called solid objects like planets, asteroids, and comets are subject to the same quantum mechanical tunneling that black holes are, and will eventually deplete away into nothing. Everything in creation will be a cold, cold gas of subatomic particles, jiggling away at just a few fractions of a degree above absolute zero. No part of this gas will be hotter than any other, and no life of any conceivable kind will be able to exist.

A ROCK AND A HARD PLACE

Every year my extended family makes a pilgrimage to Robin Hood's Bay on the North Yorkshire coast. One of the main attractions—apart from ice cream and fish and chips—is the fossil hunting. Edged with clay cliffs, the pebbled beach in the bay is a treasure trove of ammonites, devil's toenails, and sharks' teeth. The process is always the same. For the first twenty minutes or so you find nothing. Then you find a single fossil—usually not a very good one. Then suddenly there are fossils everywhere. Virtually every stone you pick up contains some half-submerged prehistoric creature, preserved in exquisite detail.

Seen from the perspective of entropy, there isn't a great deal of difference, really, between fossils and holiday snaps. If the

* That's calculated for a black hole of twenty trillion solar masses. The largest black hole currently known is NGC 4889, which weighs in at twenty-one billion solar masses.

Second Law tells us anything, it's that meaning has no permanence. That coiled black ammonite may feel ancient, but the whole history of life on Earth is fleeting, like a drop of spray flung high into the air by a crashing waterfall. We life-forms are a glorious curiosity, another way for a star to cool, and for the universe to oxidize carbon. We are a means to an end.

If that sounds bleak, it's really not meant to. If it stirs anything in you, hopefully it's a raging thirst for what the universe wants to deny you: knowledge. Or, rather, you are a manifestation of the universe's wildest wish, namely, to awake and know itself. We have learned something profound about life. Wherever we find it, and whatever its building blocks, it will require a constant source of energy. It will use that energy to organize itself, at the expense of the entropy of its surroundings. And it will be far from equilibrium, because equilibrium means death.

CHAPTER SIX

HUMANS

In which the author stakes his footling reputation on one particular hypothesis of how life got its start, and forces himself through the evolutionary bottlenecks that impede the flow from microorganism to Microsoft.

The booking hall at central London's Euston Station will never be the same again. I am crossing it now, as I hurry to catch my train, my head buzzing with ideas. Not my ideas, I hasten to add. They are the ideas of Nick Lane, the Provost's Venture Research Fellow at University College London. Though if you ask Lane—and I just have—he will say some are the ideas of one Mike Russell, now of NASA's Jet Propulsion Laboratory in Pasadena, who first proposed them some twenty years ago.

I came to see Lane because I am on a quest. To try and figure out how likely intelligent aliens are, I need to know how likely I am. We've seen that, from its very beginning, the universe was "trying for a baby," in that nature is fine-tuned to produce atoms. But it's one thing to try and another to conceive. How likely was it that a small, wet, rocky planet orbiting a humdrum star in a spiral galaxy would be the birthplace of single-celled life? And how and why did that single-celled life evolve into a technologically advanced civilization of intelligent apes?

I'LL HAVE THE PRIMORDIAL SOUP

The classic picture of what scientists call abiogenesis is most often attributed to Charles Darwin. Though he avoided the subject in his public work, in a letter of 1871 to his close friend the botanist Joseph Hooker he made his true feelings clear:*

> It is often said that all the conditions for the first production of a living organism are present, which could ever have been present. But if (and Oh! what a big if!) we could conceive in some warm little pond, with all sorts of ammonia and phosphoric salts, light, heat, electricity, etc., present, that a protein compound was chemically formed ready to undergo still more complex changes, at the present day such matter would be instantly devoured or absorbed, which would not have been the case before living creatures were formed.

Leaving aside the shocking punctuation, what Darwin seems to be saying is that, given the right conditions, something living can spontaneously emerge from something non-living

* Darwin was such a secretive soul that his letters to Joseph Hooker form the backbone of all Darwin scholarship, and you can find a link to them at http://www.darwinproject.ac.uk/darwin-hooker-letters.

simply by chance. In essence, that has been the non-religious view ever since Aristotle, and it remains our belief today, though there has been a great deal of toing and froing over exactly what those conditions might be.

In the 1920s, the Russian biologist Alexander Oparin and the British polymath J. B. S. Haldane independently refined Darwin's conjecture into what is known as the "primordial soup" theory. The gist is that once you have the right chemical elements and a source of energy, sooner or later a self-replicating molecule will emerge which is then capable of undergoing natural selection, in turn producing life.

So what might those chemical elements be? All earthly life, as you probably know, is based on one most extraordinary element: carbon. Carbon is a party animal, eager to bond with a variety of other elements—and also with itself—to form long-chain molecules and rings of almost infinite variety. In organisms, we generally find it in the company of five "usual suspects": hydrogen, nitrogen, oxygen, phosphorus, and sulfur.*

Two configurations can rightly be thought of as the "building blocks" of life. One is called a nucleobase, which makes up both RNA and DNA, and also the very important energy-carrying molecule called ATP. The other is called an amino acid, the building block of proteins. I say building block; neither nucleobases nor amino acids are particularly simple, a point I have tried to emphasize by means of my characteristically poor drawings on the next page.

* Often given the mnemonic CHNOPS. That's if you can call a string of letters a mnemonic.

Amino Acids:

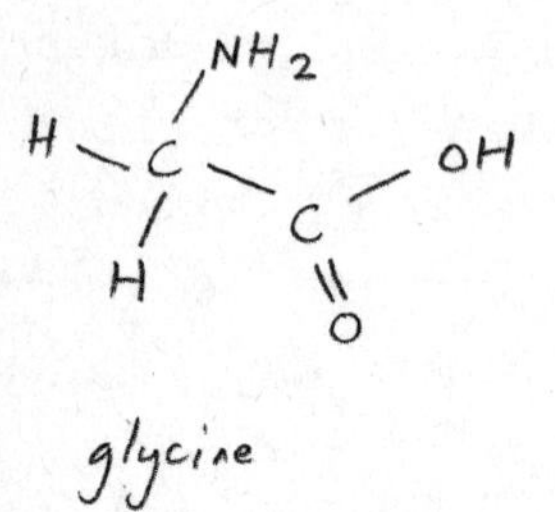

glycine

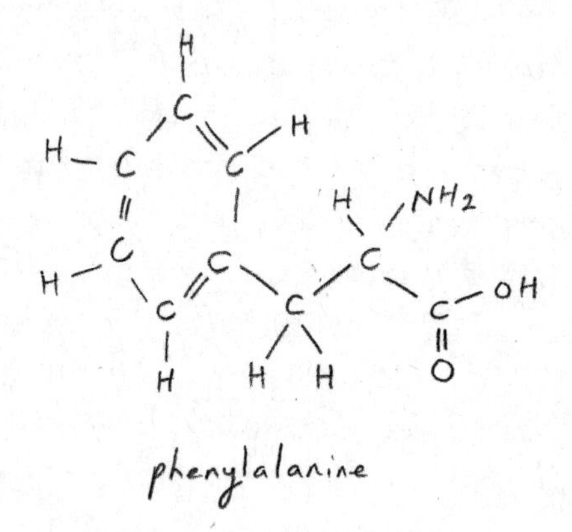

phenylalanine

Nucleobases:

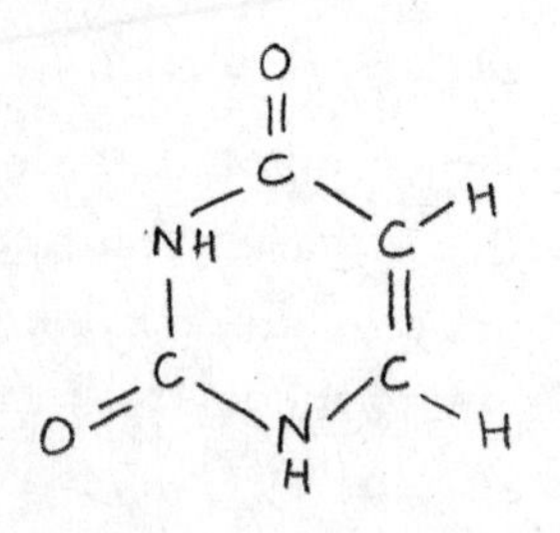

uracil

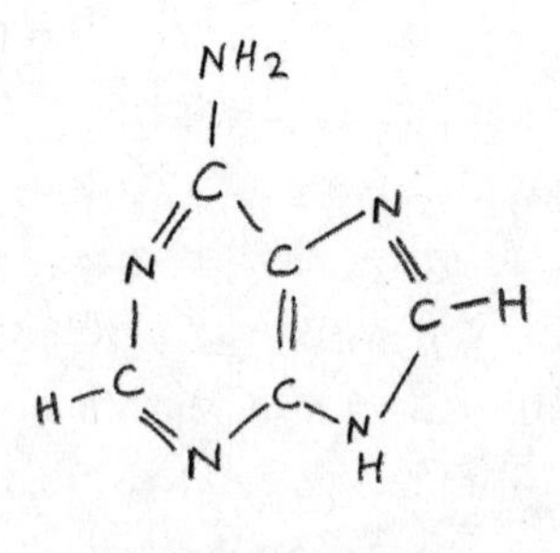

adenine

Right. So the classic picture of the primordial soup theory in action looks something like this. It's a foul night, and a bolt of lightning strikes in the skies above a broiling sea, in an atmosphere pumped full of noxious gases. Emboldened by this spark of energy, the gases react to form building blocks like nucleobases and amino acids. These building blocks then rain down into the ocean, where they fuse together to make the very first self-replicating molecules. A little packet of them gets trapped in an oily film, and the first cell is formed. Life is off to the races.*

LACKING IN ENERGY AND CONCENTRATION

Back in 1953, a Nobel Prize-winning chemist at the University of Chicago, Harold Urey, decided to put the primordial soup theory to the test. In a flask, Urey's PhD student Stanley Miller put the gases they believed to have been knocking around on the early Earth—ammonia, methane, hydrogen, and water vapor—and passed an electric current through them. The results were extraordinary. There in the bottom of the flask were amino acids, the building blocks of proteins.

What's more, in the years since what is now known as the Miller–Urey experiment, we have found amino acids and nucleobases in some far-flung places: on comets, meteorites,

* I feel like I should point out that there are tribalisms within the primordial soup theory. One favors self-replicating molecules first; one proteins; and one cell membranes. Sooner or later, of course, you are going to need all three.

and even in interstellar space.* That's significant, because for its first billion or so years the Earth was pummeled by comets and meteorites, with a particular flurry of blows taking place roughly 3.9 billion years ago during what is known as the Late Heavy Bombardment. The primordial seas could very well have been a soup of amino acids and nucleobases, some earthly and some alien. Could a lightning strike have kick-started life as we know it?

In the minds of many contemporary evolutionary biologists—and this sketch comedian—there are two reasons why the answer to this question is a polite "not really." The first is what is called the concentration problem. For a series of chance encounters to produce something as complex as a self-replicating molecule you need a lot of the right kind of molecules in the right place, which is why chemists tend to do their experiments in test tubes rather than oceans. A broiling primordial sea is simply not the place you can expect to find delicate organic chemistry.

As a workaround, some evolutionary biologists have suggested that Darwin had it right, and the vital reactions took place in a pond, where—thanks to evaporation—the "soup" was much thicker. Which sounds promising, until you consider the fact that, as already mentioned, the early Earth was being bombarded by comets. The exact time when life began is subject to intense debate, but, to be brief, we have fossil evidence of so-called "microbial mats" at around 3.5 billion years

* To qualify that sweeping statement: To date, samples of amino acids have been found in Antarctic meteorites, and their spectra have been observed in comets and interstellar space. Nucleobases have also been found in Antarctic meteorites, and the spectrum of methanamide, their chemical precursor, has been identified in comets. The spectra of amino acids have been seen in interstellar space, but not—to my knowledge, at least—that of nucleobases or their precursors.

ago,[*] and chemical evidence of cellular life from 3.7 billion years ago.[†] As a result, most pundits would be happy to set a date for "prototype" life at around four billion years ago. The really peculiar thing about that is it predates the Late Heavy Bombardment. Whatever this "prototype" life was, it appears to have survived a cataclysm.

But for me all that fades into insignificance when it comes to the second problem: energy. As we learned in the last chapter, to sustain life we need a ready supply of information-rich energy. The "wham, bam, thank you, ma'am" of lightning just won't do the trick; life needs a lover with a slow hand. And when I say slow, I mean *slow*; for natural selection to do its thing, we need an energy supply that lasts for tens, hundreds, maybe even thousands of years. Thankfully, in a flash of inspiration over two decades ago, the aforementioned Mike Russell pictured what he considered to be just the right place. It's time to meet your maker: an alkaline hydrothermal vent.

LIFE SPRINGS ETERNAL

"And what," you may ask, "is one of those?" The simplest answer would be a hot spring on the seabed, caused by a chemical reaction between sea water and a common mineral, olivine. Olivine—a greenish crystal made of iron, magnesium, silicon, and oxygen—reacts with sea water to form serpentinite. And the striking thing about the booking hall here at Euston is that

* Microbial mat fossils 3.5 billion years old have been found at the Dresser Formation in the Pilbara region of Western Australia. Microbial mats are communities of bacteria.

† The Isua Greenstone Belt in southwestern Greenland, dated at 3.7 billion years old, has carbon in it that seems to be of organic origin.

it is paved with serpentinite marble, a green stone with white serpent-like marbling, a fact that the fatalist in me can't help feel is significant.*

Anyway, back to the plot. When sea water reacts with olivine to form serpentinite, it releases a great deal of heat, and this hot, mineral-rich fluid then rises as alkaline springs on the seabed. When it meets the cold sea water at the bottom of the ocean, the minerals precipitate out, like the limescale in a kettle, producing vents of porous white limestone. One famous example is the so-called Lost City in the middle of the North Atlantic, a ghostly hoard of some thirty gnarled chimney stacks, the tallest of which is some twenty stories high.†

It's in the microscopic pores of this limestone, Mike Russell realized, that life may have gotten its start. The clue came from the way that all single-celled organisms store energy. Essentially, they act like tiny electric batteries, pumping protons across their membranes so that their insides become less positively charged than their outsides.‡ We call this a proton gradient. If energy is required for a chemical reaction somewhere within the cell, a proton is allowed to fire back through the membrane. The energy of this proton is then harnessed to

* OK, cards on the table. I am not entirely sure that what builders call "serpentinite marble" has anything to do with serpentinite the mineral other than a similarity in appearance, which is to say both are green with a serpent-like pattern. In fact, I'm not altogether sure "serpentinite marble" has got anything to do with marble. It's probably made in a factory from brick dust. Frankly, when you're looking for synchronicities you take what you're given.

† In fact, the only example. Discovered in 2000, close to the Mid-Atlantic Ridge, the joint between the Eurasian and North American plates, at 30° north.

‡ The acidity or alkalinity of a liquid is defined by its pH, being the logarithm to the base ten of the number of hydrogen ions (protons) per cubic meter.

create a molecule of energy-carrying ATP,[*] which then carries it to wherever it is needed.

Which begs the question, "Why bother?" Why go to all the trouble of storing energy in a proton gradient, instead of just making ATP straight away? Maybe, thought Mike Russell, it's a hangover from an earlier stage of life. McDonald's sells hamburgers in 119 countries, but banks its profits in dollars because its first store was in San Bernardino, California. Maybe single-celled organisms bank in protons because that's the way it was done back in the 'hood. And one place you are sure to find a proton gradient is in an alkaline hydrothermal vent.

Why? Because "acidity" and "alkalinity" are just "proton concentration" by another name. Acidic fluids have a lot of protons; alkaline fluids have few. The present-day oceans are slightly alkaline,[†] but thanks to a much higher concentration of atmospheric carbon dioxide, back in the day they would have been much more acidic.[‡]

* Once it has given up its energy, ATP is converted to ADP, or adenosine diphosphate. It's not as complex as it sounds. ATP is basically a blocky carbon molecule with three phosphate groups getting on each other's nerves, so much so that one is happy to whizz off energetically, leaving the other two behind.

† Thanks to anthropogenic carbon dioxide, this is changing, and the world's oceans are becoming more acidic. In fact, their pH is changing ten to a hundred times faster than at any time in the past fifty million years. This will have positive and negative effects on marine life; we are all hoping that the positive outweighs the negative.

‡ Ah yes—I forgot to mention. In the years since the Miller–Urey experiment, we've learned that, far from being full of ammonia, hydrogen, and methane, the early atmosphere was full of oxidized gases. How do we know this? Because of zircon crystals. The oldest are dated at 4.4 billion years old, and their chemical composition indicates that they were formed at moderate temperatures in water, in an atmosphere full of nitrogen, sulfur dioxide, and carbon dioxide. No oxygen of course; that came roughly 2.3 billion years ago, as a by-product of oxygenic photosynthesis. More about that in a moment.

EXAMINE YOUR PORES

With that in mind, let's zoom in on one of the pores in the limestone chimney of a primordial alkaline hydrothermal vent and see what's going on. Amazingly, it's full of microscopic bubbles. Each one acts like a tiny battery. Inside, we've got warm alkaline fluid. Immediately outside, we've got acidic sea water. And betwixt the two we have a thin gel-like membrane made of iron sulfide, with a proton gradient across it.

Could such a bubble be the place that some prototype of life set up shop, using a natural proton gradient to drive chemical reactions? As we saw in the last chapter, to organize matter we need to be able to do work. Thanks to the proton gradient across its membrane, a bubble in an alkaline hydrothermal vent has energy on tap. What's more, its confined space is also a great place to concentrate chemicals. The concentration problem and the energy problem have been solved in one fell swoop.

And those aren't the only things these kinds of vents have going for them, because they are also rich in another vital constituent of living cells: transition metals. Cast your mind back to the periodic table, and you will remember that the middle of it is made up of a block of colorful, dense, mildly reactive metals with dependable names like iron, nickel, copper, and zinc. Ever wondered why these are recommended as part of a balanced diet? Crucially, it's because we find them embedded within a bewildering number of proteins.*

* Roughly a third of all proteins contain an embedded transition metal.

Why might that be? Well, the thing about transition metals is that they make great catalysts.* Essentially they are wealthy philanthropists, with more electrons than they rightly know what to do with, and are happy to donate a few to the needy, safe in the knowledge that they will regain them somewhere further down the line. Alcohol dehydrogenase, for example, the enzyme in the liver that breaks down alcohol, contains a socking great zinc ion right in the middle of it, as do some three hundred other known enzymes. So far as we can make out, transition metals were the first catalysts, and were later enslaved by enzymes.

Membranes, proton gradients, and transition metals: All of them make a good case for alkaline hydrothermal vents as the cradle of life. Add to that the fact that they are tucked away on the seabed, out of reach of the 700 million year asteroid bombardment that rattled the newborn Earth, and I hope I've got your attention. So how was the trick done? How do you get from something non-living to something living?

MAKE MINE A SINGLE-CELLED ORGANISM

Knowing that life is a symphony written in long-chain carbon molecules, it's clear that we need a source of carbon. What

* Chemists refer to an inorganic element's "oxidation state," being the number of electrons a neutral atom has either gained or lost. Iron is particularly obliging, and is happy to gain one or two electrons, or lose as many as six. We then say it is capable of oxidation states −2 to +6. For organic molecules, we define oxidation state slightly differently. C-C bonds are neutral, C-H bonds reduce the oxidation state by −1, and non-carbon atoms increase it by +1. The carbon in carbon dioxide (CO_2), for example, is fully oxidized at +2; the carbon in methane (CH_4) is fully reduced at −4. In an oxidizing atmosphere such as on Earth, methane is therefore chemically unstable compared to carbon dioxide. In a reducing atmosphere—in other words, one rich in hydrogen—it would be the other way around.

better place to get it than from the carbon dioxide dissolved in the primordial ocean? In long-chain molecules we most often find carbon bonded to hydrogen, so we are going to have to find a source of that as well. What about water? We've got plenty of that at hand.

No dice. It's possible to get carbon dioxide to react with water to produce long-chain carbon molecules and free oxygen, but it's far from easy. Plants do it, but they use sunlight, harnessed by a convoluted chemical pathway called oxygenic photosynthesis. That particular party trick didn't emerge until something like 2.8 billion years ago (bya) at the very earliest.* No, water won't do. So where are we going to get our hydrogen from?

This is where another feature of alkaline hydrothermal vents comes to the fore; serpentinization produces lots of dissolved hydrogen. Carbon dioxide and hydrogen will react to make methane and other long-chain molecules, but you need both energy and a catalyst to get things going, much in the same way that you need a match and a fire lighter to get a decent fire to take in a grate. As we know, not only do our bubbles have energy on tap, but their membranes are made from iron sulfide, a catalyst which is perfect for the job.

One of my favorite Monty Python sketches is from *The Life of Brian*. "All right, all right," says a revolutionary John Cleese, "but apart from better sanitation and medicine and education and irrigation and public health and roads and a freshwater system and paths and public order . . . what have the Romans ever done for us?" After my chat with Nick Lane, I feel

* We currently think that oxygenic photosynthesis evolved around 2.7 bya in a single-celled organism called cyanobacteria. By 2.3 bya the build-up of oxygen in the atmosphere became significant, producing the Great Oxidation Event.

the same way. Apart from the protection from meteorites, the membranes, the transition metals, the proton gradients, the iron sulfide catalysts, the dissolved hydrogen, and the carbon dioxide . . . what have alkaline hydrothermal vents ever done for us?

PUSHING CARBON UPHILL

Of course it's a long way from a few smallish carbon molecules in an iron-sulfide bubble to a single-celled life-form. Unlike the primordial soup theory, however, which produces life like a rabbit out of a hat, our vent-based proto-life can make its way there in stages. One of Mike Russell's triumphs has been to show that the iron sulfide membrane of an individual bubble is permeable. That means that small molecules can escape, but larger ones are trapped, ready to undergo further reactions.

In the broadest of terms, the reaction of carbon dioxide with hydrogen would first produce small carbon molecules like methane, formate, and acetate,* all of which would be allowed in and out of the membrane. In the next stage, these small molecules would react together to form medium-sized molecules such as amino acids and nucleobases. These would be trapped by the membrane, the proverbial fish in a barrel for further reactions which would then produce larger and larger molecules.

A crucial step would have been the creation of the first long-chain carbon molecule that was capable of making copies of

* I know you're gagging for some chemical formulae here, so methane is CH_4, formate is HCO_2-, and acetate is CH_3CO_2-.

itself, or, as we say in the jargon, self-replicating. In present-day organisms, this role is played by DNA, but it's unlikely to have been the molecule of choice for the very first life. For a start, DNA has a complex double-stranded structure, being a sort of "twisted ladder," with "sides" made of ribose and phosphate groups, and "rungs" made of pairs of nucleobases.* In fact you can think of a DNA molecule as more or less being two RNA molecules fused together,† a simple fact that has led many to suppose that RNA came first and later evolved into DNA.

Supporting this is the fact that DNA is essentially passive. To decode it, translate it into a recipe for amino acids, then assemble those amino acids into proteins requires RNA. Add this to the fact that RNA is able to write code into DNA, and catalyze reactions, and you start to build a picture of a busy chef who has decided to write his favorite recipes in a cookbook. Meaning that RNA is the chef, and DNA is the cookbook. All of which hints at an earlier epoch of life, predating DNA, when RNA ruled the kitchen. Evolutionary biologists call this the "RNA World"; life, but not as we know it.

The grail of researchers like Lane is to create RNA from scratch in a model alkaline hydrothermal vent. At the climax of my visit, he leads me into a pristine lab, where a glass cylinder the size of a bricklayer's thermos flask sits on a bench top,

* This rigid structure makes it very stable. One of my favorite science facts is that real, working DNA has successfully been recovered from the toe bone of a 130,000-year-old Neanderthal found in a Siberian cave, enabling the sequencing of the Neanderthal genome, and making *Jurassic Park* seem like a real possibility. More about the Neanderthals later.

† There are two main differences. Where RNA employs the nucleobase uracil, DNA uses its close relative thymine, and the ribose groups of DNA have had an oxygen atom removed when compared with those in RNA. Hence the names RiboNucleic Acid and DeoxyriboNucleic Acid.

trailing wires and surrounded by electronic monitors. It looks like a lava lamp on life support. I peer through the glass, not really sure what I'm expecting to see. Frankenstein's Molecule, perhaps? Could it be possible that, inside this sterile-looking experiment, new life is taking its first shuffling steps?

A JUMBO IN A JUNKYARD

One of the most famous critiques of the primordial soup theory was made by Fred Hoyle. In his 1981 book *The Intelligent Universe*, he professed bemusement that something as complex as a single-celled organism could come about by chance. With characteristic Yorkshire phlegm, Hoyle put it this way:

> A junkyard contains all the bits and pieces of a Boeing 747, dismembered and in disarray. A whirlwind happens to blow through the yard. What is the chance that after its passage a fully assembled 747, ready to fly, will be found standing there? So small as to be negligible, even if a tornado were to blow through enough junkyards to fill the whole Universe.

Can a proton gradient across the membrane of an iron sulfide bubble trapped in an alkaline hydrothermal vent achieve what a tornado in a junkyard cannot? After my encounter with Nick Lane, I'm in the "yes" camp. Most importantly, unlike the tornado, the vent doesn't have to make the entire cell in one go; it can do it in incremental steps. First, it just needs to do something simple, like take one molecule of carbon dioxide and react it with hydrogen to make methane. Next it synthesizes the building blocks: amino acids and nucleobases. What's more, any large molecules that form can't ever leave, trapped as they are within gel-like bubbles of iron sulfide, encouraging them to form ever longer chains: handy stuff like proteins, and nucleic acids.

And here's the crunch. Far from being a shot in the dark, life is a slam dunk. We should expect to find it anywhere there's a hydrothermal vent bubbling alkaline vent fluid into an acidic ocean, and such vents are a feature of all newborn, volcanic, wet, rocky planets. Far from being a statistical fluke, life is just the chemical pathway by which carbon dioxide reacts with hydrogen to form methane. Or, as Mike Russell succinctly puts it: "The meaning of life is to hydrogenate carbon dioxide."

FIRST TO THE PARTY

Can that be true? Certainly life got started very quickly on Earth. To see just how quickly, let's remind ourselves of how planetary systems form. First, a shockwave from a supernova creates pockets of high density in surrounding gas and dust. These pockets then collapse under gravity to form clusters of new stars. As each new solar system condenses, it spins faster and faster, flattening into a disk, much in the same way that a skater spins faster as she draws in her arms.

Out in the disk, planets begin to clump together under gravity. The temperature of the protostar determines what type of planet forms where. Close to the star is the rock line, where the temperature is cool enough to allow rock to solidify. Here's where we find small rocky planets like Mercury, Venus, Earth, and Mars. Further out is the snowline, beyond which water, methane, and ammonia all freeze, and we find giant ice planets like Uranus and Neptune. In between are the giant gas planets, like Jupiter and Saturn.*

Eventually, the temperature and pressure of the protostar become so great that it "switches on," and begins nuclear fusion. A blast of charged particles strafes the newborn solar system, blasting away the remaining gas and dust and leaving naked planets. For the first time, their home star lights the horizon. By this time the solar system is barely fifty million years old. Another fifty million years on, and the Earth is much like it is today, with a carbon-dioxide rich atmosphere, little land, and an acidic ocean.

And here's the kicker. Fast-forward a mere 300 million years, and the organism that we call LUCA, the Last Universal Common Ancestor of all life on Earth, is eking out an existence in an alkaline hydrothermal vent. It uses the proton gradient between vent fluid and sea water to hydrogenate carbon dioxide—meaning to replace one or both of its oxygen atoms with hydrogen atoms—releasing energy.

As we know, LUCA wasn't the first life. It was the product of millions of years of evolution, one tier of which had probably been RNA-based. Other kinds of worlds almost certainly predated these RNA worlds, but their self-replicating molecules are lost to us. However you dice it, 300 million years was not a great deal of time to produce something as complex as LUCA. Life isn't rare, at least not in its proto-cellular form. It works straight out of the box.

* Rocky planets tend to be small both because there's not much rock to go around, and because their orbits are small so there's a shorter path over which to hoover it up. The gas planets like Jupiter are in the sweet spot, where there's lots of material and a large orbit. The ice planets have even larger orbits, of course, but not so much material, so tend to end up somewhere in the middle size-wise. Beyond the ice planets, orbits are enormous but material is even harder to come by, hence dwarf planets like Pluto.

Yet LUCA, as I've hinted, hadn't yet left the safety of the vent. To do that, it needed to develop a membrane capable of generating its own proton gradient. Was that a roadblock on the path to complex life? It would seem not. This may come as a surprise, but there's growing evidence that LUCA left the vent not once, but twice.

POPPING THE EVOLUTIONARY HOOD

Ever since Darwin sketched his first "tree of life," expressing his idea that all life on Earth has evolved from a common ancestor, biologists have been arguing endlessly over which species begat which. Formally known as taxonomy, the guiding principle of this somewhat fraught discipline was to try and group organisms according to their physical traits.

The name of the game was to divide creatures into groups that shared the same characteristics, and then to rank those groups in some sort of evolutionary order. On one level, of course, this makes complete sense. The grand sweep of evolution can be more or less summarized as a progression from the simple to the complex, so you'd think that you'd be able to sift and sort organisms into some kind of time line. At one end, you'd have the simple stuff, like single-celled bacteria, and at the other you'd have the complex, multicellular things such as woolly mammoths. Get into the fine detail, however, and it's a different matter.

For a start, the more closely related two species are, the greater the similarity in their outward appearance and the harder it is to rank them. A difficult job isn't made any easier by the fact that traits can just as easily evolve out as evolve in. While the grand sweep may be toward complexity, on a shorter timescale there is a great deal of ebb and flow. Humans are a classic example. Neanderthal man, who lived alongside us in northern Europe only 39,000 years ago, appears to have had a bigger brain than we have, and may have been more intelligent than we are. If an alien taxonomist arrives on a barren Earth some million years hence, and finds a human and a Neanderthal skull, he could be forgiven for assuming that the Neanderthal version was the more recently evolved of the two.

And if that weren't bad enough, there's convergent evolution to deal with. As we shall see in much more detail in Chapter Seven, there are some traits which crop up time and time again, as nature evolves similar solutions to similar problems. Eyes, for example, have evolved independently scores of times. Both humans and octopuses, for example, have camera eyes. If anything, octopus eyes are slightly better designed as they lack a blind spot. Unfortunately for the taxonomist, this means that organisms that look alike aren't necessarily close relatives.

All of these complications combined to make taxonomy one of the most frustrating, controversial, and internecine disciplines ever created in science.* Once it became possible to examine the structure of DNA, however, all that changed. As we've already seen, DNA is a ladder-shaped molecule twisted into a helix, where the sides of the ladder are formed by alternating sugar and phosphate groups, and the rungs are made of the four bases, guanine, adenine, thymine, cytosine.

* It's worth knowing that before DNA sequencing, life was divided into five kingdoms: animals, plants, fungi, protists, and prokaryotes. Within these various kingdoms, there were divisions into phylum, class, order, family, and, finally, genus and species. As *Homo sapiens*, for example, we sit within the kingdom of animals, in the phylum of vertebrates, in the class of mammals, in the order of primates, in the family of the great apes, in the genus of humans, in the species of modern humans.

The exact sequence of those bases stores all the information needed to create the organism from scratch, be it an amoeba or an ostrich.

That sounds complicated, but the way that DNA functions can be grasped simply by renaming the four bases G, A, T, C. To cut a long story short, the bases form a four-letter alphabet that can be used to make up two kinds of DNA. The first type, called "coding regions," holds the recipes for all the proteins the host organism is made up of. Somewhere in your DNA there will be a coding region—or gene—with the recipe for hemoglobin, for example, which is the oxygen-carrying protein in the red corpuscles of your blood.

The second type of DNA, called "non-coding regions," are a little more mysterious. They far outnumber the coding regions—about 98 percent of your DNA is non-coding—and control how the coding regions are switched on and off, as well as acting as a kind of a junkyard where bits of code can be stored that might come in handy at a later date.

The entirety of an organism's DNA is known as its genome, and, since the invention of DNA sequencing by Fred Sanger in 1977, it has been possible to transcribe the bases for an entire organism. It all boils down to this. Whereas heredity used to be a matter of opinion, now it is a matter of fact. Effectively we can flip the hood of an organism and read off its genetic code, then compare it with the code of another organism. We can therefore see directly how the two are related, and construct a tree of life not from an organism's outward appearance, but from its genome. This new discipline is called phylogenetics, and it has transformed biology.

CHOOSE YOUR DOMAIN NAME

One of the first big surprises of phylogenetics was the discovery by the American biologist Carl Woese in the late seventies that a previously undiscovered branch of life had been hiding in plain sight. Sadly, I'm not talking about Bigfoot, which I'm fairly sure even a taxonomist would reveal to be a man from the Washington State tourist board dressed in a furry suit.* The branch of life that Woese discovered was altogether more modest in size. In fact, it was microscopic.

In a nutshell, Woese discovered that the kingdom everyone had been calling bacteria was actually two completely different kinds of single-celled organism. Although both creatures looked similar under the microscope, when it came to their genetic make-up they were about as different as it is possible to be. What's more, neither was clearly the ancestor of the other; both appeared equally ancient.

At the time it was thought that there were only two kinds of cell on Earth: those with a nucleus, named eukaryotes—Greek for "true kernel"—and those without, named prokaryotes, as in "before kernel."† As the names suggest, it was believed that the prokarya had preceded the eukarya. Woese's discovery was that prokarya were actually of two types, as distinct from one another as they were from eukarya. He proposed a whole new classification for living organisms, dividing them into three domains: the eukarya, the bacteria, and the archaea.

* Is it just me who finds it interesting that the US state with the highest number of Bigfoot sightings is the same one where flying saucers were first spotted?

† Going back to the five kingdoms: animals, plants, fungi, and protists are all eukaryotes. In case you're wondering, protists are single-celled eukaryotes, the most famous example being the amoeba.

One of the striking differences between the bacteria and the archaea is in the structure of their cell membranes. In both cases, the building block is a lipid with a water-loving phosphate molecule at one end, but, in the case of bacteria, the lipid is a fatty acid, whereas in the archaea it's an isoprene.* If life began in a primordial soup, how could this have come about? If there are two kinds of single-celled organisms with different cell membranes, aren't we asking for a junkyard tornado to assemble a jumbo jet not once, but twice?

Alkaline hydrothermal vents, of course, provide us with a possible answer. LUCA lived in a vent, and it needed to evolve a membrane in order to leave. And it did. Twice. One iteration gave rise to the bacteria and the other to the archaea. Independently, each evolved its own proton pump, a nanomachine in its cell membrane capable of recreating the proton gradient that had powered the metabolism of LUCA. Both domains still use this mechanism today, storing energy by pumping protons across their membranes, then allowing them back through in order to generate ATP.

So far, so good. Given an alkaline hydrothermal vent, LUCA is easy to make. So are membranes; so easy they evolved twice. Is this the case all the way down the line? Is every step along the path from single-celled life to technologically advanced civilizations the evolutionary equivalent of a cascading line of dominoes? If you are hoping the answer is yes, I am here to disappoint you. The creation of intelligent life appears to have hinged on one extraordinary event. If you're new to biology, all I can say is you are in for a shock.

* Fatty acids are basically zigzag chains of hydrogenated carbon. Isoprenes are, too, but they also have side branches, also made of hydrogenated carbon atoms.

A PLAGUE ON BOTH YOUR HOUSES

To summarize where we've gotten to so far, we've learned that, far from being the statistical equivalent of a Boeing assembled by a junkyard tornado, the last common ancestor of all life on Earth arose swiftly by a series of high-probability steps. Our best guess is that it set up shop in iron sulfide bubbles, sheltered in the limestone chimney of an alkaline hydrothermal vent. It set sail into the ocean with a brand new membrane on at least two occasions, and our present-day bacteria and archaea are the direct descendants of these two rival species. Crucially, what happened next was . . . nothing.

The sad fact for fans of intelligent life is that the bacteria and the archaea have remained single-celled and, well, dumb, from their inception right up until the present day. For four billion years they have resolutely avoided evolving into anything remotely resembling a complex multicellular organism, let alone a technologically advanced intelligent one.

In truth, they haven't really needed to. Whatever we humans might want to believe, the world belongs to archaea and bacteria. They are, without question, the most successful organisms on the planet, bar none. Even in our own bodies, they outnumber our cells ten to one. We find them in the ocean, the atmosphere, salt lakes, even in the reactors of nuclear power plants. Almost anywhere there is a source of energy, bacteria and archaea have found a way to exploit it, but never in order to become more complex. All their evolution has been biochemical; structurally, they remain simple bags of chemicals, reproducing, feeding, excreting waste, dying, and doing little else.

Not that they haven't left their mark on the planet. One of their most striking contributions has been to excrete oxygen into the atmosphere, a gas completely absent from the primordial Earth. We find the first evidence of free oxygen at around 2.3 billion years ago, over a billion and a half years after LUCA's bubble burst and it left the vent. Nicknamed the Great Oxidation Event, this marks the bacteria's discovery of a sophisticated biochemical pathway called oxygenic photosynthesis, the harnessing of sunlight to rip electrons from water and force them on to carbon dioxide, fixing carbon and releasing oxygen as a waste product.*

Oxygen, as we all know, is reactive, and once it entered the atmosphere it rapidly set about bonding with anything it could get its needy little orbitals on: All too soon it was rusting metals, oxidizing salts, and converting atmospheric methane into carbon dioxide. Methane is a potent greenhouse gas, and removing it had a profound effect on climate. In fact, it's believed that falling methane levels were a significant factor in triggering the Huronian glaciation, when the global temperature dropped dramatically, causing runaway growth of the polar ice caps to the extent that the entire planet froze over in what's known as a Snowball Earth.

And that might have been that, but for one extraordinary event that was to change the entire trajectory of life on Earth. Without it, we'd still be living in a world of microorganisms. There would be no animals, no plants, no fungi, and no amoebae. Few living things would be visible to the naked eye other than the odd bacterial colony. So far as we can tell, this extraordinary event happened only once, and it created a new kind of

* Other kinds of photosynthesis came first. In fact, one of the first indications we have of life on Earth are so-called "banded iron" formations, dated at 3.2 billion years old, caused by an early kind of photosynthesis that used iron instead of water.

cell, the eukaryote. Unlike the bacteria and the archaea, eukaryotic cells are highly organized and have a nucleus. Much of the origin of these extraordinary cells remains a mystery, but this much we know: They were created by the enslavement of a bacterium by a hungry archaeon.

MIGHTY MITOCHONDRIA

There are two kinds of prokaryotes in this world, those that create their own food from inorganics, called autotrophs, and those that feed by eating other cells, called heterotrophs. Either way, the goal is to end up with glucose, which can then be burned to create energy in a process called respiration. Usually that means shoving it into something called the Krebs cycle, a repeating chain of biochemical reactions which pump protons across the cell membrane before allowing them back through to generate ATP.

Why am I telling you this? Because something like one and a half to two billion years ago, a heterotrophic archaeon—that's singular for archaea, by the way—ate an autotrophic bacterium. Or at least it tried to. It engulfed it and attempted to digest it, no doubt intending to break it down into sugars that it could then shove into its very own Krebs cycle. Thankfully for us it failed.

Instead, the bacterium survived. In fact, it more than survived; it thrived. Together, the two organisms negotiated a delicate pact. The bacterium became a permanent fixture within the archaeon, forming what biologists call an endosymbiosis. In return for shelter and a ready supply of glucose, the bacterium used its Krebs cycle to supply the archaeon with ATP. In effect, it became a power plant for its host. Why was all of this so groundbreaking? The reason, as ever, is to do with energy and information.

Because bacteria and archaea store energy by pumping protons across their membrane, they have a problem. The bigger they get, the less membrane they have relative to their mass, and the less efficient they get. You can't have more than one membrane, but you can have as many slave bacteria as you like. By outsourcing respiration to an army of supplicants, the archaeon was able to generate an extraordinary amount of energy. A whole new level of complexity became possible. The bacterium became a mitochondrion and the eukaryotic cell was born.

As I hope you can see from my drawing on the next page, they were a world apart from their prokaryotic counterparts. Where bacteria and archaea are one-horse towns, the eukarya are sparkling citadels, full of eye-catching new structures. We've already mentioned the power plant that is the mitochondrion, a stripped-down autotrophic bacterium whose job it is to supply energy to the cell; and the copyright library, in the form of the nucleus, that could now house an almost limitless quantity of DNA. In addition, these eukaryotes have a factory where RNA can assemble proteins, called the endoplasmic reticulum, and a UPS service called the Golgi apparatus which can package up those proteins for export. There's a road network in the shape of the cytoskeleton, a web of pathways throughout the cell along which metabolites can be transported, and a waste-processing plant in the form of a lysosome. Life 2.0 had arrived.

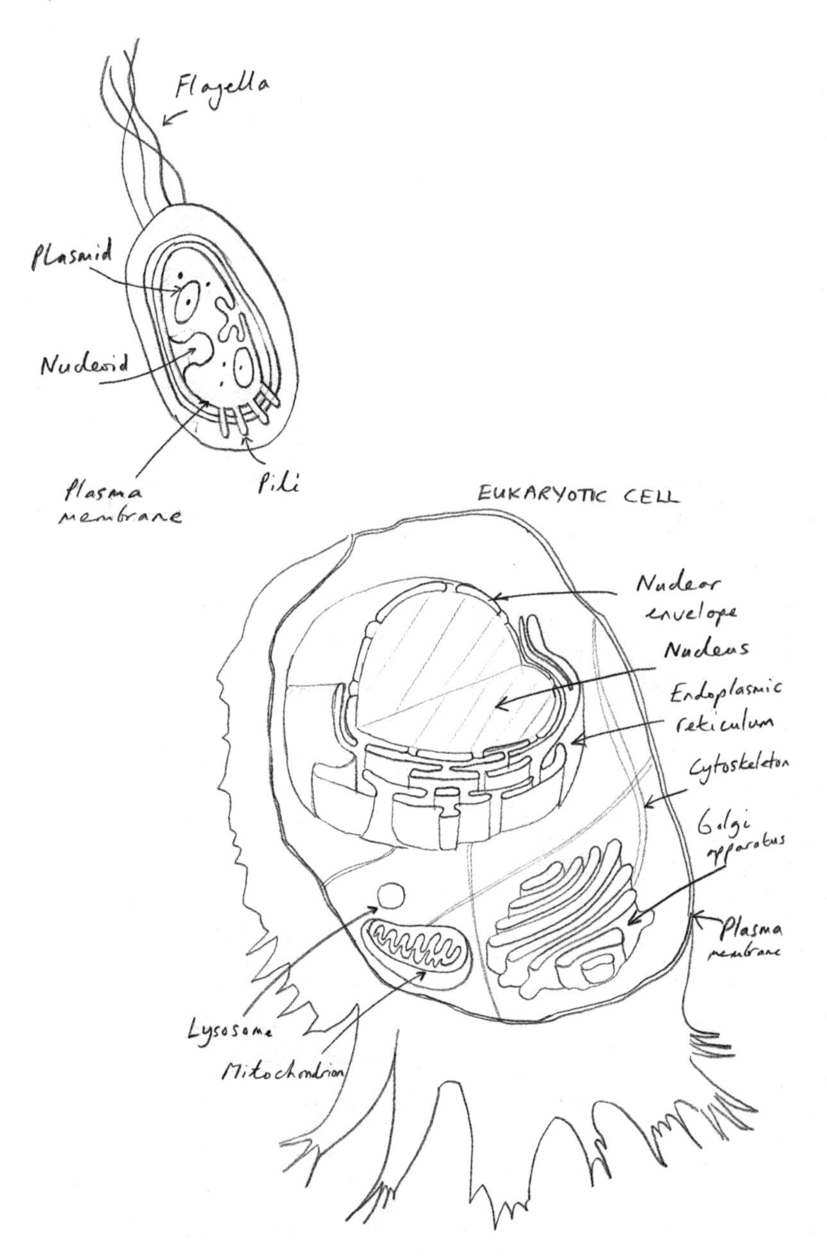
PROKARYOTIC CELL
Flagella
Plasmid
Nucleoid
Plasma membrane
Pili
EUKARYOTIC CELL
Nuclear envelope
Nucleus
Endoplasmic reticulum
Cytoskeleton
Golgi apparatus
Plasma membrane
Lysosome
Mitochondrion

GIVE ME SOME OXYGEN

The stage was now set for complex life. When the last of the Snowball Earths turned to slush around 635 million years ago, the first multicellular creatures made a tentative exploration of the newly warm shallow oceans.* We call this period the Ediacaran, and it saw the rise of some truly bizarre new life-forms as well as a dramatic increase in the level of oxygen. If you want to catch a glimpse of beings that really look alien, you need to Google the Ediacaran biota.

That name "biota"—meaning "living part of a biosystem"—is well chosen, because in many cases you'd be hard-pressed to say whether the Ediacaran fossils we have are made of sponges, plants, fungi, jellyfish, or something else entirely. The earliest appear to have been large disc-like creatures, rooted to the ocean floor at considerable depths. They don't have mouths or limbs, and we can't tell if they had internal organs. Our best guess is that they lived alongside microbial mats, absorbing nutrients through their skin.

Unfamiliar as they are, however, these strange creatures represent a crucial step toward intelligent life. Not only does size bring security—it's hard for a heterotrophic microorganism to swallow a sponge—but it also brings greater energy efficiency. Add to that the fact that burning glucose with oxygen produces much more ATP than fermenting it, and you can

* We used to think the increase in oxygen "released the brakes" on evolution, and enabled the evolution of complex life. We now believe it was the other way around; the evolution of the Ediacaran biota helped oxygenate the oceans and sea bed through filter-feeding and burrowing.

start to see how much more energy was becoming available to drive complexity.*

The end of the Ediacaran saw a boom in complex life not seen before or since. Almost overnight, a new breed of exotic creatures swept the globe. We call them the metazoans, or animals. Fittingly, we call this extraordinary radiation the Cambrian Explosion, and its resonance can be seen in all intelligent life today. Again, their key innovation was to do with energy. I'll call a spade a spade: life evolved a mouth, a gut, and an anus.

LET'S HAVE US A BILATERAL TRIPLOBLAST!

High in the Canadian Rockies, in British Columbia, there's a small limestone quarry shot through with a seam known as the Burgess Shale. This is the site of arguably the most important fossil find in history, the paleontological equivalent of Pompeii. Five hundred and forty-one million years ago, right at the beginning of the Cambrian, a mudslide into shallow water engulfed a profusion of bewilderingly diverse life-forms. The fossils they left behind are exquisitely preserved, as if a trawler's net had been lowered into the prehistoric ocean and hauled on to the deck.

* Because there were relatively low oxygen levels (roughly 1 percent atmospheric) when the eukaryotic cell came into being, it seems likely that the very first mitochondria fermented glucose rather than burning it in oxygen. The progenitor of endosymbiont theory, Lynne Margolis (ex-wife of Carl Sagan, by the way), had a different view. She believed that the first mitochondria were aerobic (oxygen-burning) rather than anaerobic (fermenting), which is why the host archaeon liked them so much, given that the oceans were filling with free oxygen. Who's right? As my father used to say, you pays your money and takes your choice.

That said, not one member of this once-in-a-lifetime catch looks like anything you would want to deep-fry and eat with chips. Opabinia, for example, is like a slug in a ball gown with a single ominous pincer. Aysheaia looks like a roll of linoleum with a mouth at one end, and Marrella resembles nothing so much as an overcreative trout lure. Yet, strange as they first appear, some of these long-extinct creatures share a basic body plan which we humans have inherited. In short, they are bilateral triploblasts* and they are the proud owners of a digestive tract.

That last bit is important, because—as you might have guessed—a digestive tract is yet another way to up the energy stakes. What better way to make a living than by shoving whole organisms in at one end, shredding them with teeth, digesting them with enzymes in the gut, then ferrying the resultant glucose to your mitochondria to generate swathes of valuable energy? Not to mention the rather satisfying moment when you egest everything you don't want in a lazy dropping.

No one is exactly sure what became of the sightless and mouthless Ediacarans, but it's a fair guess that many of them spent their final hours in the gut of a bilateral triploblast. With the ability to hunt and eat, the stage was now set for this particular brand of multicellular life to reap even more energy from the oceans, increasing its complexity along the way. Most of you will know the romance which follows. Like all the best stories, it divides neatly into three acts. In the first, life is established on land. The second sees the rise of the dinosaurs. And the third, in which we are lucky enough to be living, sees the ascendance of mammals.

* Bilateral, as you might guess, means their left side mirrors their right side. The triploblast bit refers to the fact that they have a body cavity arranged about their digestive tract.

THE AGE OF FISH AND PLANTS

The Paleozoic Era, which runs from the Cambrian Explosion to the Great Dying at the end of the Permian, spans roughly half the lifetime of complex life. Although there were no more Snowball Earths, the planet still endured two Ice Ages.* The main events were the evolution of land plants and the subsequent evolution of fish into amphibians, and then into four-legged land animals, known in the trade as tetrapods. By the end of the Permian, the sixth and final period of the Paleozoic—see my handy guide to geological periods on pages 193–194—there were trilobites and fish galore on the ocean shelves, and horrendous snaggle-toothed ancestral mammals called Gorgonopsids roaming the forests, the largest of which was the size of a grizzly bear.

Not that you are ever likely to meet one, because, as the name suggests, the Great Dying is the largest of the known Big Five extinctions.† When you hear Al Gore talk about global warming, this is exactly the kind of nightmare scenario he is worrying about. At the time the continents were all joining together in one huge geological love-in called Pangaea, and fissure-like volcanoes produced a carpet of lava the size of continental Europe. The colossal amounts of carbon dioxide released caused a runaway greenhouse effect, and the mean global temperature rose by some six degrees.

* The Andean-Saharan lasted thirty million years, and spanned the Ordovician–Silurian extinction. The Karoo Ice Age lasted roughly sixty million years, and took place during the Carboniferous.

† I know you want to know, so these are: the Ordovician–Silurian (the former pronounced "Ordo-vee-shan"), the Late Devonian, the Permian–Triassic, the Triassic–Jurassic, and the Cretaceous–Paleogene.

Ocean currents, as you may know, rely on ice at the poles to sink cold oxygenated water, which then wells up in the tropics. When the polar ice melts, as it did during the Great Dying, these currents switch off and tropical waters become less oxygenated. That means curtains for oxygen-loving marine life. Not only that, but the center of Pangaea became a barren desert, eradicating swathes of newly evolved land species. All told, at the end of the Paleozoic a staggering nineteen out of twenty species went extinct. Think about that the next time you fill up with diesel.

THE AGE OF THE DINOSAURS

The following era, the Mesozoic, also ended in a mass extinction.* Thanks to the unbounded interest of children the world over, these warm geological periods are the ones we know best: the Triassic, Jurassic, and Cretaceous. The Triassic essentially saw the recovery of life after the Great Dying, only to be followed by what many believe was a meteorite impact at the Triassic–Jurassic boundary. The resulting extinction enabled the rise of the dinosaurs, and also saw the emergence of our direct ancestors, the mammals. By the time of the Cretaceous, mammals had diverged into two main groups: those which gestated their young in abdominal pouches and those with placentae. It is from placental mammals that primates, and therefore we, are descended.

* By now you may be getting the picture; geological periods are often defined by extinctions. Typically reddish oxygenated rocks like limestones and sandstones are followed by black, unoxygenated ones like slates and shales. They tend also to be named after the places where they were first discovered: The Cambrian, for example, is Latin for Wales; the Silurian is named after the Welsh tribe the Silures. The Jurassic gets its name from the Jura Mountains, close to CERN in Switzerland. And the Devonian is named after, well, Devon.

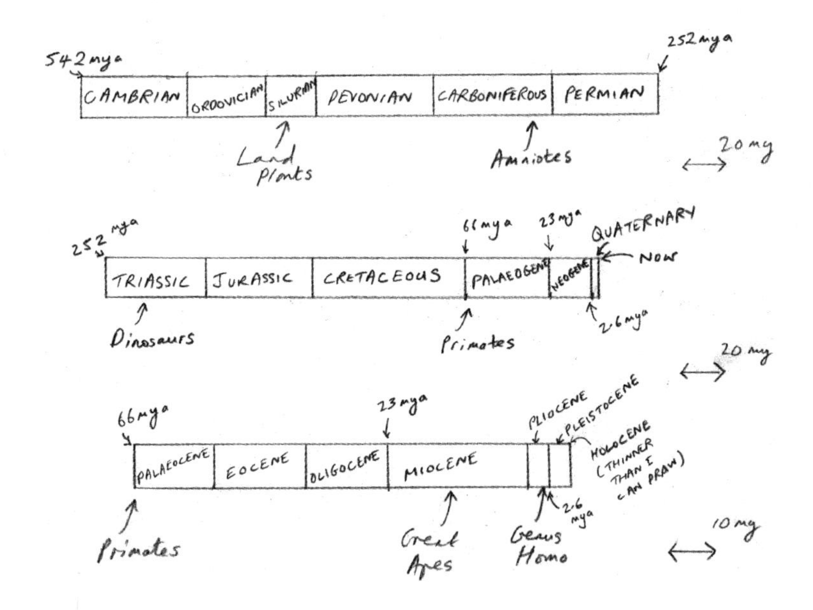
Recent Geological Time
542 mya
252 mya
CAMBRIAN
ORDOVICIAN
SILURIAN
DEVONIAN
CARBONIFEROUS
PERMIAN
Land Plants
Amniotes
20 my
252 mya
66 mya
23 mya
QUATERNARY
Now
TRIASSIC
JURASSIC
CRETACEOUS
PALAEOGENE
NEOGENE
Dinosaurs
Primates
2.6 mya
20 my
66 mya
23 mya
PLIOCENE
PLEISTOCENE
HOLOCENE (THINNER THAN I CAN DRAW)
PALAEOCENE
EOCENE
OLIGOCENE
MIOCENE
Primates
Great Apes
Genus Homo
2.6 mya
10 my

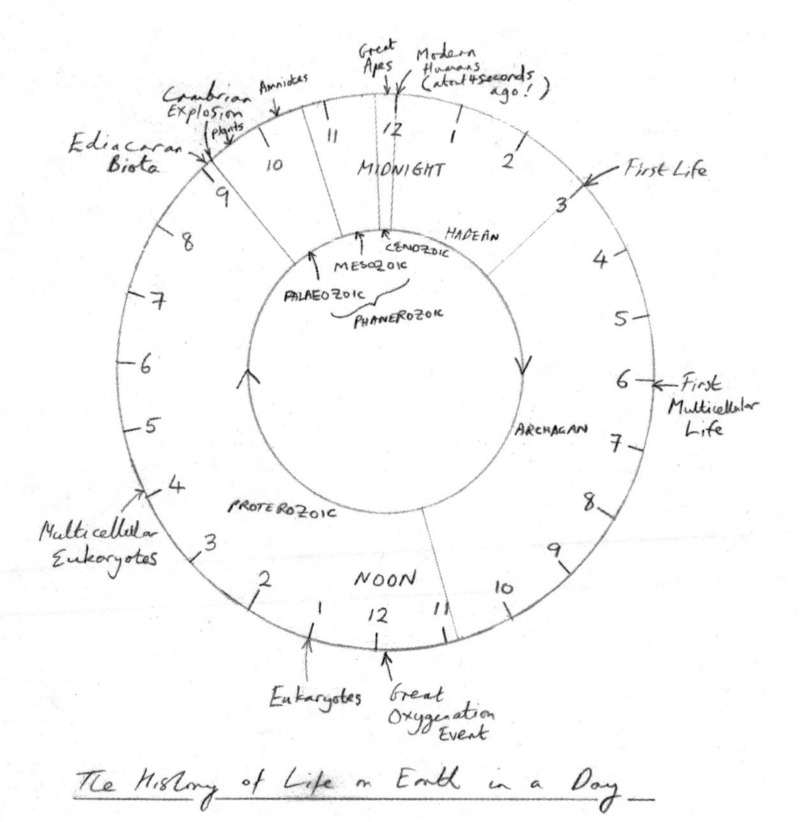
Great Apes
Modern Humans (about 4 seconds ago!)
Amniotes
Cambrian Explosion
plants
Ediacaran Biota
MIDNIGHT
First Life
HADEAN
CENOZOIC
MESOZOIC
PALAEOZOIC
PHANEROZOIC
First Multicellular Life
ARCHAEAN
PROTEROZOIC
Multicellular Eukaryotes
NOON
Eukaryotes
Great Oxygenation Event
The History of Life on Earth in a Day

We now find ourselves at the boundary of our present era, the Cenozoic. As already mentioned, the Mesozoic ended with a bang some sixty-six million years ago, when a giant space rock some six miles in diameter slammed into the Yucatán Peninsula in Mexico, releasing a billion times the energy of the atomic bomb that fell on Nagasaki. When random catastrophes like that occur, it tends to be the large predators at the top of the food chain that suffer. In this case that was the dinosaurs, of which only the birds survived.* As we all know, our ancestors the placental mammals were all too happy to step into the breach.

THE AGE OF MAMMALS

Evolution, as you can see, often proceeds by fits and starts. The first step of the cycle is a radiation, where all sorts of genetic experimentation goes on as organisms adapt to new niches. The next is the dominance of a few particular forms: the land plants, say, in the Silurian and Devonian.† Next comes an extinction, where the vast majority of species are wiped out on an utterly random basis. For the Gorgonopsids, that reckoning was the Permian extinction. The cycle is then free to begin again. And, true to form, a radiation of reptiles in the Triassic saw the rise of the dinosaurs in the Jurassic and Cretaceous.

After the greenhouse world of the dinosaurs, the Cenozoic has generally seen a slow decline in global temperature, leading to the present Ice Age, the Quaternary, which began 2.6

* Other large creatures also managed to sneak through what is called the K-Pg extinction event, such as the crocodiles.

† It's possible that the rapid expansion of land plants sucked up so much carbon dioxide that they caused the Karoo Ice Age in the Carboniferous.

million years ago. I know what you're thinking; if this is an Ice Age, you hadn't noticed, but an important feature of the Quaternary has been a gradual back and forth of polar ice. Epochs where the ice advances, known as glaciations, are interspersed with ones called interglacials where the ice retreats. As you might have guessed, we are in an interglacial right now, called the Holocene, where temperatures have been remarkably stable at roughly present-day levels for some 11,000 years.

But back to my point. It's thought that, as the polar ice formed, the planet became drier, and the African forests began to dwindle, giving way to savannahs.* Forests, as we all know, are the natural habitat of apes. Could that have been the selection pressure that encouraged our ancestors to leave the trees? Whatever the reason, some seven million years ago they began to diverge from the ancestors of chimps. Interestingly, the adaptation that helped them wasn't one that made them smarter. That came much later. They diverged because they could walk on two feet.

What good did that do them? Well, not much, so far as we can make out, for several million years. The fossil evidence tells us that for the following four million years they gradually got better at walking upright, but that was about it. We find Ardepithecus shuffling across the forest floor some five million years ago, and Australopithecus sauntering upright on the savannah about a million years later, but they weren't the

* After an initial hot flush, the Cenozoic has seen a mighty cooling. We think the Antarctic first developed glaciers around thirty-four million years ago, then began to cool dramatically once it became isolated from South America, roughly twenty-three million years ago, eventually becoming covered in ice fourteen million years ago. The Arctic took a little longer, and iced over about 3.2 million years ago.

smartest customers, with individual brain capacities roundabout the same as that of their cousins the chimps.

Clearly they were doing something right, because just under two million years ago we find a radiation of hominins, or "human-like" species, all living cheek-by-jowl in Africa.* Top billing goes to *Homo erectus*, with a brain capacity twice that of a chimpanzee, increased body size, and smaller teeth. Not only was this species the first true hunter-gatherer, it was also the first tourist, with some members leaving Africa to take up residence in Asia. The "smaller teeth" bit is particularly interesting, because it may hint at another link between energy and complexity. In short, there are some indications that *Homo erectus* could cook.

Although we have yet to find solid evidence that it could control fire, the fact that *Homo erectus* has small teeth, small guts, and slept on the ground rather than in the trees are all intriguing clues. After all, cooking breaks down the long-chain carbon molecules in food, making it easier to chew and digest, dispensing with the need for a robust gut and dentition. It also means you can extract more energy from that vital adaptation, the gut. You can then expend that energy making your brain more complex, which in turn makes you a better hunter, creating a virtuous cycle.

There's another thing about cooking, of course. It's also a group activity, set around a campfire, which must have encouraged all sorts of group bonding. As we shall see in the next chapter, intelligence is by no means confined to humans, but all the animals in which it occurs are social. It also encourages parenthood. Baby chimps are pretty much on their own

* Also around at the time were *Homo habilis* and *Homo ergaster*, as well as *Paranthropus robustus* and *Paranthropus bosei*.

once weaned, but cooking is technical, and not the kind of thing you'd ask a baby to do unless you were really in a rush. Cooking therefore encourages the dependency of children on their parents, buying time for childhood, a time of imaginative and creative development.

PULLING MUSSELS FROM A SHELL

To be fair, there's currently a great deal of disagreement about what *Homo erectus* could or couldn't do, and conclusive proof of tall stories by the campfire is a long way off. The conventional view is that they changed little in nearly two million years, using the same basic stone tools right up until their extinction some 140,000 years ago. In a book about communicating with aliens, however, I can't help but mention a recent and very controversial discovery.

This is also worth a Google. In scientific terms it's a scratch on a mussel shell, but it would take a bard to do it justice. Recently dated at 500,000 years old, it comes from a treasure trove of *Homo erectus* fossils found on the Indonesian island of Java by a Dutch surgeon named Eugene Dubois in 1891. Something—or someone—has carved a zigzag pattern on it, and, when you see it, it's hard not to feel a deeply human connection. I defy anyone to look at those markings, made by an ape nearly half a million years ago, and not to imagine that either it knew what beauty is, or sought order in a confusing world, or at the very least was bored and looking for something to do.

The next time in the fossil record we find anything like these patterns, they are scratched in ochre by anatomically modern *Homo sapiens*.* The site they come from, the Blombos Cave on the southern Cape coast, has been one of the richest sources of early human artifacts ever discovered. Dated at 100,000 years old, these criss-cross etchings speak to the existence of beings just like ourselves, capable of imagination, creativity, and abstract thought. Something crucial is happening here. Information is no longer simply being stored in DNA. It is being stored in a network of brains. The name of that network, of course, is culture.

GONE WALKABOUT

After a genesis in Africa some 200,000 years ago, *Homo sapiens* rapidly spread across the Levant, heading for Asia and Europe. With a seemingly solitary leap, its brain size had increased by over a third in comparison to *Homo erectus*, and was now over three times the size of that of a chimpanzee. If art is a proxy for language, as many paleoanthropologists believe, the indication from the Blombos Cave is that even these very first *Homo sapiens* were singing "I don't know but I've been told" as they walked.

Or maybe they just chatted about the weather. Either way, 45,000 years ago they reached Europe, where they encountered the Neanderthals, a separate hominin line that had already been living there for some 150,000 years. A talented bunch with their own toolmaking and burial traditions, as already mentioned the Neanderthals had bigger brains than

* We aren't completely sure how we are related to *Homo erectus*, but one of the most widely accepted hypotheses has it that the African branch of *Homo erectus* evolved into *Homo heidelbergensis*, which then spread to Europe and Asia, where it evolved into *Homo neanderthalis*. African *Homo heidelbergensis* then evolved into *Homo sapiens*.

us and were arguably better adapted to the cold climate. Whatever happened, it doesn't seem to have involved much cooperation, because within 5,000 years the Neanderthals were extinct. Never let it be said that humanity doesn't have a dark side.

It's around this time that we detect what is often called the "Great Leap Forward": a step change in human culture. There's evidence of elaborate burial rituals, of wearing animal skins as clothing, and using pit-traps to hunt prey. By the time we reach 40,000 years ago, you can hardly move in Europe or Asia without stumbling across all manner of cave paintings, jewelery, fishing hooks, and flutes.

The invention of farming, roughly 11,500 years ago at the end of the last glacial maximum, secured a growing network of villages with a reliable energy supply. A few millennia later, around 3600 BC (5,600 years ago), came the founding of Sumer, the world's first civilization, and, by 3100 BC, Sumerian had become its first written language. Now information could be stored not just in the brain, or in a network of brains, but as hard copy. And progress continues to be relentless. Today, a dwindling supply of energy from fossil fuels supports a worldwide population of seven billion souls, connected by a digital network that contains just about every single bit of information amassed by humanity to date.

THE UNBEARABLE LIGHTNESS OF BEING

So that's why we've never heard from aliens, right? The whole of human evolution hinges on one key event, the creation of the eukaryotic cell, without which the Earth would still, even four billion years later, be nothing but a petri dish for bacteria and archaea. There's nothing out there for us to talk to, because, although single-celled life is common, complex life is rare, and intelligent life rarer still. The series of flukes that led to intelligence in humans hardly makes communicable civilizations look like a dead cert.

So what are the chances that I should be here? Let's assume for a moment that the time in Earth's history when a given organism evolves is a good guide to its probability. This is a bit like rolling a die once a minute. After six minutes, all things being equal, you'd expect to have rolled at least one six. Since the Earth has been rolling the die for roughly four million years, and assuming the rate of mutation is roughly constant, we can calculate the probability of the various stages of life. Single-celled life has been there from the first roll of the die, so its probability is 1. Eukaryotic cells emerged halfway through Earth's history, so their probability is ½, or 1 in 2. And we humans evolved 200,000 years before the present, so our probability is 200,000/4,000,000,000, or 1 in 20,000.

Of course that calculation is riddled with so many assumptions it's little more than a curiosity. The biggest is that the Earth is somehow typical of all life-bearing planets. To really get a feel for the probability of complex life, we need detailed knowledge of a large sample of life-bearing planets, something we are unlikely to have for at least a decade or two. Another is that only complex organisms with our own particular genetic make-up—that's to say bipedal hominins—possess the kind of intelligence necessary to evolve technology and communicate. But it doesn't look encouraging, does it? It suggests that simple life is common, but intelligent life is rare. From time to time a light comes on, but everywhere else is in darkness. Our project is in vain.

Well, not quite. Because, as we are about to see in the next chapter, it turns out that we humans are not quite as unique as we might think. There are other intelligent species on the planet, and an argument can be made that intelligence is just as common an adaptation as land-dwelling or flight. As we are about to see, there's a good chance that if we rewound the tape and played evolution out again, we'd find ourselves on a planet of the apes, dolphins, or crows.

Not only that, but by focusing so sharply on the fine detail of evolution here on Earth, we've neglected the bigger picture. Yes, we are here because of a series of biological and climatic flukes. But we are also here because we had the time to evolve. We happen to live on a stable planet in a stable solar system, orbiting a quiet, long-lived star. When we look out into the galaxy, hoping to see our neighbors waving back at us, we are assuming that planets and solar systems like ours have existed since the Milky Way was founded, some thirteen billion years ago. But what if that's just not the case? What if life started on Earth at the same time that life could start anywhere in the galaxy? What if we are one of a whole host of civilizations that are just waking up? In order to climb evolution's ladder to complexity, we need to avoid its biggest snake: gamma ray bursts.

HERE BE DRAGONS

In 1963, following the Nuclear Test Ban Treaty, the US launched the Vela satellite, essentially to make sure that what was then the USSR was playing ball. These high-flying satellites were tuned to detect the distinctive pulses of visible light, radio waves, x-rays, and gamma rays emitted by nuclear weapons tests. Instead, they found something else: bursts of pure gamma rays coming from outside the solar system.

To begin with it was assumed that these bursts must be coming from somewhere within the Milky Way, but by the mid-1990s it was realized that they are in fact coming from distant galaxies, many of which are almost halfway across the observable universe. To be as bright as they appear in our sky, having traveled billions of light-years to reach us, means that they must be extraordinarily intense. One particularly sobering fact is that a typical gamma ray burst (GRB) contains as much energy as the Sun radiates in its entire lifetime, condensed into a pulse lasting just a few seconds.

We still aren't exactly sure what causes these blistering infernos. The shorter bursts are thought to be caused by the collision of binary neutron stars. The longer, more powerful bursts are thought to result from the gravitational collapse of colossal stars, known as hypernovae. One thing is certain, however. You don't want to be anywhere within 10,000 light-years of one, or you'll be fried. The near side of any Earthlike planet would be toast, and the far side would then get blasted by a shower of secondary radiation. The ozone layer would be obliterated, and any remaining life would then be decimated by the Sun's ultraviolet radiation.

We now have a lot of data on GRBs, and in 2014 two astronomers named Tsvi Piran, of the Hebrew University of Jerusalem, and Raul Jiménez, of the University of Barcelona, crunched the numbers to find out what the risks are. Their findings make extremely interesting reading. For a start, they calculate that there's a 90 percent chance that the Earth has been fried by at least one GRB at some time during the last 4.6 billion years, and a 50 percent chance that it has been hit within the last half-billion. Could one of the Big Five mass extinctions have been due to a GRB?*

But it's their conclusions about the suitability of the universe to past life that really give you pause for thought. For a start, only 10 percent of all galaxies have few enough GRBs to support life, and, even then, you'd better be nowhere near the center, where they tend to blow up more frequently. The Earth's position, some 25,000 light-years from the center of the Milky Way—which is, as you might have guessed, a member of that lucky 10 percent—is now looking like prime real estate. But, most importantly, they calculate that life would have been impossible on any planet, in any kind of galaxy, anywhere in the universe before five billion years ago.

What does that mean? The bottom line: We can trace two imaginary spheres around the Earth. The first is five billion light-years in radius, and within it we can expect to find single-celled life. Beyond that sphere the universe is barren, because any galaxy older than that is still being fried by GRBs. The second sphere is a billion light-years in radius. If life on Earth is typical—and that's a fairly big "if"—then within it we can expect to find technologically advanced societies. After all, it took four billion years for humankind to develop radio technology, and if that represents some kind of average, then the very oldest communicable societies will be a billion years ahead of us at most.

And finally, it raises a question. What if the Earth isn't typical, and the rise of our own technological society has been inordinately rapid? What if, on average, it takes six billion years to invent radio technology, rather than our own four billion?

* A prime candidate is the Ordovician–Silurian, for which there seems to be no convincing climate event or meteor strike.

In that case, we have a chilling solution to the Fermi paradox: We are alone. And if it takes, say, an average of five billion years to evolve a communicable society? Well, then, it's all to play for. Maybe the galaxy isn't dead. It's asleep. And round about now it's going to start waking up.

CHAPTER SEVEN

ALIENS

In which the author conjures real aliens, and learns that life is a servant of two masters. One, consciousness, hopes that the universe lasts forever. The second, cosmos, seeks an untimely end. No prizes for guessing which one is winning.

At first he assumed it had to be some kind of hoax. As Assistant Keeper of the Department of Natural History of the Modern Curiosities of the British Museum, George Shaw was often confronted with biological oddities, and a depressing number proved to be illegitimate. Accompanying the pelt was a sketch by its donor, Captain John Hunter, governor of the newly founded colony in New South Wales, purporting to show the animal when alive. Could the drawing be a fake, too? Even by Antipodean standards, the creature that confronted him was bafflingly bizarre.

Where to start? Its tail was flat, like a beaver's; its body was like that of an otter, or maybe a mole. Or, come to think of it, a seal; yes, he thought, that fur resembled nothing so much as a seal pelt. Like a seal, each of its four legs was home to a webbed foot. Yet search as he might, he could find nothing on the creature's abdomen that in any way resembled a nipple. As proclaimed by the Swedish zoologist Carl Linnaeus in his definitive *Systema Naturae*, the class of Mammalia was reserved for "animals that suckle their young by means of lactiferous teats". But if it wasn't a mammal, what was it? Could it possibly belong within the class of Amphibia, many of whose members "appear to live promiscuously on land or in water"? Yet who had ever heard of an amphibian with fur?

And then there was the business of the creature's rear ankles. On each was a pronounced spur, much as you might see on a champion cockerel at a country fair. And that was the least striking of the creature's avian characteristics.

To be blunt, it had the head of a duck. Or, at least, the mandible of a duck; there was something distinctly fishy about its minute eyes, so small that one had to hunt to find them amidst the tufts of fur. Shaw was well acquainted with the so-called "mermaids" of London society; freakish fakes wherein charlatans had procured monkeys' torsos and grafted them to the tails of fish. Had some Eastern mountebank sewn the bill of a duck on to the body of a mole or beaver?

He examined the base of the beak, surrounded as it was by a circular flap of somewhat leathery skin. Was the flap there to hide some surreptitious stitching? Try as he might, he could find no traces of thread that might prove the lie. Perplexed, he took a shallow dish and doused the entire pelt in water to see if he might loosen whatever glue was holding it all in place. It refused to fall apart. A horrible truth began to dawn on him. He was going to have to classify this monstrosity by naming its genus and species. But what was he looking at? A bird, a mammal, a fish, or a reptile?

A WALK ON THE WILD SIDE

If there is communicable alien life out there in the galaxy, what is it like? The possibilities seem endless. Life on Earth is bewilderingly diverse. We find it everywhere, both microbial and multicellular, eating every conceivable type of food and inhabiting a myriad physical forms. Surely no one could have predicted the existence of the duck-billed platypus from first principles, no matter how much they knew about cellular biology, genetics, paleoclimatology, and the ecology of east Australian rivers? And if we can't second-guess life on our own planet, what can we ever hope to deduce about aliens?

Surprisingly, the situation isn't nearly as hopeless as you might think. As we are about to see, there's a wealth of information about life on Earthlike planets to be gleaned from life-as-we-know-it, and an abundance of informed speculation to draw on with regard to life-as-we-don't. Riddles such as the duck-billed platypus have given up their secrets, and the lessons we've learned are particularly instructive when it comes to alien-hunting. Not only are we about to meet a menagerie of exotic, exquisitely alien life-forms inhabiting all manner of weird locations, but when it comes to rocky wet planets it turns out the Earth isn't the only game in town.

LOOKING IN THE LILY POND

As we know, the Kepler Space Telescope tells us that roughly one in five Sun-like stars has an Earth-sized planet in its habitable zone; that is to say, at the right distance to have liquid water on its surface. On average, that means the nearest such planet could be as little as twelve light-years away. According to Einstein, that means that even if we find it tomorrow, and invent speed-of-light space travel the day after, it's still going to take over a decade to get there. And if we fail to invent speed-of-light travel, with present technology the journey is going to take tens of thousands of years.

I tend to be optimistic about these kinds of things. Knowing a little about humankind, I can't help thinking that once we have found our nearest Eden we will want to get there as quickly as possible, and necessity, as they say, is the mother of invention.* But even if we conjure a technology that can drive our ships at, say, 10 percent of the speed of light, the adventure of breathing the air on a planet similar to the Earth has to be several generations away at best.

So what do we do in the meantime? Are there any clues that we can glean from life on Earth as to what life on other Earths might be like? Happily, there are.

All life on our planet may share a common ancestor, but it has often inhabited parallel worlds. In some cases, these worlds have been evolving independently for tens of millions of years, with startling consequences. They are, in short, the closest thing to setting foot on an Earthlike planet that we will experience in at least three generations. I'm talking, of course, about some of the greatest wonders of the globe: islands.

* By then the father of invention may be global warming, but that's a topic for another book.

CONTINENTAL DRIFT

As we discussed in Chapter Six, at the time of the Permian extinction some 250 million years ago, all the present-day continents were joined together into one giant landmass, or supercontinent, known as Pangaea. Thanks both to its size and a runaway greenhouse effect, the center of Pangaea became a dune-filled desert. We can still see its legacy today in the form of colossal deposits of sandstone; in fact, my own county of Cheshire in the UK is home to a sandstone ridge that is a relic of the Permian.*

The Permian, as you may know, marks the end of the Paleozoic Era, when plants and animals colonized the land. The Mesozoic Era, which followed immediately after, saw Pangaea begin to break up. The first big rift came toward the beginning of the Jurassic, when Pangaea split into two: the northern lump, called Laurasia, was made up of North America, Greenland and Eurasia; while that in the south, called Gondwana, was made up of South America, Africa, Antarctica, India, and Australia.†

By the beginning of the Cretaceous Period, roughly 140 million years ago, Gondwana was starting to break up. One by one, continents broke off the southern land mass and headed north to join Laurasia. Africa was the first to leave, splitting from South America to create the famous jigsaw puzzle by which the right hip of South America fits neatly into Africa's left. India followed soon after, making a beeline for Asia, while Africa headed north toward Europe. In the middle of the Cretaceous, about eighty

* See my handy guide to geological time on pages 193 and 194.

† The name means "Land of the Gonds"; the Gonds were an Indian tribe. The old name for Gondwana was "Gondwanaland," meaning "Land of the land of the Gonds."

million years ago, it was New Zealand's turn to take the plunge, while Madagascar split off from a still-migrating India.

Australia and South America were the last to leave, remaining joined to Antarctica until well into our present era, the Cenozoic. Australia departed first, in the Paleogene Period, roughly forty-five million years ago. South America and Antarctica then hung on until the beginning of the Neogene Period, roughly twenty-three million years ago, at which point the Drake Passage opened up and the forests of Antarctica were replaced by permanent snow.

For the next twenty million years, South America migrated north, until eventually, at the beginning of our present Quaternary Period, an upwelling of volcanic rock created the Isthmus of Panama, and it became joined with North America. Finally, the globe had taken its present form, and a permanent ice cap began to form in the Arctic. The transformation from the tropical hothouse of the Jurassic, Cretaceous, and Paleogene into the icehouse of the Neogene and Quaternary was complete.*

What all this means is that from the early Cretaceous onward, Gondwanan life effectively became marooned on a series of islands and continents, each of them like a tiny Earth. Fascinatingly, these alternate worlds paint a very different picture of how evolution might have turned out. On

* While we're on it, a quick refresher on recent geological time periods. The Cenozoic Era spans the sixty-six million years since the impact that wiped out the dinosaurs at the end of the Mesozoic Era, and is subdivided into the Paleogene, Neogene, and Quaternary Periods. The Paleogene was a hothouse, and the Neogene and Quaternary have been progressively colder. We divide the Paleogene into the Paleocene, Eocene, and Oligocene Epochs, and the Neogene into the Miocene and Pliocene Epochs. The Quaternary is divided into the Pleistocene and Holocene Epochs. Roughly speaking, primates evolved at the beginning of the Paleocene, and hominins at the end of the Pliocene. Anatomically modern humans emerged in the Pleistocene, and human civilization in the Holocene.

North America, for example, placental mammals came out on top, with marsupial mammals forced to extinction; on South America, however, the reverse was the case, and marsupial mammals dominated. In Australia, it was marsupial mammals that made the cut while the placentals perished; in New Zealand, no mammals survived at all.*

IF IT WALKS LIKE A DUCK . . . IT'S NOT NECESSARILY A DUCK

Madagascar is a case in point. Located in the Indian Ocean, just off the coast of East Africa, it is home to a fascinating array of mammals, none of which were present on the island when it rifted from India. As far as primates go it has no monkeys or apes, and no dogs or cats among its carnivores. All of the mammal species it does have—and it has a great many—are the descendants of creatures that either swam, floated, or flew across from Africa and India.†

At some point round about sixty million years ago, a family of lemurs appears to have made the trip from East Africa. Not that they booked a package tour, of course; our best guess is that they were clinging to a tree during a hurricane when they got washed out to sea. A bunch of tenrecs—small, shrew-like mammals—followed in their wake roughly thirty million years ago, and other rafts bobbed up on the shore around

* The last known New Zealand land mammal died out in the Miocene around 16mya. Interestingly, several species of bat have repopulated New Zealand, where they have started to fill the vacant niche of ground-dwelling shrews.

† I should point out that, like New Zealand, the fossil record of Madagascar shows that it was once home to native mammals that died out. The recently (2014) discovered 66 million-year-old-fossil of *Vintana sertichi*, a strange-looking creature the size of a modern-day groundhog, is a case in point.

twenty million years ago, carrying rodents and carnivorans, the mongoose-like ancestors of carnivorous mammals.

It's the lemurs that most people have heard of; present-day Madagascar is home to nearly one hundred different species, all descended from that first vagabond crew of early primates that crossed the Mozambique Channel on a fast-moving ocean current. The remarkable thing about lemurs is not only are they box-office catnip, as demonstrated by the fabulous *Madagascar* movies, but they are a striking visual echo of what the first primates must have looked like. Gaze into the eyes of a lemur, and you can almost imagine you are eyeballing one of our shrew-like ancestors in the tropical rainforest of the Paleocene.

But while it's lemurs that get the glory, as far as our story goes it's the carnivorans that are the real stars. Do yourself a favor and Google a creature called a fossa. What pops up is something that looks like a cougar; so like a cougar, in fact, that many nineteenth-century taxonomists placed it in the cat family. The fossa has a head like a cat, a body like a cat, and a tail like a cat. It climbs trees and can semi-retract its claws, just like a cat. Yet, intriguingly, a fossa isn't a cat. Instead, DNA studies have shown it to be a direct descendant of the African carnivorans that washed up in the Miocene some twenty million years ago.

Let's think about that for a moment. There have never been any jungle cats on Madagascar, yet the fossa looks just like one. How can that be? We'll never know for sure, but the smart money says that the fossa is a rather striking example of what we call convergent evolution. The same way of making a living—chasing small mammals through tropical trees—has produced two animals that are remarkably similar in appearance, despite the fact that they are only distantly related.

The fossa is one example, but Madagascar is home to many others. One of my favorites is the tenrec, several of which have evolved to look exactly like hedgehogs, despite being only distant relatives. Not that examples of convergence are limited to Madagascar; Australia, New Zealand, and South America are all full of cases where unrelated creatures have ended up looking remarkably alike.

Where we find placental flying squirrels and moles in North America, for example, we find marsupial flying squirrels and moles in Australia, even though their last common ancestor was probably doing its best trying to avoid being trodden underfoot in the Cretaceous jungle. The saber-toothed placental tiger of Ice Age North America had a much earlier doppelgänger in the saber-toothed marsupial tiger of South America.* Even more striking, in my humble opinion, is the outward resemblance of the Australian thorny devil lizard, *Moloch horridus*, to the North American desert horned lizard, *Phrynosoma platyrhinos*. Both are desert-dwelling ant-eating lizards with blotchy markings and protected by exaggerated spines, and yet they are about as unrelated as it is possible for two lizards to be.

Such whole-animal convergences are striking, but they are only part of the story. Partial convergences are even more common, where creatures have very different body plans but similar body parts or behaviors. Flight, for example, has evolved at least four times, in insects, pterosaurs, birds, and bats, while, as we have already learned, camera eyes have

* The classic example of the placental saber-toothed tiger is *Smilodon fatalis*, which first appeared in North America around 1.6 million years ago during the Pleistocene; along with the woolly mammoth and the dire wolf it belongs to an elite group of large mammals known as the Pleistocene megafauna. The marsupial tiger is the older *Thylacosmilus atrox*, which first appeared in South America in the late Miocene roughly eleven million years ago.

evolved independently both in vertebrates and octopuses. The bearing of live young is estimated to have evolved over one hundred times among lizards and snakes, usually in response to cold climates, while hosts of animals we previously thought were related—flightless birds, for example—turn out to be only distant cousins.

RED IN TOOTH AND CLAW

It was just these kinds of partial convergences that so confused the European investigators of the duck-billed platypus. Following Shaw's examination in 1799, the classification of what is properly known as *Orinthorhynchus anatinus** only became more problematic when, in 1802, the surgeon and anatomist Sir Everard Home reported that it possessed a cloaca, a single opening for the alimentary, reproductive, and urinary tracts. That would seem to class the platypus among either amphibians, reptiles, or birds, and Home even theorized that it laid eggs. At this point the French anatomist Etienne Geoffroy Saint-Hilaire waded in, declaring that both the platypus and its countryman the echidna, or spiny anteater, represented an entirely new class of vertebrates. He named this new class the monotremes, from the Latin for "single opening."

Not that his fellow naturalists took much notice, preferring to join the mammal vs reptile fracas. The pendulum swung even further in the direction of reptiles with the 1823

* Shaw gave it the name *Platypus anatinus*, with *Platypus* meaning flat-footed and *anatinus* meaning duck-like. Sadly the genus *Platypus* turned out to have been already taken by a wood-boring beetle. Meanwhile, the German naturalist Johann Friedrich Blumenbach had independently named the same creature *Orinthorhynchus paradoxcus* (bird-snout, paradoxical). Compromise was then reached with *Orinthorhynchus anatinus*.

discovery by the German anatomist Johann Friedrich Meckel that the platypus' spur was venomous. Venom, of course, is associated with reptiles like snakes and lizards. Three years later, however, Meckel published a paper that proved beyond doubt that the platypus had mammary glands, propelling the pendulum hard in the opposite direction. The entire platypus debate now hinged on one crucial question: Did it lay eggs?

The Aborigines were adamant that it did, but so rigid was the belief that mammals only gave birth to live young that, to begin with, few academics took the idea seriously. Even as late as 1884, the *Sydney Morning Herald* declared that any evidence to the affirmative must be "examined and reported on by scientists in whom the world has faith, then all the scientific world will stand convinced and will believe where they have not seen."

That same year, a tyro Scottish zoologist named William Hay Caldwell decided to blow his entire academic grant on traveling to Australia to settle the matter once and for all. An ecologically sensitive program of shooting and dissecting platypuses had been in progress since 1834, overseen by the Australian naturalist George Bennett, the curator of the Australian Museum, Sydney, New South Wales, but the obnoxious Caldwell adopted what can only be described as a slash-and-burn approach.

In the Australian winter of 1884, assisted by an army of local Aborigines, he set up camp on the banks of the Burnett River in northern Queensland, and began slaughtering all the platypuses he could find. On August 24, following a three-month period in which he had destroyed more than seventy platypus, he shot a female which had not only just laid an egg, but also had one in her uterus, ready to be laid. His triumphant telegram, "monotremes oviparous, ovum meroblastic," was seen as the last word on the matter. Not only did the platypus lay eggs, it said, but those eggs were reptilian. The platypus was officially a conundrum.

THE EVOLUTION OF MAMMALS

If the study of islands and continents that have been isolated for tens of millions of years is the next best thing to finding an Earthlike planet, then to my mind at least the discovery of the duck-billed platypus is the next best thing to capturing an alien. And thanks to the Platypus Genome Project of 2008, we now have a much clearer idea of how this extraordinary creature fits into the grand sweep of evolution.

Part of the answer, as you might have guessed, is that the platypus is a descendant of a third group of mammals, the monotremes, which sits alongside the marsupials and placentals. The fossil record of monotremes is patchy, but suggests a radiation in the late Triassic or early Jurassic, eventually becoming extinct everywhere except in Australia, where they continue to thrive.

So far as we can tell from DNA studies, the last common ancestor of monotremes, marsupials, and placentals probably lived some time during the Triassic, and appears to have been a furry egg-laying creature that suckled its young. Throughout hundreds of millions of years of evolution, the platypus has continued to lay eggs, while the other two surviving branches of the mammal family tree—marsupials and placental mammals—have instead evolved the ability to give birth to live young.*

* We used to say that the monotremes were therefore more "primitive" than the marsupials and placentals, but, of course, that isn't the case at all. Monotremes have undergone just as much evolution as marsupials, for example; they continue to lay eggs because in Australia that's what works.

So much for eggs. What about the platypus' other extraordinary traits, like its venomous spur and its duck bill? Fascinatingly, this is where convergent evolution comes in. Platypus venom is remarkably similar to reptile venom, but turns out to have evolved completely independently. The platypus' bill provides even more of a surprise. Everyone knew the platypus had to be doing something clever to be able to catch half its body weight in insect larvae at the bottom of muddy streams in the dead of night with its ears, eyes, and nostrils closed; that something turned out to be what's known as electroreception.

Electroreception is very much de rigueur in fish, but almost unheard of in mammals. It turns out that the platypus' extraordinary bill, as well as mimicking shovelers such as the duck, is home to a vast array of receptors. As it swims, it swings its head from side to side, detecting the minute electric fields given off by its prey. In other words, despite having their last common ancestor way back in the Devonian, fish and platypus are both capable of sensing electric fields. Their evolutionary journeys couldn't have been more different, but given the same selection pressures—trying to make a living in murky water—the final destination was the same.

So what does all this tell us about aliens? Well, for starters it tells us that just because a planet is Earthlike, four billion years of evolution isn't necessarily going to produce anything remotely human. After all, on New Zealand and Australia placental mammals didn't even make the cut, let alone the subdivision that we belong to, the primates.* And even if primates do evolve, and subsequently give rise to ground-dwelling,

* The picture painted by phylogenetics shows placentals evolving first in Africa, and then spreading to Asia, North America, and eventually—once it joined via the Isthmus of Panama—South America.

bipedal apes, there's no guarantee that they will fare any better. The Madagascan lemurs, for example, convergently evolved several "ape" species, but they all went extinct.*

On the other hand, thanks to convergence, what comes around goes around. The islands and remote continents of the Earth show us that certain adaptations arise again and again: when it comes to life on planets like our own, we should expect to find the same notes, just not necessarily in the same order. Things like wings, eyes and teeth will be common on Earthlike planets throughout the galaxy, even though the creatures that possess them may be as unfamiliar to us as the duck-billed platypus was to nineteenth-century naturalists. These are, after all, the solutions that work time and time again, and toward which evolution will always stumble. And here's the rub. Intelligence, the vital commodity we need to find if we are to communicate with aliens, turns out to be a convergent trait.

FEATHERED APES

"I always wanted to fly," says Nicky Clayton, as she spins her Audi TT around in a Cambridge side street, evading yet another civic roadblock. "That's why I love dance." I nod enthusiastically. That explains why my bag is currently rolling around in the trunk next to an entire wardrobe of lacy costumes and what can only be described as killer heels. Seconds later we are heading toward her laboratory in Madingley, on

* The diversity of subfossil lemurs is astounding, and includes giant ape-like lemurs such as *Megaladapis*, as well as orangutan-sized forms, and large ground-dwelling Gorilla-sized species. They also convergently hit upon "sloths" (see *Archaeoindris* and others) with a wide array of sloth-like subfossil lemurs and giant Aye-ayes.

the outskirts of Cambridge, and I briefly wonder whether the front of the car is going to achieve aerodynamic lift and grant her wish.

The building we arrive at has the understated hush of a local tennis club, with pavilion-style wooden huts surrounded by lawns and meshed enclosures. Instead of the thwock of brushed cotton on catgut, however, the air tinkles with bird-song. That's because, by night, Clayton is a dancer, but by day she is Professor of Comparative Cognition in the Department of Psychology at the University of Cambridge. What's more, she has made a considerable name for herself studying the intelligence of a previously overlooked species, the crow.*

Through a series of ingenious experiments, Clayton and her coworkers—her husband Nathan Emery is one of her collaborators—have demonstrated that, far from being "bird brains," when it comes to intelligence, crows show a great deal of similarity to apes. Crows are foragers, for example, and love to hide food. In one famous experiment, Clayton and her team devised a sort of "crow motel," where western scrub jays spent the night in one of two bedrooms. In one, breakfast was served in the morning; in the other, no such luck. After a stay of six nights, alternating between the two bedrooms, the jays were unexpectedly given nuts in the evening. Much as you or I might do, the crows hid food in the bedroom that came without breakfast, just in case.

* I know what you're thinking: Crows can't be that smart because they have tiny brains. Crucially, however, it's not size that matters; it's the ratio of size to body mass. A crow's brain may be the size of a walnut, but its body is extremely light, giving it a so-called encephalization quotient, or EQ, on a par with apes. In fact, the western scrub jay, a particularly smart species of crow, has an EQ on a par with early hominins such as Australopithecus.

That result is interesting, because it has been mirrored in similar experiments with apes. In one test, for example, chimpanzees and orangutans were shown how to use a plastic hose to suck fruit juice from a container. Later, in a separate room, they were given a choice of four objects, one of which was the hose. Knowing that they would later encounter the container, the apes cunningly selected the hose. Apes and crows, in other words, aren't imprisoned in the here and now; they are capable of imagining their future and of planning for it.

Not only that, but Nicky's team has also demonstrated that crows are capable of designing tools, as well as reasoning, problem-solving, empathizing, and even deliberately deceiving one another. Other experimenters have shown that all of these abilities are shared by apes, though in my opinion, when it comes to tool-making, it's the crows that have the edge. And all this despite the fact that crows and apes have completely different kinds of brains, as you might expect with two species whose last common ancestor was an amniote* that roamed the Carboniferous forests some 300 million years ago.

So what caused the intelligence of crows and apes to converge? Clayton makes some intriguing suggestions. Both animals are highly social, and, as we all know, to get ahead in society you need an ability to play politics. That requires brainpower, and it may be that group living is one of the drivers toward animal smarts. Second, both apes and crows are foragers, living off seasonal foods that are tricky to identify, hard to get at and distributed over a wide area. In this case, it's

* Meaning something that was capable of laying hard-shelled eggs on land. Amniotes then split into the synapsids and sauropsids. Synapsids eventually gave rise to mammals and sauropsids to reptiles and birds.

not just the early bird that catches the worm, but the one that's gifted enough to drop stones into a pitcher of water.*

And third—and to my mind, most significantly—both chimpanzees and crows first appeared between ten and five million years ago, a time of rapid climate variation as the Earth limbered up for the present Ice Age. Interestingly, that's also when our own hominin line diverged from the common ancestor it shares with *Pan*, the genus to which chimpanzees and bonobos belong. One way to get around a changing climate, of course, is to use your intelligence to help you find food and shelter. Could what Clayton calls the "clever club" of apes, crows, parrots, dolphins, and elephants be a direct result of unpredictable weather?

For those of us seeking to contact aliens, the implications of all this are profound. Just as we can expect the inhabitants of Earthlike planets to fly, bite, and see by the light of their home star, we should also expect them to be smart. Because despite what we might want to believe, intelligence is not a uniquely human trait; we share it with crows, dolphins, elephants, and no doubt a whole zoo of other terrestrial species yet to be investigated.† Its basic components—things like reasoning, problem-solving, imagination, memory, and mental time travel—crop up time and again.

* I know that sounds like I'm culling my data from Aesop's Fables, but Sarah Jelbert at the University of Auckland actually did this experiment. In fact, if you put "crow and pitcher" into *New Scientist*'s YouTube channel you can watch a version for yourself.

† Just for starters, and in no particular order: octopuses, dogs, cats, rats, whales, parrots, and pigs. The problem-solving abilities and tool use of octopuses is particularly significant, because they are about as far removed in animal evolution from the other examples as it is possible to be.

WAR OF THE WORLDS

So we have half an answer to the question of what an intelligent alien might look like. First, it will have a complex brain. Whether that brain is centralized, as it is in birds, mammals, and dolphins, or decentralized as in octopuses, we can be less sure. It will have acute senses, whether attuned to light, sound, heat, chemicals, or electric fields. It will most likely eat a wide variety of foods, be extremely dexterous with physical objects, communicative, and social. Large portions of its long life will be spent learning from its society, parents, and peers. And it will quite possibly have evolved in an ever-changing terrain, where it used its intelligence to stay one step ahead of the game.

As to whether it has six legs, fur, feathers, or a two-inch-thick coating of mucus, all bets are off. Certainly, if it resembles any of the family groupings we recognize on Earth—vertebrates, or invertebrates, for example—that will be a fluke. There is nothing special about the order in which species have diverged here on Earth; if we re-ran evolution it might just as well be a ray-finned fish rather than a lobe-finned fish that gave rise to the first amphibians, for example, and we might have inherited six digits per hand rather than five.

If it comes to that, of course, if we re-ran Earth evolution we might not even get as far as the lobe-finned fishes. There are considerable hurdles to jump on the track to complex life as we know it, with one of the hardest to clear being the creation of the eukaryotic cell and its energy-giving mitochondria. Another is the evolution of oxygenic photosynthesis, whereby chlorophyll is used to harness light energy to rip electrons from water and stuff them on to carbon dioxide to make sugars. Fail to make it past either of those and all you get is an ocean full of bacteria, much as we had for the first two billion years of life on Earth.

As we shall see in a moment, there are reasonable arguments as to why the eukaryotic cell might be convergent, and how we might get by with ordinary photosynthesis rather than the souped-up, oxygenic kind, but however you slice it, the sixties *Star Trek* trope of landing on an opulently flowered Eden inhabited by nubile blondes and beaux with negligible body hair is very much an outside bet. Even though it ticks all the above boxes for brain size, dexterity, and communication, the intelligent alien on the other end of the phone might look more like a crab, or a spider, or an octopus; even more likely, it will resemble none of the above, but for a sucker here and an eye there.

Of course, it takes a lot more than just intelligence to make a communicable alien. Humans are smart, but not that smart. What really sets us apart is civilization; what made that possible was the invention of farming. As we entered the Holocene, and the snows retreated, everyone got their hoes out. Or to be precise, they got the halters out; the hoes came shortly after. Farming, too, appears to be highly convergent, emerging roughly 11,500 years ago in the stable climate of the Holocene in Southeast Asia, the Levant, the Fertile Crescent, South America, and Europe. Oh yes, and roughly fifty million years ago in the Attine ants of the Amazon rainforest.

ALIEN ANT INVASION

Not that they were farming rice and rye, of course; the preferred crop of the Attines is fungus. One particularly sophisticated group, the leafcutter ants, live in civilizations over five million strong, with precise divisions of labor. Their underground nests reach enormous size; one excavated in Brazil in 2012 was more than five hundred square feet across and more than twenty-five feet deep. A network of tunnels provides access and ventilation to large underground chambers where the fungus is fed with leaf mulch; the fungus breaks down the cellulose in the leaves so that both it, the ants, and their larvae can digest it. As the fungus grows, the ants prune it, fertilize it, and even treat it with antibiotics when it becomes infected with other parasitic fungi.

I'm sure by now you are getting the picture: There is almost nothing about humans that is unique. There are plenty of other creatures out there that walk on two legs, give birth to live young, or farm for a living. Ants, of course, aren't great shakes in the cognitive department, which is why they aren't planning a flyby of Europa—not on this planet, anyway. On the one hand, that's a little disappointing; we humans love to feel important, and it's a bit disconcerting to discover that a six-legged creature in the Amazon jungle mastered the antibiotic before we did, but when it comes to our quest to find other communicable life in the galaxy it should give us great heart.

All the important stuff like intelligence, language, tool use, farming, and civilization is convergent; it has evolved before, in countless other species. And while no one was waiting for humans to be the ones that got their act together first, by chance that is exactly what happened. True, evolution has not been conducted with us in mind; when the first archaeon left the vent, it wasn't with the express intention of one day wandering about Soho with a copy of *Time Out*. But the ratchet of natural selection has spent the last four billion years layering complexity upon complexity; it was just a matter of time before one species lucked out.*

THE INFORMATION SUPERHIGHWAY

And the precise way in which we lucked out, of course, was that we came up with a method for storing information outside our physical bodies. As previously mentioned, the first phonetic writing system is believed to have emerged in Sumer in Mesopotamia around 3100 BC. As Claude Shannon would have noted, this marks a profound change. Before writing, DNA was the only way nature had of making a "hard copy"; afterward, anything anyone had a mind to write down could be preserved for generations.

From Mesopotamia, the technology of writing spread quickly west to Ancient Egypt, and then around the Mediterranean, courtesy of the Phoenicians. The Ancient Greeks adopted the Phoenician alphabet, and bequeathed it to the Romans. We have no way of knowing whether there were cavemen as wise as Socrates, but thanks to writing, even 2,400 years later we are intimately acquainted with his every thought.

Information, let's not forget, is a physical thing. While the Sumerians had scored their marks in wet clay, the Egyptians

* That's not to say every species on Earth has become more complex, obviously. The vast majority of life on the planet is still single-celled, in the form of archaea and bacteria. Organisms can lose complexity as well as gain it, or even stay much the same. Like many cave-dwelling species, the Hawaiian Kaua'i cave wolf spider has lost its eyes, for example, and the coelacanth is doing a pretty good impression of other coelacanths that lived 400 million years ago. It's average complexity that increases over time, which is why we would be so surprised to find a fossilized ichthyosaur with a brain-to-body ratio bigger than a bottlenose dolphin, or an Attine ant farm with elevators and designated parking spaces.

had something far handier—papyrus—from which they were able to make the first books and eventually to found the first library in roughly 300 BC in Alexandria. Writing not only meant that information could be preserved; it could also be copied. Reportedly, when a ship cast anchor in the port of Alexandria, any written works on board would be confiscated on behalf of the library and judiciously copied.

As seekers of alien civilizations, we would do well to note that besides the Mediterranean and Middle East, writing appears to have independently evolved in both South America and China, so we can be reasonably sure that it, too, is convergent. The Chinese, in fact, not only invented modern paper, but also the first woodblock printing. They were also the first to invent movable type,* but for some reason—the vast number of characters in Chinese script, perhaps—the idea appears not to have caught on.

In AD 751, the western expansion of the Tang Dynasty was halted by the Abbasid Caliphate at the Battle of Talas, and the Chinese prisoners of war revealed the secrets of papermaking to their captors. By AD 794, paper was being manufactured in Baghdad. Throughout the Middle Ages, Islamic scholars built upon the classic Ancient Greek texts, laying the foundations of modern science and mathematics.† Thanks to the Crusades, Muslim scholarship and papermaking technology eventually began to reach Western Europe, and by the late thirteenth century first Italy, and then France and Germany, became centers for papermaking.

* In 1041 a Chinese alchemist named Pi Sheng invented movable clay type, and in 1313 a Chinese magistrate named Wang Chen invented movable woodblock type.

† It is the Islamic mathematician al-Khwarizmi (c. 780–850) who gives us the word "algebra," being the Anglicized form of "al-jabr," one of the mathematical operations he used to solve quadratic equations.

Another leap then came with the invention of the printing press by Gutenberg in 1450; as information flooded across Europe, it suddenly switched from being a backward region to being the proud home of scientific greats such as Leibniz, Kepler, and Newton. Their discoveries in turn ignited the industrial revolution of the mid-nineteenth century with its mechanization of farming, manufacturing, and the production of energy. Finally, the present digital age has seen every bit of information in existence made available online, orbiting the planet in an impenetrable electromagnetic swarm.

As humans, we find it difficult to see our technology as part of nature, as much the product of natural selection as, say, the anthill or the shell of a snail. Our snaking motorways and dotted radio masts appear to us to be a very different thing from the veins of a leaf, or nodes of a spider's web, but of course they are fundamentally the same. To see something like the internet as a natural extension of the human organism feels absurd, but it's important to ask the question: What is all this information for?

Take the evolutionary biologist's view, and the answer comes back with resounding clarity: It's not "for" anything. Pieces of information that are good at getting themselves copied will eventually come to dominate the world, whether they are the gene for blue eyes or the code for a cute cat video. One fruitful strategy is to improve the fitness of their host. The information in your iPad improves the chances that you will have offspring, just as that in your DNA does. Admittedly, it's a bit hard to see the direct effect that reading a good Dick Francis novel has on your capacity to bear children,* but, believe me,

* It keeps you sane in a crazy world, thereby enabling you to avoid traffic and eventually go to bed with someone.

that may well be why he's there. Zoom out and look at the big picture and the facts are undeniable. Since the invention of writing, human population has grown with exponential fury.*

Throughout this book we have tried to figure out what intelligent life on Earth can tell us about intelligent life on other planets. The first vital step was the creation of the cell; the next was photosynthesis. The evolution of the eukaryote enabled the leap to multicellular life; that was followed by the evolution of animals, or metazoans. Finally, the shift from DNA to silicon as a means of storing information enabled the accomplishments of human civilization to transcend anything that might be possible for an individual human. Writing is convergent, and with writing comes an explosion of information that transforms a tribe into an accomplished technological civilization. If there are other Earths out there, there could well be aliens like us to talk to.

But what about planets that aren't like the Earth? Might we find intelligent communicable life on them? And, if we do, what would those aliens look like? It's time to cast our net a little wider, and fish our neighboring solar systems for life-as-we-know-it. And, finally, we shall need to cast our net a little wider still, and fish for the ultimate prize: life-as-we-don't. Could there be intelligent beings out there whose chemistry is based on an element other than carbon? Does life even need chemistry at all? Could it take root in a dust cloud, or on the surface of a star? And how would we know it if it did?

* Human population is estimated to have been less than ten million at the beginning of the Holocene. In 1804 it reached one billion, and two billion in 1925.

SUPER-SIZE ME

If we've learned one thing in the course of our journey together, it's that here on Earth the vital steps that led to the emergence of communicable life took a great deal of time, a total of roughly four billion years.* The problem is that while the lifetime of Sun-like stars is something like ten billion years, after half that time they start to run out of hydrogen and heat up. Bit by bit, over the next 500 million years the Sun will vaporize the world's oceans and turn our planet into a smog-choked desert. Put simply, the window of opportunity for stars like the Sun to develop creatures like ourselves is a little bit on the snug side.

Fine, you might think: Let's find the longest-lived stars out there and look for Earth-sized planets in their habitable zones. But there's a problem. The longest-lived stars are the smallest, dimmest ones, known as red dwarves. Red dwarves have lifetimes of up to trillions of years, but not only are they far less stable than the Sun, spitting out all sorts of nasty radiation, their habitable zones sit much closer in. That means any wet rocky planets they happen to have will be right in the firing line.†

* In broad strokes, cellular life emerged from the vent around four billion years ago, evolved oxygenic photosynthesis around three billion years ago, evolved the eukaryotic cell around two billion years ago, multicellularity around 1.5 billion years ago, then complex multicellularity around 575 million years ago. Plants moved on to land in the Ordovician roughly 480 million years ago, amphibians followed them in the Devonian 375 million years ago, and 2009 saw the release of the alpha version of Minecraft.

† There might be a way around this. Since red dwarves are relatively lower in mass, a rocky planet in the habitable zone may well be in tidal lock with the parent star. If one hemisphere is always facing away from the star, it would be shielded from radiation doom, and could have energy transferred to its radiation-shielded night side via the atmosphere. Alternatively, a giant planet might generate a large protective magnetic field that could shield itself or its moons from the worst effects.

Happily for us, however, there's an alternative. There's an intermediate size of star, the so-called orange dwarves,* that are perfect for nurturing life. Not only do they have long lifetimes, in excess of fifteen billion years, but they are stable, like the Sun. In fact, they even emit much less harmful ultraviolet light than Sun-like stars; here on Earth we only managed to get decent protection from UV after the formation of the ozone layer, following the Great Oxidation Event roughly 2.3 billion years ago. Perfect. So we should look for Earth-sized planets in their habitable zones, right? Wrong.

Unfortunately, when it comes to long lifetimes, small rocky planets like the Earth won't cut it. One of the most crucial ingredients for carbon-based life, as we know, is plate tectonics. Not only does outgassing from volcanoes keep the atmosphere full of carbon dioxide, and therefore keep geologically active planets from freezing over, but, according to our pet theory, it's in volcanic vents on the sea floor that life gets its start. No volcanoes, no life. And to be volcanic, of course, the center of a planet needs to be hot. The problem is, small rocky planets like the Earth will cool down long before an orange dwarf burns out.

This is where super-Earths come in. Being several times more massive than the Earth, they keep their heat longer and can easily remain volcanically active for the long lifetime of an orange dwarf star. Along with a molten core, of course, comes a magnetic field, and protection from harmful cosmic rays and solar flares. Some, such as René Heller of the Institute of Astrophysics, Göttingen, even go so far as to call such planets

* It was a while ago, so here's a brief reminder that we label the hydrogen-burning stars as follows, brightest to dimmest; O, B, A, F (yellow/white dwarf), G (yellow dwarf), K (orange dwarf), and M (red dwarf).

"super-habitable," better than Earth for the evolution of life. And the kicker is that, unlike diminutive Earth-sized planets, they will show up well in the next generation of telescopes.

INTELLIGENT LIFE 2.0

So what would life on a super-Earth in the habitable zone of an orange dwarf star look like? Just as in the case of an Earth-sized planet, while it is impossible to predict the precise route that evolution will take to get there we can make some informed guesses as to where it will end up. And thanks to the fact that it can run for nearly twice as long as it can here on Earth, that may be a very interesting place indeed.

A CLOCKWORK ORANGE

Let's start with what physics can tell us about a larger-than-life Earth planet. No prizes for guessing that a bigger Earth means stronger gravity, but what effect will that have on the land, oceans, and atmosphere? Interestingly, quite a lot. Calculations show that plate tectonics kicks off when planets are roughly Earth-sized, and switches off when they reach around five times Earth mass. At the lower bound, you get deep oceans and large continents; by the time you reach twice the Earth's mass, stronger gravity produces shallow oceans with island archipelagos.

That's exciting, because shallow oceans and islands, as we know, are great for biodiversity. Even today, we find the interiors of large continents and the depths of our deepest oceans more sparsely populated, while islands and lagoons teem with life. But we are getting ahead of ourselves. Would we expect life on a super-Earth to follow the same path it did here on Earth, with single-celled life emerging in the oceans, evolving complexity, then invading the land?

Again, in my view, the path may be different, but the destination remains the same. According to our pet hypothesis, single-celled life started in the vent at least twice: once as bacteria and a second time as archaea. Anything that happened more than once is a prime candidate for convergence, so we can assume that single-celled life is a given. Both bacteria and archaea then navigated the next step; namely, they weaned themselves off the electrical energy of the vent, and learned to harness light energy, also known as photosynthesis.

Of course, this early photosynthesis involved using light energy to rip electrons from chemicals that surrounded the vent: Hydrogen sulfide, for example, was a favorite; iron was another. Those electrons were then forced on to molecules of carbon dioxide, creating sugars. Eventually, in one type of cell only, the cyanobacteria, it gave way to a much more complicated, but altogether more successful process: oxygenic photosynthesis. Here, light energy was used to rip electrons from water, also creating sugars, but this time releasing oxygen as a waste product.

Primitive photosynthesis, in other words, emerged several times; oxygenic photosynthesis only once. To my mind, that means we have to surrender to the possibility that other Earths and super-Earths will not be flooded with highly combustible oxygen. But that needn't cramp our style. After all, free oxygen was a huge problem for primitive life. Not only was it toxic, but it oxidized atmospheric methane, removing a vital greenhouse gas from the atmosphere and triggering a Snowball Earth in the form of the Huronian glaciation.

Eventually, of course, life adapted, and eukaryotes which could burn oxygen in their mitochondria came to rule the Earth. But couldn't eukaryotes have evolved anyway, without the oxygen problem? Endosymbiosis, which created the eukaryotic cell, has happened more than once; to take a notable example, just as we think an archaeon engulfed and enslaved a bacterium to produce the eukaryotic cell, so we believe that a eukaryotic cell engulfed a cyanobacterium to produce the plant cell.

The eukaryotic cell is the McLaren P1 supercar to the prokaryote's Mazda MX-5; once it appears on the scene, the brakes to complex multicellularity are well and truly off. Here on Earth, complexity has arisen in animals, fungi, plants, and algae, and it's a fair bet it will emerge on our super-Earth, too.* Which complex forms will first populate the oceans, however, is anyone's guess. The strange creatures of the Ediacaran are probably our best rule of thumb, but who knows how things will progress from there?

All we can rely on, really, is the same extraordinary adaptive power that we find here on Earth. If some single-celled organism does manage to crack oxygenic photosynthesis, no doubt plant-like forms will invade the land just as they did here. They might not be green, of course. No one is entirely sure why chlorophyll absorbs mostly blue and red light: It may just have been the nearest pigment at hand, or it may be optimized in some way to the kinds of light photons that make it through the atmosphere. Whatever the answer, there's no reason that alien chlorophylls would have to be green. To our

* Of course, in terms of complexity, plants are streets ahead of algae and fungi, and animals are light-years ahead of plants.

eyes, the plants on an orange dwarf planet could just as easily appear yellow, or even black.

And if oxygenic photosynthesis fails to emerge, well, no doubt life will find another way. Complex multicellular organisms that use sunlight to synthesize sugars—a weird cross between a plant and a fungus, perhaps, or a bioluminescent slime mold—will eventually populate the oceans and then the land. Once they do, other complex creatures that are capable of eating them will follow.

Again, none of their body plans would necessarily be recognizable, but weapons, armor, jointed limbs, eyes, mouths, and brains have evolved so many times here on Earth it's reasonable to assume they will put in a regular appearance. Maybe the thicker atmosphere will encourage flight, and the air will swarm with all manner of airborne creatures; maybe the many island habitats will produce a greater diversity of terrestrial life than here on Earth. We can't be sure. Eventually, however, one or more intelligent species will develop civilization, and that's when the real fun will start.

BACK TO THE FUTURE

The possibility that super-Earths are even more habitable than our own planet brings one thing into sharp focus: any alien civilizations we do find are likely to be much older than our own. That makes it hard to know what to look for. It's hard enough to imagine where human civilization will be in five years, let alone an alien civilization in 500 million. How would such planets show up in our telescopes?

Kepler 62 is a perfect example. Discovered on April 18, 2013, it's an orange dwarf star roughly 1,200 light-years away in the constellation of Lyra. Not only might it be as much as eleven billion years old, but it also has two super-Earths in its habitable zone. Assuming that the Earth is average, and it takes four billion years to raise a technological civilization, that means intelligent communicable alien life could have existed on either or both of those planets for a billion years.* Where do we start? What do we look for?

We certainly have no way of predicting what one-billion-year-old technology would look like. But given what we know about life here on Earth, we can still make a shopping list of things to look out for. From the Second Law of Thermodynamics we know that all life, no matter how intelligent, dissipates energy as heat. Heat, as you know, is essentially infrared light. So all we have to do is point one of our infrared space telescopes at a nearby orange dwarf with super-Earths in its habitable zone and see if they are giving off a splurge of infrared light, right?

Sadly not. Even though orange dwarves are far dimmer than our own Sun, they are still way too bright for even the next generation of infrared telescopes such as the James Webb to be able to pick out the heat signature of a civilization. We might have more luck picking out what's known as a Dyson sphere—a civilization which has blacked out its home star with light-absorbing satellites—but there are lots of darkish things in the sky which emit in the infrared, such as protostars surrounded by dust, making their existence very hard to prove.†

* Not 6.5 billion years, remember, because of GRBs. Anything older than five billion years will be toast.

† Nevertheless, Fermilab runs a program on the IRAS infrared satellite which attempts to do just this. So far they have identified seventeen "ambiguous" candidates, four of which they class as "amusing but still questionable."

As I hinted earlier, however, it might be within the capabilities of the James Webb, and indeed the forthcoming ground-based European Extremely Large Telescope (E-ELT)* to pick up enough infrared from transiting super-Earths to be able to work out what they have in their atmospheres. We could look for the distinctive spectra of oxygen and methane, for example, both of which we know are produced by life here on Earth, or pollutants such as chlorofluorocarbons (CFCs). But to really examine the infrared given off by nearby super-Earths we are going to need something game-changing.

And that something might just be NASA's New Worlds mission. Currently at the drawing-board stage, New Worlds is an ingenious way to solve the problem of life-bearing planets being very dim compared to their home stars. Essentially, a space telescope is launched together with an occulter, a colossal starshade which can be maneuvered to block out the light of the home star so that the planets show up. That way we can get a clear look at both the infrared and the visible light that such planets give off, and maybe even resolve surface features like oceans and land masses. Imagine the excitement if the images show networks of light, just as you might see on the dark side of Earth.

CLOSER TO HOME

So much for Earths and super-Earths. What other types of planet might harbor life? The first place to look, obviously, is our own solar system, where currently there are two leading candidates: Europa and Titan. Europa, you'll remember,

* Planned for 2022.

is one of the four moons of Jupiter picked out by Galileo in 1610, photographed by the *Pioneer* and *Voyager* probes in the 1970s, and last visited by NASA's *Galileo* spacecraft in 2003.*

Crucially, the Galileo mission discovered that Europa is an icehouse world, with a warm saltwater ocean contained within a crust of ice. There are clays sitting on top of the ice, suggesting a recent collision with an asteroid or comet; as we know, comets and asteroids also carry organic materials, so there's a real chance Europa's ocean is seeded with long-chain carbon molecules. Not only that, but the Hubble Space Telescope has recently spotted plumes of water vapor billowing out from Europa's south pole, suggesting that we might not have to dig through the ice to sample its ocean.

There are two missions slated to launch to Europa in the early 2020s. The ESA's Jupiter Icy Moon Explorer (*JUICE*) will arrive around 2030, though NASA's as-yet-untitled Europa mission might get there first if it manages to piggyback their new Space Launch System. As the name suggests, *JUICE* will also fly by Ganymede and Callisto; the NASA mission, on the other hand, will focus specifically on Europa. By the time they're done, we'll know exactly what temperature the ocean is, how deep, and how salty. Even more excitingly, the NASA mission may yet attempt to fly through one of Europa's water plumes with its mouth open to look for interesting chemistry.

If it's looking positive, no doubt a landing mission will soon follow; anything from a probe that touches down near the south pole to sample surface water from the plume, to a

* The *Galileo* spacecraft was launched in 1989, arrived in 1995, and spent eight years in the Jovian system. It was deliberately crashed into Jupiter to avoid contaminating its moons with Earth life.

submarine capable of burrowing down through the ice into Europa's ocean. What might the life-forms down there look like? The short answer is we have no idea. Assuming we don't come face to face with the Europan equivalent of the Kraken, the smart money would be on some form of microbial life. Beyond that, we would be looking for complex carbon molecules—things like amino acids and sugars—but, of course, Europan life might use a completely different suite of carbon molecules from the ones we find here on Earth. Even if we find it, we might not immediately recognize it.

Titan is another enticing prospect, although a new mission to follow up on the ESA's Huygens probe is still several decades away. Titan, you'll remember, is truly an alien world. Its climate and terrain are remarkably similar to our own, except methane takes the place of water, and water ice takes the place of rock. Methane clouds float through its thick nitrogen atmosphere, showering methane rain into lakes and dumping methane snow in its mountains.

Here, the most exciting mission would be one where a submersible is landed on one of Titan's methane oceans, and explores the depths for signs of life. Could its methane lakes provide the perfect solvent for silicon chemistry? Silicon belongs to the same chemical group as carbon, and is also capable of forming long-chain molecules. Might it be possible that complex life could evolve on Titan, only based on silicon rather than carbon?

The astronomer Maggie Aderin-Pocock thinks so; in a 2014 article she proposed that such aliens might take the form of enormous floating jellyfish, buoyed by enormous bladders, sucking in atmospheric nutrients through giant mouths, and communicating with one another via pulses of light. Oh yes—and their bottoms would be orange to camouflage them in the hazy Titanian skies. But if you think that's weird, you need to know about Vadim Tsytovich and his living clouds of dust.

SEARCHING FOR THE CORKSCREW

Might clouds of interstellar dust be alive and have seeded the first life on Earth? That's the intriguing proposition of the veteran Russian plasma physicist Vadim N. Tsytovich. In 2007 he published a speculative paper in the *New Journal of Physics* in which he described how plasmas—that's clouds of charged particles to you and me—might actually organize dust grains into self-replicating helical structures that tick every box for living things.

Plasma, as you may know, is the fourth state of matter after solid, liquid, and gas; essentially, it's what you get when gas molecules break down to form charged particles, or ions. A classic example would be a neon sign, a glass tube of neon gas that becomes a light-emitting plasma when it has electricity passed through it. Another would be an electric spark, where free electrons from cosmic rays or background radiation are accelerated by an electric field, and collide with air molecules causing them to form a plasma. Electricity then flows through the plasma, producing light, heat, and sound.

Although we don't tend to come across them that often on Earth, plasmas are actually one of the most abundant forms of matter in the universe. Not only are there vast filaments of plasma out in intergalactic space, but we also find them in interstellar molecular clouds, in the proto-planetary disks that surround young stars and the upper atmosphere. In all these cases they are mixed with dust, which they organize into fascinating structures known as "plasma crystals."

In modeling such crystals, Tsytovich discovered that they have some remarkably lifelike properties. Essentially, under the right conditions they are capable of forming double-helical structures that are highly reminiscent of DNA. Not only does the width and length of the helices change, providing a way of encoding information, but in some simulations they would divide into two, effectively reproducing themselves.

We have yet to create such structures in the laboratory, but Tsytovich is convinced that the helical dust structures within plasmas exhibit all the properties of living matter. After all, they feed on the energy of the plasma, they reproduce, and they evolve into permanent complex structures. Could they in some sense be alive? And, for that matter, where exactly is the boundary between living and non-living things? As Fred Hoyle suggested in his 1957 novel *The Black Cloud*, could there be intelligent beings made up of nothing more than electrically charged dust?

COOKING UP SOME ANSWERS

Last night I was preparing some pasta for my middle son. I took a small pan of boiling water, gave it a glug of olive oil, and chucked in a handful of pasta tubes. Ten minutes later I lifted the lid to check on it. All the tubes were standing up on end, crowded cheek-by-jowl on one side of the pan, surrounded by boiling water. From a jumbled mess in the bottom of the pan, they had been jostled into a highly ordered state.* Why should that be?

* The olive oil may be the crucial ingredient here. Omit it—and fail to stir—and your penne will limpet to the bottom of the pan in a congealed mess.

We see this sort of thing all the time in nature, but as yet we don't quite have the physics to describe it. Our thermodynamics—our theory of how energy and information are related—has really only been figured out for systems that are in balance, or, as a physicist would say, equilibrium. But most systems in the real world aren't like that. Supply a thing with energy, and more often than not it organizes itself. Thump a tank of water with a regular beat and you will produce a pattern of ripples. They may look pretty, but their deeper purpose is to dissipate the energy of your thumps as quickly as possible. In the same way, when I heat the bottom of the pan, the pasta organizes itself because that way it can dissipate the heat more rapidly into the room than if it were jumbled up.

There is some very interesting work being done in this area,* and it indicates that something very similar is going on with living systems. The gravitational energy of the Sun needs dissipating. The Sun becomes layered, and initiates nuclear fusion, radiating highly organized light. That light then enters the biosphere, where carbon molecules become ordered into life in order to dissipate the energy of that light. Life exists to speed the heat death of the universe.

Our problem as organisms, of course, is that we don't see the bigger picture. We see our existence as a battle, an attempt to maintain order in a world that demands chaos. But we are missing the point. In our struggle to stay alive, we are serving the universe because we are dissipating energy. The universe doesn't want those fossil fuels in the ground, it wants them burned. What better way to do that than to throw up an intelligent civilization with an insatiable thirst for energy?

* I'm thinking specifically of Jeremy England of MIT and his 2013 paper "Statistical physics of self-replication," available from all good search engines.

Following that logic, intelligent life should truly be everywhere. Wherever there is an energy source we should expect to find that matter organizes around it, helping it to dissipate its energy as fast as possible. In the case of a star, one way is to form a solar system. Within the gas, dust, and planets of that solar system, matter will organize itself into spheres, weather patterns and life. The reason we developed intelligence and technology is the same one that caused bacteria to develop photosynthesis: It's a great way to dissipate the energy of the solar system.

There is nothing fundamental that separates life from non-life. It is all just matter. We are ripples. Seen this way, the aliens are everywhere. Some are no more than the smooth regular pebbles on the beach, or the bubbles of carbon dioxide in your fizzy drink. Still others are molds and fungi. Others are quantity surveyors. And others are helical crystals in clouds of dust.

CHAPTER EIGHT

MESSAGES

In which the author quarries for a Rosetta Stone, and blasts the skies with Big Data.

> *With what meditations did Bloom accompany his demonstration to his companion of various constellations?*
>
> *Meditations of evolution increasingly vaster: of the moon invisible in incipient lunation, approaching perigee: of the infinite lattiginous scintillating uncondensed milky way, discernible by daylight by an observer placed at the lower end of a cylindrical vertical shaft 5000 ft deep sunk from the surface toward the center of the earth: of Sirius (alpha in Canis Major) 10 lightyears (57,000,000,000,000 miles) distant and in volume 900 times the dimension of our planet: of Arcturus: of the precession of equinoxes: of Orion with belt*

and sextuple sun theta and nebula in which 100 of our solar systems could be contained: of moribund and of nascent new stars such as Nova in 1901: of our system plunging toward the constellation of Hercules: of the parallax or parallactic drift of socalled fixed stars, in reality evermoving wanderers from immeasurably remote eons to infinitely remote futures in comparison with which the years, threescore and ten, of allotted human life formed a parenthesis of infinitesimal brevity.

JAMES JOYCE, *Ulysses*

On July 20, 2015, at the Royal Society in London, the Russian internet billionaire Yuri Milner called a press conference. Accompanied by luminaries such as Stephen Hawking, Frank Drake, and Martin Rees, as well as by Carl Sagan's widow and co-collaborator on Sounds of Earth, Ann Druyan, Milner announced a game-changing new SETI initiative called *Breakthrough Listen*. Over the next decade, mankind will search the nearest million stars and the nearest hundred galaxies for signals. If ET is calling, we are about to pick up the phone.

In essence, what Milner has done is provide SETI with a big pot of cash* to buy time on three of the world's most powerful telescopes. The Green Bank Telescope in West Virginia and the Parkes Radio Telescope in New South Wales will search for radio signals,† while the Lick Telescope in California will

* $100 million.

† Green Bank is, of course, the telescope on which Frank Drake conducted the first ever SETI search, when, in 1961, he pointed it at our two nearest Sun-like stars, Tau Ceti and Epsilon Eridani.

hunt for optical laser transmissions. Up until now, it has been very hard for SETI to afford more than one day a year on this kind of kit, which is just one of the reasons why they built the Allen array. Now they will be getting thousands of hours a year, vastly increasing the speed, scope, and resolution of their search.

SETI has been conducting searches for visible light transmissions—known in the trade as "optical SETI"—for over a decade, but the recent step increase in our own use of lasers makes the *Breakthrough Listen* search seem all the more timely. In January 2014, NASA used a laser to broadcast an image of the *Mona Lisa* to the Lunar Reconnaissance Orbiter (LRO), a robotic craft which is currently making a survey for future Moon landings. That was swiftly followed by the International Space Station's Otical PAyload for Lasercomm Science (OPALS), where NASA proved they were able to send data much faster over laser than they are currently able to with radio waves. Could it be that in looking for radio transmissions we are way behind the times? Maybe ET has moved on, and communicates with its satellites and lunar base stations using lasers?

Adding the Lick Observatory to the search means we can ramp it up for just this kind of signal. One of the disadvantages of radio waves is that they spread out over a wide area, dissipating their power; lasers, on the other hand, are much more directional. The snag is that to detect a laser beam you need to be close to its line of sight. There's also the risk that laser signals would be heavily encrypted; after all, anyone broadcasting with radio waves is happy to be overheard, but if you are using a laser there's a chance you are trying to keep your communications secure. As ever with SETI, the slim chance of success is outweighed by the extraordinary advantages that detection would bring. What if we suddenly find ourselves plugged into the galactic internet? Imagine how many videos of Richard Dawkins arguing with creationists we could watch then.

As well as *Breakthrough Listen*, Milner also teased a second, equally enticing prospect called *Breakthrough Message*. Although full details have yet to be announced, the gist is that a million dollars in prize money awaits anyone who can create a digital message that will "represent humanity and planet Earth." As Ann Druyan explained, "the *Breakthrough Message* competition is designed to spark the imaginations of millions, and to generate conversation about who we really are in the universe and what it is that we wish to share about the nature of being alive on Earth."

And if all that's not enough to get your head spinning, try this: All of the data will be available online, as will any code that SETI writes to crunch it. If you want to get in on the act by analyzing the data yourself, or by pimping *Breakthrough Listen*'s software, you are most welcome. Or if that seems a bit hands on, you can just download the SETI@home app and allow it to piggyback your laptop's spare processing power, becoming yet another node in the world's largest supercomputer. As a man who made his money via social networks like Facebook, Milner is clearly intent on making *Breakthrough Listen* a club that everyone wants to join.

ANOTHER EARTH

The timing of Milner's announcement couldn't have been better, coming just days after the Kepler Space Telescope found Kepler 452b, the planet they are calling "Earth 2.0." It would probably be more accurate to call it "super-Earth 2.0," because it's actually 60 percent bigger in diameter than our home planet, and would have twice the surface gravity. If it has water, it would have all the exciting life-friendly features of super-Earths that we mentioned in the last chapter, with Indonesian-type volcanic archipelagoes surrounded by shallow seas.

In that case, of course, we were talking about super-Earths orbiting close to orange stars; Kepler 452b, on the other hand, is orbiting in the habitable zone of a yellow star like our own Sun.* Like all of the Kepler planets, it's quite some way away; 1,400 light-years away, in fact.† In other words, if we did intercept a message, they would have to have sent it 1,400 years ago, just as if they were to receive one from us tomorrow we would have had to have sent it in ad 615, fifteen years before the Prophet Muhammad's conquest of Mecca.

On the other hand, the star that Kepler 452b is orbiting is six billion years old. Again, let's assume that the Earth is typical, and, on average, intelligent communicable life arises on Earthlike planets after four billion years. Remembering that, thanks to gamma ray bursts, no life would have been possible on any planet before five billion years ago, we can conclude that life on Kepler 452b could have a billion-year head start. What will life on Earth look like after another billion years of evolution?

* Kepler 452b makes an orbit every 385 days.

† Kepler is pointed at a distant but dense portion of the starfield between the constellations of Lyra and Cygnus, where it monitors the brightness of around 150,000 stars, all of which are between several hundred and several thousand light-years away.

I'm not one for predicting the future, so let's do our usual trick of falling back on what we already know. One billion years of evolution on Earth takes us back before the Cambrian explosion, to a time when life was, by and large, single-celled. No trilobites, no sponge-like Ediacarans, just the odd brightly colored microbial mat. It's an unsettling thought, but whatever we find on Earth in another billion years might easily be as different from us as we are from bacteria. Martin Rees proposes that such creatures won't even be carbon-based lifeforms as we know and love them, but the robots of a long-dead civilization.

Whatever the case, it's becoming clear that, when it comes to communicating with aliens, there's a timing problem. They may be out there, but are their signals reaching us right now, and are we capable of understanding them? When Frank Drake formulated his famous equation to calculate the number of communicable civilizations, a crucial factor was the length of time for which a civilization is detectable. But there's a world of difference between detectable and decipherable. We might be able to receive radio broadcasts from Kepler 452b—albeit via some extraordinarily poky transmitters—but how would we ever be able to decode them?

Let's imagine for a moment that *Breakthrough Listen* is wildly successful, and in seven years' time, after searching 752,656 stars, we finally detect a laser signal from an Earth-sized planet orbiting a Sun-like star 355 light-years away.* Just like in the movie *Contact*, there's a recognizable call signal—prime numbers, perhaps, or the digits of π—followed

* The website for the Institute for Computational Cosmology at Durham University gives 250,000 stars within 250 light-years, so I've assumed even density of stars to get roughly 750,000 stars within 360 light-years.

by a short broadcast. How will we know if that broadcast contains a message, rather than just being random noise? And if it does contain a message, how will we translate it?

Strangely enough, this isn't the first time that scholars have confronted such a question. For centuries, some of the greatest minds in Europe struggled to understand the sacred texts of a distant people, convinced that they might contain wisdom that would speed the technological and spiritual progress of mankind. So many had tried and failed that whoever managed to crack the code was guaranteed immortal glory. Against the odds, one young linguist succeeded, becoming a French national hero. We are about to meet the brilliant Jean-François Champollion.

THE PAST IS AN ALIEN COUNTRY

At the time of the French Revolution, if you were seeking answers to the big questions in life beside the Bible there was really only one place to look; the secrets of the Ancient Egyptians. Not only was the Hebrew story entwined with that of the pharaohs, but there were hints that Egyptian civilization had equally divine roots. There was a catch, however. While a few Ancient Egyptian artifacts had found their way to Europe during the time of the Roman Empire, Egypt itself had long been out of bounds.* And even more tantalizingly, the script on those artifacts, known as hieroglyphs—from the Greek for "sacred writing"—had been indecipherable for the best part of fourteen centuries.

Thanks to Napoleon Bonaparte, that was about to change. In an attempt to emulate his hero Alexander the Great, he decided

* Since the Islamic conquest of Egypt in AD 641, in fact.

to colonize Egypt, with the added intention of digging a canal through Suez that might renew France's interest in India. On July 1, 1798, after evading Nelson's fleet in the Mediterranean, he landed near Alexandria with over 400 ships and 38,000 men.* Among them were the cream of French intellectual society, such as the mathematician Joseph Fourier and the naturalist Etienne Geoffroy Saint-Hilaire,† spread among seventeen different ships for safe-keeping. While Napoleon ultimately failed to subjugate Egypt, it was these so-called savants who were to bring back the real prize: the Ancient Egyptian artifact known as the Rosetta Stone.

THE VALLEY OF THE KINGS

After Alexander the Great had conquered Egypt in 331 BC, the country had been ruled by the Ptolemaic Dynasty, a rather decadent line of Macedonian noblemen who had eventually assumed the role of pharaohs. It was Ptolemys I to III who built the world's first library at Alexandria, for example, and Ptolemy XIV who married his sister, the famous Cleopatra VII, only to be cuckolded by Julius Caesar.‡ Following the Islamic conquest, Egypt was ruled by a succession of Islamic Caliphates

* Precise figures are hard to come by, but, roughly speaking: 30,000 infantry, 3,000 cavalry, and 3,000 artillery and engineers, plus Napoleon's personal bodyguard of 380. Of the rest, 300 were women and 167 were savants. As well as 400 transport ships, Napoleon had thirteen ships of the line and seven frigates. Nelson had thirteen ships of the line but—following a storm in the Med—no frigates.

† Whom we last met courtesy of our mutual friend the duck-billed platypus.

‡ I'm not saying that building a library is decadent; it's more the incest that I'm aiming for. That said, the practice of brothers marrying sisters—and worse—was common in the upper echelons of Ancient Egyptian society.

and Sultanates, the last of which was that of the Mamelukes, which lasted from 1250 to 1517.

At that point the Ottomans took over, controlling the country from Constantinople but retaining the Mamelukes as an aristocratic ruling class. By the time of Napoleon, however, the Mameluke leaders Ibrahim Bey and Murad Bey had become increasingly powerful, disrupting trade and generally doing little to ingratiate themselves with the Ottoman Sultanate. It was this power vacuum that Napoleon intended to exploit, presumably relying on the fact that France's erstwhile allies the Ottomans were cheesed off enough with the Mamelukes to remain neutral.

To begin with, he was largely successful, easily capturing Alexandria and Rosetta on the coast, then defeating the Mameluke forces of Murad Bey at Shubra Khit on the Nile on July 13, 1798, and taking Cairo following the Battle of the Pyramids on July 21. It wasn't to last. Barely two weeks later, on August 1, Nelson finally caught up with the French fleet at Aboukir Bay (otherwise known as the Battle of the Nile) and delivered one of the most crushing naval victories in recorded history. No British ships were lost, while eleven of Napoleon's thirteen ships of the line were sunk. Napoleon had his colony, but they were a long way from home.

WHAT'S IN A NAME?

While Napoleon battled both the British and the Mamelukes, his savants were conducting just as earnest a campaign on the cultural front. Their first Ancient Egyptian find, at Alexandria, were two obelisks, one standing and one fallen, both of which were covered in hieroglyphs. Nicknamed

"Cleopatra's Needles" by the savants,* each contained a high proportion of what Napoleon's soldiers called *cartouches*, or "gun cartridges"; namely, groups of hieroglyphs encircled with an oval and abutted by a line. For example, toward the top of Cleopatra's Needle in London you can clearly make out the following:

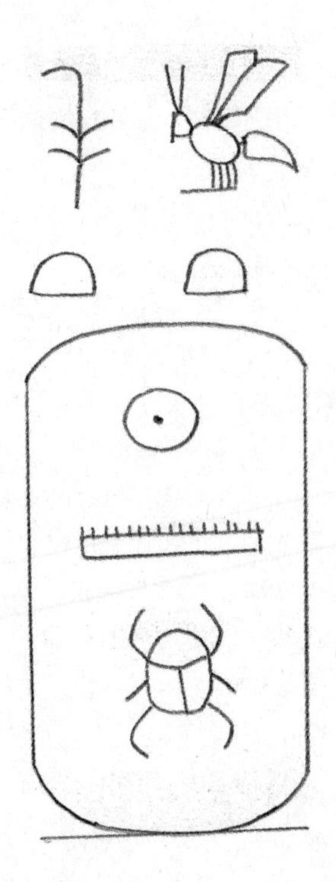

* Or, rather, les Aiguilles de Cléopâtre. The supine one now resides on the Thames Embankment in London, the other in New York's Central Park. Both were gifts from the Egyptian government decades after Napoleon's departure from Egypt. The connection with Cleopatra is that she designed the Caesarium, or "Palace of the Caesars," at Alexandria, and after her death the Emperor Augustus transported the needles from outside the Temple of the Sun in the city of Heliopolis, near modern Cairo. They were originally commissioned in 1450 BC by Pharaoh Thutmose III to honor the sun god Ra.

To put you out of your misery, this is the so-called throne name of the 18th Dynasty Pharaoh in whose honor the obelisks were commissioned, the great military hero Thutmose III, who ruled Egypt from 1479 to 1425 BC.* In appropriating the needles for Caesar's monument, Augustus certainly knew what message he was sending out to the Egyptians.

Thanks to Augustus and others, who brought many obelisks back to Rome as trophies, cartouches had been known to Western scholars for centuries. The belief at the time was that they were made up of what we call ideograms, that is to say, pictures with a symbolic meaning. The sedge ideogram, for example, that sits above the cartouche for Thutmose III, represents Upper Egypt. The bee, on the other hand, signifies Lower Egypt. The semicircle, for reasons best known to the Egyptian scribes, represents "Lord" or "King." Putting it all together, you get "Lord of Upper and Lower Egypt"; or maybe "He who will unite Egypt."

Part of the problem about ideograms is that you can argue about them all day. Some scholars have suggested, for example, that the sedge, being constantly renewed, represents eternal life, while the bee signifies mortality. It may be that as far as the Ancient Egyptians were concerned, those four hieroglyphs contain all of those meanings and more. The cartouches of Alexandria, however, were nothing compared to the riches that awaited the savants in Upper Egypt.

* The 18th Dynasty, by the way, is the one that belongs to another famous pharaoh, Tutankhamun, who reigned from (roughly) 1332 to 1323 BC.

THE FOUNDATION STONE

If Napoleon was disheartened by the defeat of the French fleet at the Battle of the Nile, he wasn't letting on. Apparently undaunted, on August 22, 1798 he inaugurated the Egyptian Institute of Arts and Sciences, and installed its member savants in one of the Mameluke palaces of Cairo. The work of the Institute was to be published a few years later as the *Description de l'Egypte*, and launched a wave of Egyptomania in Napoleonic France every bit as feverish as the one that had once gripped Ancient Rome.

While Napoleon's generals attempted in vain to engage Murad Bey in the deserts of Upper Egypt, one particular savant accompanied them in the hope of discovering further Ancient Egyptian monuments. Dominique Vivant, Baron Denon was an artist who had been big news at the court of Louis XV, but had managed to avoid proscription, eventually befriending Napoleon via Josephine's Paris salon. At Dendera he encountered the extraordinary Temple of Hathor, with its every wall and ceiling covered in Ancient Egyptian inscriptions. They contained far fewer cartouches, and a profusion of new, as yet unknown hieroglyphs.

Returning to Cairo in the middle of August 1799, Denon impressed upon his fellow savants the vital urgency of translating the hieroglyphs. Any lingering doubt as to whether these elegant drawings were literal or symbolic was gone; all the wisdom of Ancient Egypt was suddenly theirs for the taking. More savants—Fourier included—were dispatched to Dendera to make further copies, while others attempted to decipher those in Denon's drawings. Yet with no way of knowing the subject matter, there was no way to begin a translation. Just days later, all that changed with the surprise arrival in Cairo of a large lump of dark grey granite.

TONGUE AND GROOVE

As luck would have it, a mere month earlier, on July 19, 1799, a group of French soldiers had been hard at work strengthening the defenses of Fort Rashid, an Ottoman-built outpost at the mouth of the Nile near Rosetta. As the British now controlled the Mediterranean, this was to be a vital link in the defense of the Nile, and hence the entire colony. An ancient wall was being demolished when a soldier called D'Hautpoul spotted a grey slab with some sort of inscription on the side. The stone was passed up the chain of command to an officer named Michel Ange Lancret, who had been recently elected to join the Institute of Egypt.

On examining the stone, Lancret saw that the inscription was made up of three different scripts. One was Ancient Greek, one Egyptian hieroglyph, and the third he failed to recognize. Working on the Greek script, he was able to see that the text was a fairly workaday decree by the priests of Memphis detailing the good deeds of Ptolemy V and how precisely he was to be honored. Dull though it might appear, Lancret immediately grasped its immense importance. If the text of all three scripts was the same, the savants finally had the means of decoding Ancient Egyptian.

IDIOT SAVANTS

Sadly, no sooner did the savants have the stone than it slipped through their grasp. Following the French defeat at the Battle of the Nile, the Ottomans got off the fence and sided with the British against Napoleon. With his hold on the colony looking increasingly shaky, Napoleon first repelled Ottoman forces in Syria, and then marched to the coast, where on July 25, 1799 he led a decisive defeat of the Ottomans at the Battle of Aboukir.*

While exchanging prisoners following the battle, Napoleon learned from the British that the political situation in France had deteriorated, and that the Directoire Exécutif were facing a potential coup d'état. Sensing his time was at hand, Napoleon hastened to Alexandria and set sail for France on August 22, leaving the colonists in the lurch. Back in France, and buoyed by his recent victory over the Ottomans, Napoleon was able to stage a coup of his own, and by November 9 he was installed as the First Consul; by 1804 he was Emperor.

Needless to say, Egypt soon fell to the Ottomans and the British, and the savants were placed in the undignified position of having to barter their way back to France. One of the many treasures they were forced to surrender was the Rosetta Stone, which was acquired by the British and transported to the British Museum in London, where it has been on display more or less ever since. The savants, of course, had their copies of the Rosetta Stone, made by applying ink directly to its surface and using it as a printing block, and which they eventually presented over three plates in the *Description de l'Egypte*.

* The very bay, in other words, where the French had lost the Battle of the Nile in 1798.

CLEOPATRA'S NOODLES

This, in case you were wondering, is the point where Jean-François Champollion comes in. A few months after his return from Egypt, the great mathematician Fourier took up residence in Grenoble. Like the other savants, Fourier styled himself as *"Un Egyptien"* and remained obsessed with all things ancient, including the hieroglyphs. When inspecting one of the local schools, he was so impressed by the linguistic abilities of a twelve-year-old student named Champollion that he invited him to his study to show him some of his Egyptian antiquities. Learning that none of the inscriptions could be understood, Champollion declared his intention to decipher the hieroglyphs.

Not that it was easy. Part of the problem, as Champollion was to discover, was that the third script on the stone, known as Demotic, was virtually unknown in academic circles. With little progress being made with the hieroglyphs themselves, the logical step seemed to be the translation of the Demotic, but that too proved a stubborn problem. A later form of Egyptian script, Coptic, was better known, and appeared to be derived from Demotic, and Champollion made it his business to become fluent in it.

Unknown to Champollion, however, he had a rival. Thomas Young was a trained physician and polymath, probably best known for proving that light is a wave by means of an experiment known as Young's Slits.* Fluent in a baffling number of

* Young showed that if a light source was obscured but for two parallel slits, the light from those slits would produce an interference pattern. This landmark experiment was subsequently modified by one of my all-time physics heroes, G.I. Taylor, to prove the quantum nature of light. Basically he turned the brightness of the lamp down until there could only be one photon of light in the system at any one time, and went on a sailing holiday. When he returned, there was an interference pattern, just as in Young's experiment, proving there was no such thing as a "single path" for a quantum object. Small things, in other words, can be in two places at once; in fact they can be in all places at once.

languages, Young was late to the Rosetta Stone but quickly made up for lost ground. Focusing on the cartouches as the means to crack the code, he made a thrilling deduction. Knowing that the Ancient Greek and the Demotic scripts contained the name of Ptolemy, he set about trying to find the relevant cartouche among the hieroglyphs.

His reasoning was simple. Although believing, as everyone did at the time, that hieroglyphs were ideogrammatic, while the Demotic script was phonetic, he couldn't help noticing that some of the Demotic letters appeared to be derived from hieroglyphs. Could it be that some of the hieroglyphs represented sounds rather than ideas? If that was true, then those sounds would almost certainly be used to spell out a foreign name like Ptolemy. Sure enough, he found a cartouche which seemed to do the trick, and assigned the following letter sounds:

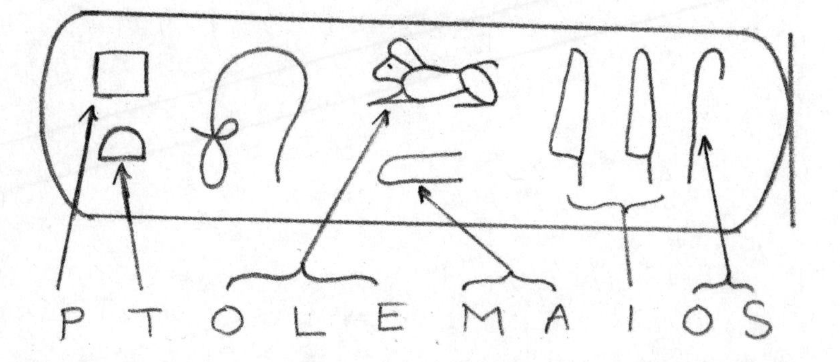

Unfortunately for Young, he remained convinced that the Egyptian names—and the rest of the hieroglyphs—were ideograms, not phonograms, and so made little further progress. Reading of Young's success, however, Champollion was inspired. Counting the characters on the Rosetta Stone, he found that 486 words of the Greek script were matched by 1,419 hieroglyphs. Grouping the hieroglyphs as best he could, he found that the total number of these "words" was roughly 180. Clearly something was amiss. Could it be that Ancient Egyptian was much more complex than anyone had imagined, and contained a mix of "picture" glyphs and "sound" glyphs?

The final key was provided not by the Rosetta Stone, but by an obelisk obtained by the British adventurer William Bankes. Discovered in the Temple of Isis on the sacred island of Philae in the Nile, near Aswan, the fallen obelisk and the base from which it had become detached had already made an appearance in the *Description de l'Egypte*. When it arrived in England in the summer of 1821, Bankes realized that a Greek inscription on the base contained the names of Ptolemy VIII and Cleopatra III, and that one of the two cartouches within the hieroglyphs matched that for Ptolemy in the Rosetta Stone. Could the other cartouche be that of Cleopatra?

Excitedly, Bankes had a lithograph printed of both the Greek and Ancient Egyptian inscriptions, and sent a copy to Young. Unable to make any progress, Young decided that the copy was inexact and abandoned any attempt at translation. Another found its way into the hands of Champollion in France, however, who checked first that the cartouche for Ptolemy matched that on the Rosetta Stone. It did. Disregarding Young's original letter sounds, Champollion came up with the following:

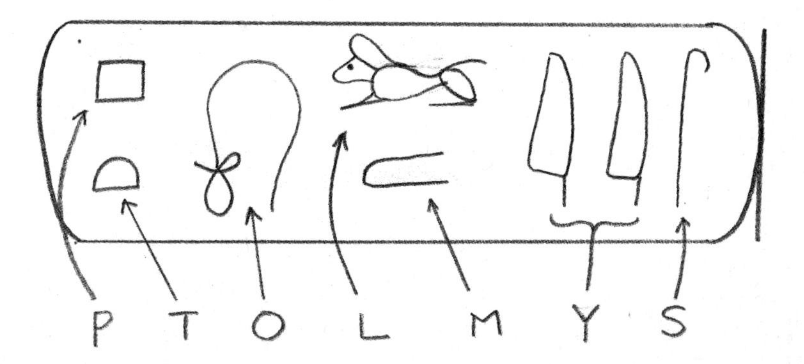

Turning to the second cartouche, he immediately recognized four of the glyphs:

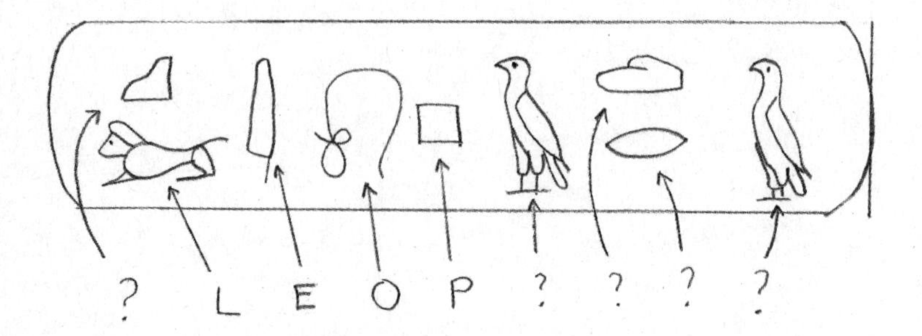

Assuming that the others must also be phonograms, and that the same sound could be represented by more than one phonogram, he assigned them the following values:*

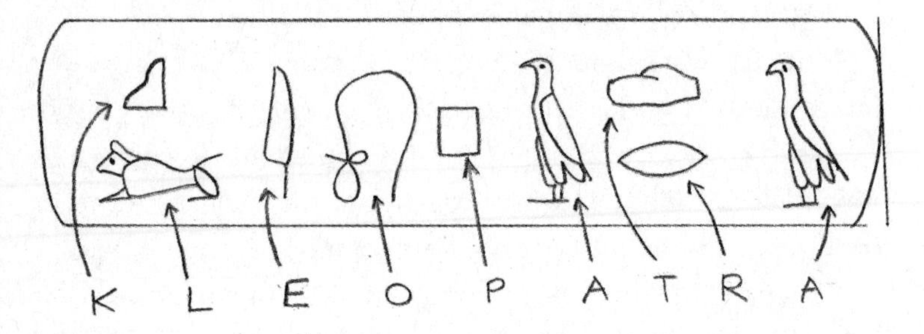

* There's an added bit of brilliance here; hieroglyphs don't usually contain vowel sounds, and neither does Demotic script. Vowel sounds came in with the Egyptian–Greek combo that is Coptic.

He then applied himself to a third cartouche:

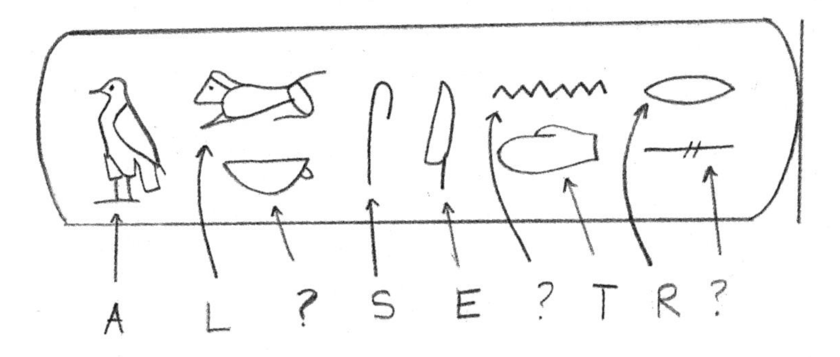

Knowing six of the glyphs, he was able to deduce the name of Alexander the Great:

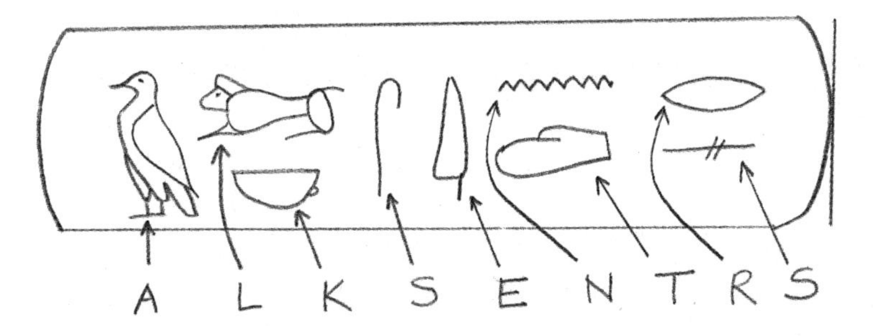

By 1824, Champollion had cracked virtually the entirety of hieroglyphics. Publishing his findings in the sensational *Précis du Système Hiéroglyphique*, he showed that there were actually three types of glyphs: ideograms, phonograms, and determinatives. It was phonograms, not ideograms, that made up the heart and soul of the Ancient Egyptian hieroglyphs; in fact, there was some interplay between the two, since the consonants that made up an ideogram could also be used as a form of phonogram.

To give one example: The ideogram for a scarab beetle, as in the case of Thutmose III's cartouche, can represent the idea "that which will be" or the three consonants "hpr." This was confusing even to the Ancient Egyptians, which is where determinatives come in. Placed after a phonogram, they let you know the sense of a word. The phonograms for chmplln, for example, might be followed by the determinative for a man 𓀀, showing that they spelled "champollion."*

WE ARE NOT ALONE

Contrary to the hopes of Champollion and others, our studies of the Ancient Egyptian texts have not made us adept in divine secrets, or shown us how life originated on Earth. Instead, they have intimately connected us with one of the first civilizations, revealing both its cruelty and its wisdom, its economic power and moral weakness. Considering that we are separated by over five thousand years of cultural evolution, we and the Ancient Egyptians are remarkably alike. The closeness of that bond is not only rewarding in its own right, but hints at another, equally close kinship with the humans that migrated to the Levant from sub-Saharan Africa. But for writing, we are all strangers.

The worldwide mania that followed the discovery of Ancient Egyptian hieroglyphics no doubt has its parallels in how the international community would react to the detection of an alien signal. There will be people who expect it to contain

* So what do the sun 𓇳, the gaming board 𓏠,and the scarab 𓆣 mean in the throne name of Thutmose III? The sun 𓇳, meaning "that which is," is easy enough, and we already know that the scarab denotes "that which will be," but what about the gaming board? Crucially, it's not an ideogram, but a phonogram, "mns." "Menes" was the first pharaoh of Egypt, flipping us back to an ideogram with the meaning "that which was." Altogether we get "present, past, future," or "everlasting." So Thutmose's throne name is "Son of Ra, He who is everlasting." Phew.

the secrets of the universe, of technologies beyond our wildest imagination: the final fulfillment of our blocked wish for unbounded energy, perhaps; time travel; maybe even an end to world poverty. What we find in it will almost certainly be something different. Whatever it is, like the deciphering of the hieroglyphs, it will dissolve many of the physical boundaries between our cultures. In short, we will not be alone.

Like the hieroglyphs, however, any message we do detect will be unbearably hard to decode. Ancient Egyptian society, being more technologically primitive than ours, might have produced a more simple form of writing, but it did not. The complexity of the hieroglyphs speaks to a different kind of literacy from that we are used to today, one where the very act of writing was a divine act, where symbolism and literalism collided as violently as they do in, say, *Ulysses* by James Joyce. Modern English, for example, readily submits to digitization, becoming a string of 1s and 0s with little lost in the way of meaning. Could such a highly visual language as Ancient Egyptian be encoded as faithfully? Our languages have become more pliant, certainly, but drained as they are of visual symbolism, have they become less poetic?

Finally, the great lesson of the hieroglyphs is the crucial part played by the Rosetta Stone. Without knowing the content of at least some part of a message, we have no way to decode it. When we talk to aliens, what will we use in its place? After all, in the case of the hieroglyphs we shared enough culture with the Ancient Egyptians to know what a scarab beetle looked like, for example, or that the Sun takes the shape of a circle. What do we do when the life-forms on the end of the phone are giant centipedes that communicate via chemical odors, or gelatinous-tendrilled air-dwelling balloons that signal using bioluminescence?

RADIO GA GA

Thankfully, there is one obvious candidate for a Rosetta Stone, and we don't need to go digging for it in the dunes of a future desert planet: the physics and mathematics required to build a radio telescope. We may not share body parts, or culture, or even the same biochemistry, but in order to send us a radio message our alien callers will at least share our penchant for radio technology. And that in turn means they will have at least the same understanding of mathematics and physics as we do.*

Just as we assume the aliens will broadcast in the least noisy part of the spectrum, we are fairly safe in assuming that they will begin their message with some form of Rosetta Stone. The most common element in the universe is hydrogen, so why not start there? They could kick off with the charge on the electron, for example, closely followed by the mass of a proton, then seal the deal with the speed of light. The problem, of course, is that to put a number to such things, intelligent creatures need a system of measurement. That in turn requires some kind of unit, and it's not that likely that the Kepler 452f-ians are using the foot, second, and pound. Which is where some rather ingenious things called dimensionless numbers come in.

A dimensionless number is one that has no units, and is therefore the same whatever measurement system you choose. You already know one: pi, being the circumference of a circle divided by its diameter. Pi crops up everywhere from General

* Granted, we are making the assumption that mathematics is universal, and not just something we cooked up to try and make sense of life on Earth. So in a sense we are funneling the number of communicable civilizations down to "those that have radio dishes and share our mathematics."

Relativity to Quantum Mechanics, and would no doubt be as handy to alien mathematicians as it is to earthbound ones.* Most people could probably tell you the value of pi to three decimal places, being 3.142, or that it is approximately equal to 22⁄7. But to physicists there is another number that is just as famous: 137.

That—or, rather, its inverse, 1⁄137—is the approximate value of what is known as the fine-structure constant, α. This enigmatic number describes the tendency for an electron to emit or absorb a photon, and while it is fundamental to a quantum mechanical description of electromagnetism—and therefore, the manufacture and operation of radio dishes—no one has the faintest idea where it comes from. And it's not alone. The gravitational coupling constant, α_G, would also be as familiar to alien physicists as the name of Cleopatra to the Ancient Egyptians. To those we can add such gems as the ratio of a proton mass to an electron mass, and the ratio of a neutron mass to a proton mass, all of which will be well known in alien worlds just as they are here.

Even with a Rosetta Stone in the form of basic physics, decoding the rest of an alien radio message will be fiendishly difficult. After all, it took some of the brightest minds on the planet twenty-three years to effect a translation of the hieroglyphs. To be successful we will need both brilliant scientists like Thomas Young, and equally gifted linguists such as Jean-François Champollion. The incipient age of interstellar communication will require a coming together not just of the sciences, but also the arts.

In the meantime, however, what do we look for? The practical answer is simple: any kind of transmission for which there

* Pi, like many dimensionless numbers in physics, is irrational, meaning it can't be neatly expressed as a fraction.

is no known natural source. Like Jocelyn Bell Burnell, we will turn up a few pulsars along the way, but we will be all the richer for that. For the savants, the Ancient Egyptian monuments of Rome hinted that somewhere out there lurked the collected wisdom of an entire civilization. In the quest for extraterrestrial intelligence, we ourselves are that hint. Assuming we do find some kind of unidentified transmitted signal, what then? Amazingly, even with no Rosetta Stone, all is not lost. And the story starts in the most unlikely of places: with *Ulysses* by James Joyce.

THE BLOOMSDAY BOOK

First serialized in the American journal *The Little Review*, James Joyce's *Ulysses* is rightly considered the masterpiece of literary modernism.* Everything about it defies convention. Its narrative, if you can call it that, might be summed up as "two men go for a walk in Dublin, and very little happens." Virtuosic, untrammeled, ruthlessly academic, and intentionally obscene, *Ulysses* is about as idiosyncratic a text as you can imagine, dispensing with all the preexisting conventions of character, speech, style, comprehensibility, and believability. It is exhilarating to read, at the same time unlocking the gilded cage of form and imprisoning you with its vaulting genius.

Yet as anarchic and free-flowing as it appears, when it comes to its underlying structure, *Ulysses* is subject to the most rigid of rules. As first noticed by the Harvard linguist George Kingsley

* *Ulysses* was serialized from March 1918 to December 1920, then published in book form in Paris in 1922. Its title is the Latin name for Odysseus, the hero of Homer's epic poem the *Odyssey*; like his editor Ezra Pound, Joyce wasn't afraid of an allusion or two.

Zipf in his 1949 masterpiece *Human Behavior and the Principle of Least Effort*, when submitted to statistical analysis, *Ulysses* is indistinguishable not only from one of the texts to which it alludes,* Homer's *Iliad*, but also to the Old English poem *Beowulf*, four Latin plays of Plautus, and the language of the Plains Cree Indians. To cut to the chase, if you take all the different words in *Ulysses*, count how often they occur and then rank them in order, you start to see something extraordinary.

To see what I mean, take a look at the chunk of *Ulysses* that appears at the beginning of this chapter. To really see the pattern, you need to analyze the text as a whole, but all we are after here is the gist. It's fairly easy to see which word crops up the most: it's the rather unprepossessing "of," which I count as appearing twenty-one times. Next is the equally un-Joycey "the," which crops up eleven times. After that we have "in" at seven times, and "and" at five times . . . by which point you may have picked up the pattern. The second most frequent word occurs half as many times as the first most frequent; the third most frequent, a third as many; the fourth, a quarter. In short, if we define the *rank* of a word to be its place in the pecking order, and its *frequency* to be the number of times it appears, we can write:

rank = constant/frequency

It surprises me all over again just telling you, so to convince us both let's take a look at the data that Zipf presents in *Human Behavior*. This is the point at which a table comes in handy, so here goes:

* The other, of course, is the *Odyssey*.

Word	Rank	Frequency
I	10	2,653
say	100	265
bag	1,000	26
orangefiery	10,000	2

Even more strangely, it doesn't end with written texts and spoken languages. Turning to data from the Sixteenth United States Census, conducted in 1940, Zipf showed that the same relationship applied for the population of cities, the number of shops within a city, and the wages of the citizens within them. In other words, the tenth most populated city had a tenth of the population of the first most populated, and the one hundredth richest man had a hundredth of the wealth of the richest. What could all of this possibly mean?

To this day, no one is exactly sure. Anyone and everyone in practically every field of the humanities has had a crack at Zipf's law, and no one explanation has yet garnered a decisive following.* As the title of his book suggests, Zipf 's own suggestion was that it is something to do with that universal human maxim "anything for an easy life." Why go out of your way to shop on the main street when the supermarket on the outside of town has everything that you need and more? Why

* OK, this is the most nerdy joke in the entire book but here goes: When we rank the various explanations of Zipf's law together with the number of academics that subscribe to that explanation, we get the only data set involving human decision-making which isn't subject to Zipf's law. Snarckle.

hire another actor when Tom Cruise is available? Why use DuckDuckGo when you've got Google?

Of course, when it comes to language, two things quickly make you lose the will to live. The first is when too many sounds are repeated, as anyone who has been on a long car journey with a three-year-old can attest. The other, as you will know from trying to speak French on holiday, is when there are too many different sounds to be able to tell them apart. Zipf saw language as a trade-off between these two extremes, a middle ground where the balance—weighted somewhere toward shorter words that are easy to say and understand—was just about perfect. Whatever the case, it seems that Zipf's law is a necessary-but-not-sufficient property of language. And most fascinatingly for our story, in 1999 three researchers named Laurance Doyle, Brenda McCowan, and Sean Hanser found that Zipf's law applies to the whistles of bottlenose dolphins.*

THE ORDER OF THE DOLPHIN

SETI has a long history of research into dolphin communication. One of the ten attendees at the first SETI conference at Green Bank in 1961 was the neuroscientist John Lilly. His book of that year, *Man and Dolphin*, had been an international bestseller, and claimed not only that dolphins were capable of complex emotions, but that they might also be capable of speaking human languages.

* You can find a fascinating discussion of all things Zipf in Alex Bellos' rather brilliant book *Alex Through the Looking Glass*. Thanks, Alex, for bumping into me in a coffee shop and handing me a copy at the very moment I happened to be writing this chapter.

Lilly's book had caught the attention of Frank Drake, who wanted to understand the potential challenges of communicating with other intelligent species. Lilly, with his boundless charisma and movie star looks, was an instant hit. Not only were the brains of bottlenose dolphins larger than ours, he informed the conference, but they were just as densely packed with neurons; in fact, parts of their brains looked even more complex than their human counterparts.

What's more, dolphins appeared to have their own language. Using tapes he had recorded at his new purpose-built Communication Research Institute on the island of St Thomas in the Virgin Islands, he demonstrated how bottlenose dolphins were able to communicate with one another with whistling sounds that they made using their blowholes. If he slowed the tapes down, he showed, the dolphins' squeaks and clicks even sounded like human language. Might it be possible to teach them to speak English?

Later, Frank Drake was to come to the reluctant conclusion that Lilly's work was "poor science," and that he had probably distilled hours of recordings to find those little bits that made their speech sound humanlike. At the time, however, Lilly's findings were enthralling, providing just a taste of the non-human intelligence that they were all seeking. Only Philip Morrison offered a note of skepticism, observing that dolphins, intelligent as they were, couldn't build telescopes with flippers.

On disbanding, the group decided to call themselves "The Order of the Dolphin." A few weeks later, Frank Drake received a small package in the mail from Melvin Calvin.* An identical

* For his work on the Calvin Cycle, the process by which plants photosynthesize sugars using the light energy captured by chlorophyll.

package was addressed to Struve, and Drake later learned that each conference participant had received one. Inside the box was a silver badge, a replica of an Ancient Greek coin in the shape of a leaping dolphin. The badge was a reminder that not only had they formed their own fraternity, but that in admiring the intelligence of dolphins they were honoring an academic tradition that went back to the Ancient Greeks.

THE GIRL WHO TALKED TO DOLPHINS*

The interest from Frank Drake, Carl Sagan, and others helped Lilly to secure NASA funding for his Communication Research Institute, and in 1965 he conducted one of the all-time strangest scientific experiments, when he agreed that a twenty-two-year-old researcher, Margaret Howe, should be isolated for ten weeks with a young male bottlenose dolphin called Peter. Howe had been attempting to teach Peter to speak by repeating English letters, numbers, and words and trying to get Peter to "say" them back to her, much as a mother might teach a child to speak. Howe became convinced that she would make more progress if she was in constant contact with Peter, and persuaded Lilly to allow her and the dolphin to cohabit.

The upstairs of the house was plastered and flooded so that Howe and Peter could share the same living space, with the knee-deep water being just shallow enough for Howe to wade through, and just deep enough for Peter to swim in. Howe slept on a foam mattress in the middle of one of the pools,

* I highly recommend that you check out the film of the same title by Chris Riley, he of *In the Shadow of the Moon* fame, which contains original footage of the experiment and a present-day interview with Margaret Lovatt.

worked at a desk suspended from the ceiling, and ate tinned food to minimize contact with outsiders. The days were run to a strict timetable, with Lilly giving precise instructions as to what Peter was to be taught and how it was to be documented. Howe was extremely dedicated, cropping her hair and even at one point painting the lower half of her face in thick white make-up and applying black lipstick so that Peter might more clearly see the shapes her lips made.

Dolphins are promiscuous by nature, and as the experiment went on, Peter began to show a sexual interest in Howe; by week five she records in her notes that "Peter begins having erections and has them frequently when I play with him." When she rebuffed his advances, Peter would become aggressive, using his flippers and nose to bruise her shins. When his conjugal visits with the other dolphins at the facility became so frequent as to be disruptive, Howe decided to relieve him manually so that he might focus on his lessons. Lilly, whose judgment may have been clouded by his incipient interest in LSD, noted, "I feel that we are in the midst of a new becoming; moving into a previous unknown . . ."

Gregory Bateson, the distinguished linguist who served as the Institute's director, and who was leading the research on dolphin to dolphin communication, was less than impressed. In his view, Peter was simply mimicking Margaret's speech in order to get fish, with little or no comprehension of what was being said. Carl Sagan, too, was skeptical that any progress was being made and suggested that Lilly switch to experiments which could verify whether dolphins could convey information to one another rather than trying to teach them English.

Eager for results, Lilly became increasingly desperate, and in the summer of 1966 he took the drastic step of injecting two of the dolphins with LSD to see whether that might improve their language skills. Thankfully, the drug appeared to have no effect, not even when Lilly took a pneumatic drill to the rock at the side of the pool to mess with the dolphins' super-sensitive hearing. This new twist was too much for Bateson, who left the facility. Lilly's NASA funding was withdrawn, the Institute closed down, and SETI was forever blemished by association.

JUMPING THROUGH HOOPS

The lesson of the Lilly experiments—apart from "don't give dolphins handjobs or LSD"—is a basic one. When it comes to close encounters with other species, it's all too easy to try to impose our own ideas of how they should think, feel, and behave. To us, a dolphin's lack of facial expressions, for example, might seem to indicate a lack of empathy; to a dolphin, our reluctance to stand on our heads when we greet them might appear equally inscrutable. On one count, however, Lilly does appear to have been correct: Dolphins have all the trappings of high intelligence.

For example, we know that not only are bottlenose dolphins capable of understanding signs, but the order in which they are shown those signs—in other words, the syntax—makes a difference. In the 1980s, Louis Herman and his team at the Dolphin Institute in Hawaii successfully taught captive dolphins "words" in the form of arm signals. Bottlenose dolphins were shown to be capable of understanding "sentences" of up to five "words," easily distinguishing between "take the ball to the hoop" and "take the hoop to the ball."

As demonstrated by the work of Diana Reiss, bottlenose dolphins are capable of complex manipulation, creating underwater bubble-rings with their blowholes that they can then swim through; in one case a captive dolphin formed a stream of bubbles with its blowhole that it then shaped into a ring using its tail.* They also show self-organized learning, interacting with a large poolside keyboard where they could request objects by pressing the relevant key, and exhibit self-awareness when presented with an underwater mirror, orienting their bodies to examine temporary marks made by Reiss and her fellow researchers.

What was less clear, up until the late 1990s, was whether or not bottlenose dolphins' wide range of vocalizations constituted any form of language. In the broadest of strokes, they make two main kinds of sounds with their blowholes: whistles and clicks. The clicks, generally speaking, are their equivalent of sonar, helping them to locate their prey and generally find their way around when visibility is poor.† It's the whistles—many of which can be outside the range of human hearing—where all the linguistic action appeared to be, but no one had any concrete proof.

That was until the collaboration between SETI's Laurance Doyle and UC Davis's dolphin researchers Brenda McCowan and Sean Hanser. As an astronomer at SETI, Doyle had an interest in whether the mathematical techniques that were being used to pick out planets in the Kepler data could be used to decode alien messages. In particular, he was interested in what information theory, as devised by Claude Shannon,

* Diana Reiss has some extraordinary footage of this in her TEDxBrussels talk entitled "The Dolphin in the Mirror."

† Dolphins also make another type of sound called burst-pulses, which are like a rapid series of clicks. Like whistles, there is evidence that these are also social calls.

could tell you about an alien radio source. As you may remember from Chapter Five, Shannon showed that the maximum amount of information, *H*, contained in a message of *i* letters, each with probability p_i, was simply:

$$H = -\sum p_i \log_2 p_i$$

In other words—and this is the only part you need to grasp—we can work out how much information could possibly be in a message *just by knowing the letters the message is made up of, and how often they occur.*

If that sounds incredible, it should: To my mind, it's a leap on the level of Einstein's Theory of Relativity or Schrödinger's Wave Equation. Let's think back for a moment to the example we looked at in Chapter Five, where our apple-scrumping message is made up of two "letters," "flashlight on" and "flashlight off." Crucially, we calculated that the information contained in the message was 1 bit, even though we had no idea what that information was, i.e., whether you were coming scrumping or not.

The team's idea was simple. They had no SETI signal to look at, so why not look at the messages of dolphins using information theory to see how much information they were capable of carrying? And if they were capable of carrying a lot of information, was it anything like as much as human language?

NAME, RANK, AND NUMBER

Doyle knew McCowan through the Planetary Society, a US non-profit which maintains an avid interest in SETI. One of the problems with understanding dolphins is their prodigious acoustic ability: Water is much better at transmitting sound

than air, and a dolphin can hear higher frequencies than a bat.*
To make matters even more complicated, dolphin whistles can be very short, lasting as little as a few tenths of a second, and change rapidly in pitch. As part of her PhD, McCowan had developed software that could not only accurately sample the individual whistles but could then categorize them by type. Now that there was a way of identifying the different "letters" in a dolphin message, could they be analyzed using information theory?

One of McCowan and Hanser's recent papers contained a table listing a number of different dolphin whistles and how often they had occurred, and to get things rolling the team decided to analyze them with a Zipf plot. For each of the forty-odd whistles in the table, they noted rank and frequency. Next they plotted them on a graph. Zipf 's Law, you'll remember, can be written as:

rank = constant/(frequency)

or

rank = constant x (frequency)$^{-1}$

For reasons that shall soon become clear, let's rewrite this as

rank = (frequency)$^{-1}$ x constant

* A little brown bat (*Myotis lucifugus*) can hear frequencies up to 115KHz; a common bottlenose dolphin (*Tursiops truncatus*) is good up to 150KHz. Humans hear up to 20KHz.

This kind of relationship is known in the trade as a power law, meaning simply that one variable (rank) is related to another (frequency) by a power index, in this case -1.

That's no fun to plot on a graph, so a standard trick in physics is to take the logarithm of both sides of the equation:

$$\text{Log (rank)} = \log \{ (\text{frequency})^{-1} \times \text{constant}\} \ \log (\text{rank})$$
$$= -\log (\text{frequency}) + \log (\text{constant})$$

Which is of that famous form, y = mx + c, the equation of a straight line. Plotting the logarithm of each whistle's rank against the logarithm of its frequency, the team was astonished. It came out as a dead straight line with a backwards-sloping gradient of –1.00, like this:

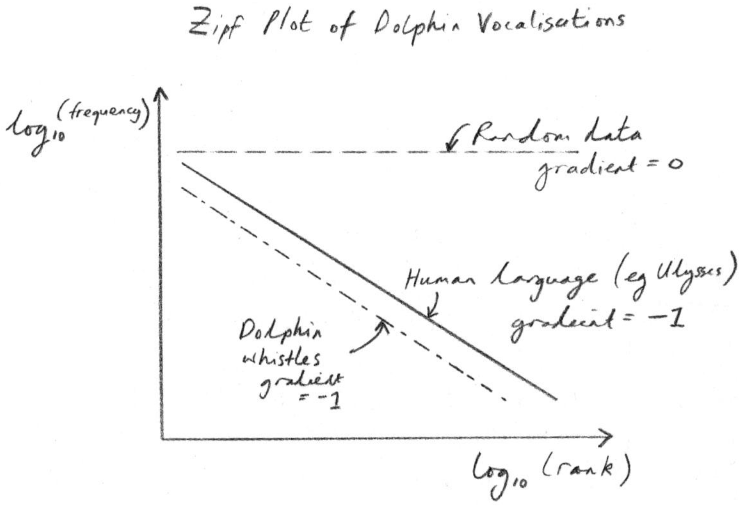

THE PLOT THICKENS

Unbelievably, dolphin whistles meet one of the basic requirements of symbolic language: Like *Ulysses*, the *Iliad*, and Plains Cree, they obey Zipf 's Law. A result like that needs checking, so next the team looked at the Zipf plots of baby dolphins. Sure enough, the line ran with a flatter slope, indicating that a much wider variety of whistles was being used; in other words, the baby dolphins were babbling. Next, they plotted the whistles of dolphins between two and eight months old. This time, the Zipf line steepened to a –1.05 slope, showing that the toddler dolphins were repeating themselves. Finally, at between nine and twelve months, teenage dolphins plotted at –1.00, just like the adults. They had finally started making sense, no doubt earnestly informing one another that their parent dolphins were total losers.

Next, the team decided to check how the calls of other species compared with bottlenose dolphins. Squirrel monkeys are highly social New World monkeys from Central and South America which make alarm calls to warn one another of predators, and Belding's ground squirrels from the western United States chirp to warn one another of danger. Adults of both species were recorded, and their vocalizations analyzed on a Zipf plot. The squirrel monkeys plotted at a less impressive but still noteworthy –0.75, and the ground squirrels at a measly –0.30. Whatever the Zipf plot was measuring, adult bottlenose dolphins and humans had a lot of it, squirrel monkeys had a bit, and ground squirrels virtually none at all.

SO COME ON, DO DOLPHINS HAVE LANGUAGE?

As we've already mentioned, a Zipf plot is a necessary but not sufficient requirement for symbolic language, and it's fairly easy to see why. *Ulysses* follows a Zipf plot, but it would be a brave literary theorist indeed that claimed that much of the complexity in Joyce's masterpiece was captured by the knowledge of how often each word occurs.

Clearly there is a lot more going on, and luckily Shannon has a fair bit to say about that, too. To write *Ulysses*, you not only need the right words, but you also need to put them in the right order. Looking at the text at the start of this chapter, for example, we can see that "of the" occurs five times; "of our" occurs three times. These two-word chunks are known as "digrams." Clearly we need a lot more text to make a decent job of it, but you can see that we could do the same job for digrams as we did for individual words, working out the probability of each for *Ulysses* as a whole and plugging them into one of Claude Shannon's—admittedly fierce—equations.

If you do that, the number that you end up with is called a message's "second order Shannon entropy." In fact you can do the same sort of job with trigrams, quadrigrams, quintigrams, and so on, producing figures for the third, fourth, fifth, and higher Shannon entropies that capture more and more of the complexity of the message. Human language, for example, has a Shannon entropy of around eight, meaning that, on average, rules of syntax connect eight words at a time.* And the dolphin whistles? They plotted at four, round about the same

* Another way of looking at this is: Thanks to grammar, if you know seven words of a phrase you can have a go at guessing the eighth. Knowing eight words, however, won't help you with the ninth.

number of symbols that Louis Herman showed dolphins were capable of handling when they learned sign language.[*, †]

TALK TO THE ANIMALS

So that's it, right? Dolphins have language, but it's not nearly as complex as ours. The answer, as you might suspect, is "not quite." Firstly, we are measuring Shannon entropy, which tells us the *potential* for a message to contain information. Like a Zipf plot, a fourth-order Shannon entropy is a necessary-but-not-sufficient condition for complex language. Dolphins' whistles might obey complicated laws of syntax without actually meaning anything, although given their highly social behavior that seems unlikely. Just as with the Ancient Egyptian hieroglyphs, until we manage to decipher bottle-nose dolphin whistles we can't be sure.

And, second, a fourth-order Shannon entropy might not be the maximum entropy of dolphin whistles. As we know, to be able to calculate higher order entropies you need more and more text.[‡] Doyle, McCowan, and Hanser only had around 10,000 whistles to work with, so fourth-order entropy was the

* For the keeny-beany Shannonites: Zipf's law gives the distribution of word probabilities in the first-order Shannon entropy. There, I've said it.

† I'm playing fast and loose with the words "letter," "symbol," and "word" here, but I'm sure you get the point: Shannon's formulae apply whether you look at the text at the level of letters, phonemes, words, or phrases, etc. That said, a higher Shannon entropy for letters will correspond to a lower one for words, e.g., first-order entropy for words of English corresponds to somewhere between fifth- and sixth-order entropy for English letters.

‡ As a rule of thumb, to calculate the nth-order Shannon entropy, you need 10n letters. In other words, for third-order entropy you need 1,000 letters; for fourth you'd need 10,000. *Ulysses* is approximately 265,000 words in length, so is capable of revealing anything up to fifth-order Shannon entropy, since $\log_{10}(265{,}000) \approx 5.4$

highest they could expect to find. You know you've reached maximum entropy when each *n-gram*—quadrigram, in this case—appears roughly the same number of times. Some dolphin quadrigrams, however, appeared a lot more often than others, which implies that if Doyle and his team had more whistle data they might have found higher orders of entropy. For all we know, dolphin whistles might have maxed out at eighth- or ninth-order, implying that—shock, horror—not only can they talk, but they are smarter than us.

HOMO SAPIENS PHONE HOME

In my humble opinion, the work of Doyle, McCowan, and Hanser has huge implications for any alien transmission we receive. Assuming we find a way to extract a message from it, one of the first things we are going to want to do is analyze it using information theory. That way, without knowing anything about what's in the message—"We buy gold," for example—we can figure out its potential to carry meaning. A fifteenth-order Shannon entropy would have our cryptographers quaking in their Birkenstocks, but at least we would know what we were dealing with.

For us to be able to determine a fifteenth-order Shannon entropy, of course, we would need a whole lot of message. We'd also want it to be as diverse as possible, incorporating every conceivable kind of medium and exploring every nook and cranny of their culture. We'd want to hear alien music, flick through alien holiday snaps, and watch alien movies. After all, without a Rosetta Stone, decoding it is going to be prodigiously difficult. All we can hope for is some random cultural overlap—that we both spend long years raising children, for example, or share a love of selfies—so that we can find the equivalent of a cartouche: a small section of code for which the meaning is clear.

In short, we want the aliens to send us their internet.

METI-PHYSICS

So, finally, if we are to send a message, what should it be? In fact, should we even signal at all? Many eminent scientists, Stephen Hawking among them, believe we shouldn't. Contact with a more technological civilization didn't turn out too well for the Plains Indians, he points out. To a certain extent, he's right; we don't know what's out there. Maybe an advanced civilization will travel here by wormhole and suck up the Pacific into a giant water tanker.

As Seth Shostak, the director of SETI, recently pointed out in a *New York Times* article, that's a concern we never used to have. As we know, the Mir Message in 1962 set the ball rolling, with a brief Morse code message to Venus. Next came the so-called Golden Plaque message on *Pioneers* 10 and 11, launched in 1972 and 1973. The following year we sent the most powerful message we have ever transmitted, the Arecibo Message, which was fired at a cluster of some 300,000 stars known as Messier 13. Depicting a stick man, a twist of DNA, and a map of the solar system, the cruddy pixilation of the Arecibo Message makes the 1970s video game *Pong* look sophisticated.

That was followed by the *Voyager 1* and *2* probes in 1977 and the famous Golden Record. On it, as we've heard, Ann Druyan and Carl Sagan curated just about every kind of information they could get their hands on: speech, whale song, classical music, rock and roll, as well as images of the Taj Mahal and the underside of a crocodile. What they excluded, of course, was anything to do with war, politics, or religion. After all, we didn't want to disappoint the aliens with our bad behavior.

Since the Golden Record, the Crimea has become the main focus for Messaging to Extraterrestrial Intelligence, or METI, as it is now known thanks to the Russian astronomer Aleksander Zaitsev. In 1999 and 2003 he supervised the sending of two messages known as the Cosmic Calls to nine nearby stars. And, not to be outdone, in 2008 NASA beamed the Beatles' song "Across the Universe" at Polaris, also known as the North Star.

Rather than continuing to send such "greeting cards," Shostak has a radical proposal, and one I heartily agree with: We should start sending "Big Data." As Laurance Doyle's work shows, we don't need to overthink this. The first thing the aliens will want to do is work out whether there might be any information in our message, and to do that they will need a lot of data. There are other good reasons, too. In sending them everything we've got, we are being as honest as we can be about who we are, warts and all. Let's not pretend we are sages, or saints. Let's be human.

We don't have to worry about overloading them. To a technological civilization more advanced than our own, the world's several hundred exabytes of stored information is going to seem like a flash drive.* At the same time, while their technology might be more advanced, let's not assume that they themselves are necessarily smarter than us. It's our ability to manipulate information that has enabled our ascendance, not our individual smarts.

* Estimated in 2007, by Priscila Lopez of UOC and Martin Hilbert of USC, to be 2.9×10^{20} bytes.

Let's send the internet. And while we're doing it, let's redouble our efforts to understand the communication systems of our fellow creatures. Until we can talk to dolphins we have little hope of being able to talk to ET. Just do one thing for me: somewhere up the front, let's stick James Joyce's *Ulysses*. Not only does it have maximum entropy to make your eyes water, but it's funny, and humor is something that has been sadly lacking in our messages so far.*

The supreme irony in all of this is: With the wacky hippy inclusiveness of the Golden Record, Carl Sagan had it about right. He and Ann Druyan sent the internet of their day; now we have to do the same with ours. After all, as with all worthwhile communication, the real message here isn't what we know about physics, or where we are in our solar system. It's the fact that we want to talk in the first place.

ARE WE ALONE?

We started this journey with a question. To answer it, first we needed to set a couple of things straight: The evidence for UFOs is weak, and the scientific credentials of SETI are strong. Next, we learned how the fundamental structure of the universe—the strengths of the interactions between its building blocks, for example—are fine-tuned to make carbon-based life a reality. All known life, we learned, is one; we are intimately related to every other organism on this planet, be it fungus, elk, or bacterium.

To get a steadier grip on the slippery issue of detectable alien signals, we summoned the Drake Equation. Crucially we learned

* Plus one of its best jokes is that it is impossible to understand.

that there were a number of factors that combine to produce the overall probability that other Earthlike worlds might be calling. First, we needed to know the rate of formation of Earthlike planets. To calculate that, we needed to know the rate of formation of Sun-like stars, the fraction of those stars that have Earthlike planets, and how many Earthlike planets they have.

Thanks to the Kepler Space Telescope, these are questions to which we have some fairly accurate answers. Incredibly, those answers are remarkably close to those that the original Order of the Dolphin guessed at. Sun-like stars are formed at the rate of around one a year, and between a fifth and half of such stars have one Earthlike planet.

Next, the Drake Equation asked the rate of formation of life-bearing planets. Again, as per the original SETI meeting, our best guess is a large number; virtually the same as the rate of formation of Earthlike planets. Our evidence for this is the early emergence of microbial life, and the fact that it appears to have taken a vital first step—namely, the evolution of a cell wall—at least twice, giving rise to both the bacteria and their seabed-fellows, the archaea.

The next step was to get a feel for the likelihood of complex life, by looking at the major transitions that made it possible. Among them all, only one appears to be a bottleneck: the evolution of the mitochondria, which happened only once in our planet's 4.5-billion-year history, and without which the eukaryotic cell would never have gotten its start. Speaking personally, it's this stepping stone across the river of chaos that keeps me awake at night. Everything before and everything after seems to follow naturally, but without the vast reserves of energy provided by an archaeon enslaving a bacterium, it's hard to see how complexity might ever arise.

Returning to the Drake Equation, this "eukaryotic bottleneck" provides our first departure from the early calculations of Drake and co. How can we put a number on the fraction of Earthlike planets that develop complex life? Granted, our investigations suggest that complex life inevitably develops intelligence, but with only one example of such life to go on, how can we conjure a number? We may not be in the dark, but we are definitely in the gloom. One in a hundred? One in a million, maybe?

To comfort us, we have the thought that there was at least one more instance of an endosymbiosis, that is to say, when one type of cell enslaved another. That example, of course, was the co-option of the cyanobacterium by a eukaryote, creating the type of cell that we find in plants. Complex life may be rare, but like our own solar system, with its idiosyncratic Jupiter and absence of super-Earths, not that rare.

Finally, we learned that we, as humans, share civilization, agriculture, language, and culture with countless other beings here on Earth. If we are going to learn to talk to aliens, first we are going to have to learn to communicate with the other terrestrial and non-terrestrial intelligences on our own planet. Huge, ancient brains are out there, wanting to play, to commune, and to teach. Are we alone? No. But it's down to us to make the first move.

FURTHER READING

All the books below are, in my humble opinion, not just great science writing, but great writing.

EXTREMOPHILES

The Voyage of the Beagle by Charles Darwin, Penguin Classics, 1989

Weird Life: The Search for Life That Is Very, Very Different from Our Own by David Toomey, W. W. Norton & Company, 2014

UFOS

Aliens: Why They Are Here by Bryan Appleyard, Scribner, 2005

The Demon Haunted World: Science as a Candle in the Dark by Carl Sagan, Ballantine Books Inc., 1997

SETI

The Eerie Silence: Renewing Our Search for Alien Intelligence by Paul Davies, Mariner, 2011

Rare Earth: Why Complex Life Is Uncommon in the Universe by Peter D. Ward and Donald Brownlee, Copernicus (2000)

UNIVERSES

Just Six Numbers: The Deep Forces That Shape the Universe by Martin Rees, Basic, 2001

The Hidden Reality: Parallel Universes and the Deep Laws of the Cosmos by Brian Greene, Vintage, 2011

LIFE

What Is Life? by Erwin Schrödinger, Cambridge University Press, 2012

Creation: How Science Is Reinventing Itself by Adam Rutherford, Current, 2014

HUMANS

Human Universe by Brian Cox and Andrew Cohen, William Collins, 2014

The Accidental Species: Misunderstandings of Human Evolution by Henry Gee, University of Chicago Press, 2013

ALIENS

What Does a Martian Look Like? The Science of Extraterrestrial Life by Jack Cohen and Ian Stewart, Ebury Press, 2004

Bird Brain by Nathan Emery, Princeton University Press, 2016

MESSAGES

The Information by James Gleick, Vintage, 2012

Cells to Civilizations: The Principles of Change That Shape Life by Enrico Coen, Princeton University Press, 2015

ACKNOWLEDGMENTS

Firstly, I would like to thank Dan Clifton, who planted the seed that eventually blossomed into this book. Dan wrote and directed my BBC Horizon documentary *One Degree* and is one of those multitalented, cross-disciplinary bods who you feel might just be an alien himself. I am hoping that the publication of this book means he will finally stop sending me web links about extraterrestrials. Just as Dan planted the seed, I am equally grateful to the green fingers of Elly James and Celia Hayley at hhb for training its wandering tendrils, and to my radiantly brilliant publisher Antonia Hodgson for encouraging it to unimaginable heights.

Praise be to three gifted researchers: Suzy McClintock, Lizzie Crouch, and Andrew Bailey tirelessly sought the most wonderful stories, the most eminent contributors, and explained bits of biology I am utterly ignorant about with the most otherworldly patience. Thanks also to the crack team at Little, Brown, including—but by no means limited to—Rhiannon Smith, Kirsteen Astor, Rachel Wilkie, Sean Garrehy, Zoe Gullen, and everyone on the sales team, especially Jennifer Wilson, Sara Talbot, Rachael Hum, and Ben Goddard. For the US edition, profound gratitude to the team at The Experiment.

I also wish to express my continued gratitude to the legendary Heather Holden-Brown and legistic Jack Munnelly at hhb.

Writing this book has been a great adventure, and I have met some outrageously gifted scientists along the way. Particular mention must go to Mazlan Othman, who was truly inspiring in her enthusiasm for all things alien. Nick Lane was kind enough to give me a basic lesson in biology—a bit like asking Einstein to fix your bike. Nicky Clayton and Nathan Lane not only opened their laboratory, but unlocked their thoughts on the potential biology of intelligent aliens. Warm thanks too to Laurance Doyle, who lucidly explained Shannon's orders of entropy, to Carlos Frenck for turning me on to GRBs, to Jocelyn Bell Burnell for patiently retelling the extraordinary story of her discovery of quasars, and to Richard Crowther, who gave me the lowdown on Near Earth Objects and the best freebee of the entire book: a model of the *Skylon C1*. And special mention must go to John Elliott, who explained the challenges of decoding alien signals.

It goes without saying that the mistakes within these pages are mine, but there would have been a great deal more of them without the help of my embarrassingly over-qualified referees. Professor Paul Alexander, Head of Astrophysics at Cambridge University, was kind enough to correct some howlers in my cosmology, and Dr. Ben Slater of the Department of Palaeobiology at Cambridge University put me right on everything from the habitability of red dwarf stars to the deep history of Earth. Professor Peter McClintock of Lancaster University brushed up my thermodynamics, while Dr. Nick Lane of University College, London vetted my evolutionary biochemistry without feeling the need to point out that most of it was cribbed from his books in the first place.

The only thing more daunting than writing a book on a subject you know very little about is being married to someone who is writing a book on a subject they know very little about: Thank you so much, Jess, for your love and encouragement. Sonny, Harrison, and Lana: This book is for you. Heartfelt thanks, as always, to my ever-supportive family: my mum, Marion, and my sisters Bronwen and Leah, my brothers-in-law Richard, Phil, and Josh, and my in-laws Stephanie and Alan Parker. They, together with friends Alexander and Hannah Armstrong, Torjus and Amelia Baalack, Pierre and Kathy Condou, Steven Cree and Kahleen Crawford, Jono and Amanda Irby, Gary and Lauren Kemp, Bruce McKay and Jonathan and Shebah Yeo have listened patiently to more guff about aliens than anyone would think humanly possible.

Finally I would like to thank Rafael Agrizzi L De Medeiros, Daniella Jackson, Sophie Lewis, and Elouise Ody at my favorite coffee shop, where most of this book was written. I promise to write the next one somewhere else.

ABOUT THE AUTHOR

BEN MILLER is, like you, a mutant ape living through an Ice Age on a ball of molten iron, circulating a supermassive black hole. He is a trained physicist, an actor, and a comedian. He is also the bestselling author of *It's Not Rocket Science*, and host of the TV show of the same name. He has hosted numerous other TV and radio documentaries on subjects as varied as temperature and the history of particle physics. He is slowly coming to terms with the idea that he may never be an astronaut.

@ActualBenMiller

Connected Marketing

Pour Virginie, ma 'Bella Bleue' — Paul Marsden

For Bridget and Beth — Justin Kirby

Connected Marketing

The Viral, Buzz and Word of Mouth Revolution

Edited by

Justin Kirby and Paul Marsden

AMSTERDAM • BOSTON • HEIDELBERG • LONDON • NEW YORK • OXFORD
PARIS • SAN DIEGO • SAN FRANCISCO • SINGAPORE • SYDNEY • TOKYO

Butterworth-Heinemann is an imprint of Elsevier

Butterworth-Heinemann is an imprint of Elsevier
Linacre House, Jordan Hill, Oxford OX2 8DP, UK
30 Corporate Drive, Suite 400, Burlington, MA 01803, USA

First edition 2006
Reprinted 2006 (twice)

British Library Cataloguing in Publication Data
A catalogue record for this book is available from the British Library

Library of Congress Cataloging-in-Publication Data
A catalog record for this book is available from the Library of Congress

ISBN–13: 978-0-7506-6634-3
ISBN–10: 0-7506-6634-X

For information on all Butterworth-Heinemann publications
visit our website at books.elsevier.com

Printed and bound in *The Netherlands*

06 07 08 09 10 10 9 8 7 6 5 4 3

Contents

Contributors

Stéphane Allard
Associate Director, Spheeris

Schuyler Brown
Co-creative director, Buzz@Euro RSCG

Idil Cakim
Director of Knowledge Development, Burson-Marsteller

Andrew Corcoran
Senior Marketing Lecturer, Lincoln Business School, University of Lincoln

Steve Curran
President, Pod Digital

Bradley Ferguson
Founder, Intrinzyk/Founder emeriti, Informative, Inc.

Justin Foxton
Founding Partner and CEO, CommentUK

Graham Goodkind
Founder and Chairman, Frank PR

Justin Kirby
Managing Director, Digital Media Communications (DMC)

Paul Marsden
London School of Economics/Associate Director, Spheeris

Liam Mulhall
Founder, Brewtopia

Greg Nyilasy
Henry W. Grady College of Journalism and Mass Communication, University of Georgia

Martin Oetting
ESCP-EAP European School of Management/MemeticMinds.com

Bernd Röthlingshöfer
Author

Sven Rusticus
CEO, Icemedia

Pete Snyder
Founder and CEO, New Media Strategies

Thomas Zorbach
Founder and CEO, vm-people

Foreword

Not long after *The Anatomy of Buzz*[1] was published in October 2000, a journalist called to interview me for a story he was writing about buzz marketing. We talked for a long time and later exchanged some emails. A few weeks passed and the story did not come out, so my publisher sent a friendly enquiry to the journalist: 'Is there a problem?' As it turned out, the journalist had a difficult time convincing his editors to run the story. 'I spent 20 minutes arguing with my editor yesterday on why this story needs to run before the whole fad is dead!' he wrote.

This was the first time I realized that what I see as a major shift in marketing, is seen by some other people as just another marketing fashion.

Today, five years later, the 'fad' is not dead. You'll find plenty of evidence that buzz marketing is still going strong in the numerous case studies in this book. In fact, in the past few years I have seen a consistent increase in interest in word of mouth, buzz and viral marketing. More companies are working on ways to use buzz, conferences and panel discussions are being held all over the world, and yes, books and articles on the subject (including the article by that journalist which ran shortly after our enquiry) are coming out regularly.

Why the continued interest? In his introduction to this book, Paul Marsden points out that 'advertising hyperclutter, media channel fragmentation and new ad-busting technology that enables people to skip and block unwanted advertising' are major challenges to the traditional marketing model. Throughout the book we are also reminded of the increasingly cynical, sceptical, and marketing-savvy consumer. Indeed, noise and scepticism are major forces that send marketers looking for alternatives. In addition, Marsden and other contributors correctly point out the significant impact technology has in boosting the power, reach and speed of consumers' word of mouth.

There is another way in which technology affects the future of connected marketing. Technology has guaranteed that marketers will constantly be reminded of the power of word of mouth. Ten years ago, a marketing manager could brush off talk about word of mouth and go back to discussing her next advertising campaign. Today she stumbles over buzz wherever she goes. She searches on Google for her company's name and comes across a discussion between two unhappy customers. As she stands in line at the coffee shop, the man behind her, cell phone in hand, talks about the movie he saw last night. On search engines such as BlogPulse.com our marketing manager can actually see how much buzz her brand has been getting on blogs in the past six months.

As word of mouth becomes something that can be measured, archived and searched, we can expect the interest in connected marketing to increase even further. With this comes a word of caution: as more vendors compete in this space, there will be more pressure to get results at all costs. As glossy proposals are put on desks, as exciting presentations are made, it is crucial that marketers continue to ask one simple question: 'Is this the right thing to do?'

In Chapter 14 of this book Schuyler Brown writes: 'Periodically, you'll hear stories about companies enlisting cheap labour to monitor and populate chatrooms and post commercially motivated messages to blogs. Ethical marketers know that this crosses the line.' I agree. Full disclosure is a prerequisite to good buzz marketing. Not only because undercover campaigns can backfire, but first and foremost because full disclosure is the right thing to do.

One reason I've been captivated by the phenomenon of word of mouth communications is the raw quality that it possesses. Just visit a few blogs and newsgroups to read what people say about a certain product and you'll see what I mean. The tone is straightforward, no-nonsense, sometimes harsh, often sarcastic. There's always a variety of opinions – for every view expressed, you find an opposing one. Companies that tried to control what's being said about them have realized very fast that it's impossible to tame the beast. Word of mouth is wild.

In some ways, this book is like the phenomenon that it describes. If you expect to hear one voice, one opinion, I'm afraid you'll be disappointed. If you expect to agree with everything you read, you probably won't (I know I haven't). What you can expect is a lively conversation about a significant trend in marketing. Rich with data, war stories and case studies, this book is an important addition to the word of mouth literature. As you'll quickly find, there's more than one way to stimulate people to talk. Which ones are the best? It is up to you to sort through the copious examples and form your opinion, much in the way you would after asking a few friends for advice about which car to buy.

If you are a marketing manager looking for ideas, the analysis and examples in this book are very likely to stimulate you to think. In this sense, the question of whether buzz marketing is a fad or not is not very relevant. Only time will tell whether the journalist was right or wrong. In the meantime, there's plenty of insight in this book that will help you come up with ideas for how to spread the word about your products and services.

Emanuel Rosen
Menlo Park, 2005

Reference

1 Rosen, Emanuel (2000) *The Anatomy of Buzz: Creating Word of Mouth Marketing*. New York: Doubleday.

Acknowledgements

The editors would like to thank: Dave Chaffey, author of *Total Email Marketing*, who coined the term 'connected marketing' and pointed the way; Michael Holgate from Brand Genetics who helped move the thinking on, Sean Gogarty from Unilever who pioneered putting the principles into practice, Emanuel Rosen for his encouragement and helpful advice; Martin Oetting for helping to organize everyone and everything; Sue Weekes for fab subbing, rewriting and boundless enthusiasm; Greta the ghost; Yazied for reading, reading and more reading; Piers Hogarth-Scott at DMC Australia for his strategic insight, particularly into local integration issues; Dr David Cannon from the London Business School for his ongoing encouragement and input; Paul Hepher; and Hervé Hannequin, planner-*extraordinaire* from WCRS for his 'inspiring revelations' into marketing trends.

We would also like to thank all the marketing teams we have been working with while preparing this book – they have played a big part in shaping the content of *Connected Marketing*. In addition to thanking the VBMA, whose members have provided helpful comments, special thanks go to Michael Davey from Ford Australia, Jeremy Thomson, Shirley Tang, Steve Jelliss, Maria McCullough and Mel Sroczynski from Mazda, Charlotte Emery, Alfredo Piedra, Guy Lawrence and John Burke from Bacardi, James Kydd, Alison Corfield and Richard Duff-Tytler and Michael Garvey from Virgin Mobile, Mark Cornell and Laure Moreau from LVMH, Andrew Christophers and the rest of the crew from Brand Genetics, and Crispin Manners, Liz Andrews and the team from the PR firm Kaizo.

Last but by no means least, we'd like to thank all the contributors to *Connected Marketing* for their time, understanding and willingness to share their expertise.

Justin Kirby and Paul Marsden
London 2005

Trademarks

Icecards is a registered trademark of Icemedia.
Talkability is a registered trademark of Frank Public Relations Ltd.
e-fluentials is a registered trademark of Burson-Marsteller.
Influentials is a service mark of NOP World.
adidas, the adidas logo and the 3-Stripe trademark are registered trademarks of the adidas-Salomon group, used with permission.

All other trademarks, service marks and registered trademarks mentioned in the text are the property of their respective owners.

Introduction and summary

Paul Marsden

London School of Economics/
Associate Director, Spheeris

The naming of New York

Swiss botanist Conrad Gesner first saw the flower that was responsible for the naming of New York whilst on a trip to the Bavarian Alps in 1559. The delicate bloom had been imported to Europe from a faraway valley between the great Yangtze River and the Central Asian Steppe via Constantinople. With petals red in colour and with a sweet, soft and subtle scent, the exotic flower was believed by Turkish traders to have divine origins - they had even named it '*lale*', an anagram in Arabic script of Allah. Gesner, however, was struck by the peculiar turban-shaped form of the petals, and taking the Turkish word for turban, *tülbend*, as inspiration he gave it a European name - '*tulipa turcarum*', or tulip for short.

The extraordinary chain of events that followed, and which ultimately led to the naming of New York, has since become a popular tale that parents relate to their children when they want them to grow up to become stockbrokers. The story recounts how news of the new, beautiful and rare flower spreads by word of mouth across Europe, piquing the interest of Dutch nobility who soon begin importing tulips and adopting them as exotic status symbols - visible signs of their good taste and wealth. The combination of limited supply and the association with the rich and famous then triggers a word of mouth epidemic of demand for tulips.

To cater for the demand, Dutch shipping companies begin importing tulip bulbs from Turkey, local farmers begin cultivating them and city merchants begin trading them. But supply cannot keep up with demand - everybody wants tulips - and this only fuels the demand. Throughout Holland, thousands of people give up their jobs to grow tulips, selling their homes and their land just to get their hands on the precious bulbs. By 1635, Holland is consumed by tulip fever, pushing tulip prices up to astronomic levels; a single Viceroy tulip bulb sells for the equivalent of US$40 000: four tons of wheat, eight tons of rye, one bed, four oxen, eight pigs, 12 sheep, one suit of clothes, two casks of wine, four tons of beer, two tons of butter, 1000 pounds of cheese and one silver drinking cup.

But just as Dutch tulip buzz reaches its feverish crescendo in 1637, word of mouth suddenly turns negative. Rumours begin to spread that tulips are no longer worth the extraordinary amount people are paying for them. In a few short weeks, this negative word of mouth triggers a precipitous and dramatic crash in tulip prices – to less than a hundredth of their previous value. Because so many people have so much money tied up in tulips, the Great Tulip Crash of 1637 virtually bankrupts the Netherlands, and for decades the country is often unable to pay for soldiers to defend its interests abroad. One such interest is the Dutch settlement of New Amsterdam, lying on the east coast of North America.

Without military defence, New Amsterdam lies open to attack, and in 1664 the English army march into the fledgling city and declare it their own without a single shot being fired, renaming it in honour of the English Duke of York. And that's how a Dutch seventeenth-century word of mouth craze for tulips resulted in the renaming – or rather naming – of New York.[1]

Introducing *Connected Marketing: The Viral, Buzz and Word of Mouth Revolution*

The tale of how a seventeenth-century word of mouth tulip craze led to the naming of New York illustrates the power of what is typically called word of mouth – product-talk between people – in both driving market demand and decimating it. Three and a half centuries after tulipomania, word of mouth remains a powerful influence on what people say, do and buy. Management consultants McKinsey & Co. estimate that two-thirds of the US economy is driven by word of mouth,[2] and recent research has scientifically proven what businesses have known for some time: word of mouth drives business growth – companies that stimulate high levels of positive word of mouth in their markets grow fast, whilst those that don't stagnate.[3]

Connected Marketing is a book about using word of mouth connections as marketing media to drive growth. It covers three emerging techniques in marketing: viral marketing, buzz marketing and word of mouth marketing. Whilst differences may exist between viral, buzz and word of mouth campaigns – although the terms are often used interchangeably – all are based on the same fundamental insight that the most powerful media open to marketers are the word of mouth connections joining everybody to everybody in any target market by no more that six links.[4] So whilst viral marketing is often the label of choice for campaigns harnessing online word of mouth connections,[5] word of mouth marketing for campaigns using traditional or 'offline' word of mouth connections, and buzz marketing for campaigns that harness both - often in combination with traditional news media coverage - they all seek to exploit connectivity between people as marketing media. To avoid getting into divisive and sterile debates on definitions, we have coined the term 'connected marketing' as an umbrella term for all viral, buzz and word of mouth marketing techniques, to include all promotional activity that uses word of mouth connections between people, whether digital or traditional, as communications media to stimulate demand.

Written by 17 experts working at the cutting edge of viral, buzz and word of mouth marketing, *Connected Marketing* introduces the range of scalable, reliable and measurable solutions for driving business growth by stimulating positive brand talk between clients, customers and consumers. Illustrated through gold-standard case studies, the topics this book covers include seeding trials, viral advertising, brand advocacy programmes, live buzz programmes, brand blogging, buzzworthy PR, influencer outreach initiatives, buzz measurement and metrics, as well as chapters on the science and future of viral, buzz and word of mouth marketing. Advocating a new 'connect and collaborate' vision of marketing, *Connected Marketing* will show you how businesses can harness connectivity between clients, customers and consumers as powerful marketing media for driving demand.

30-second primer: from marketing to connected marketing

- **Marketing**: Satisfying market needs through the commercialization of products and services in such a way that satisfies internal company needs and those of the company's investors.
- **Mass marketing**: Satisfying widespread market needs with standardized mass-produced products and services, typically promoted though standardized mass media advertising.
- **Mass media advertising**: The promotion of a company or its products and services through paid-for persuasive messages from an identified sponsor appearing in media with a large audience: newspapers, magazines, cinema films, radio, television and, increasingly, the Web.

- **Viral marketing**: The promotion of a company or its products and services through a persuasive message designed to spread, typically online, from person to person.
- **Word of mouth marketing**: The promotion of a company or its products and services through an initiative conceived and designed to get people talking positively about that company, product or service.
- **Buzz marketing**: The promotion of a company or its products and services through initiatives conceived and designed to get people *and the media* talking positively about that company, product or service.
- **Connected marketing**: Umbrella term for viral, buzz and word of mouth marketing. Any promotional activity that uses word of mouth connections between people, whether digital or traditional, as communications media to stimulate demand.

Why connected marketing, why now?

The traditional marketing model we all grew up with is obsolete
Jim Stengel, Global Marketing Officer, Procter & Gamble, 2004[6]

Mass marketing today is a mass mistake
Larry Light, Chief Marketing Officer, McDonalds, 2004[7]

For the first time the consumer is boss, which is fascinatingly frightening, scary and terrifying, because everything we used to do, everything we used to know, will no longer work
Kevin Roberts, Chief Executive, Saatchi & Saatchi, 2005[8]

Marketing today is in a state of turmoil. Industry leaders say that the traditional marketing campaigns, based on mass media advertising, are not working anymore. And the facts back them up. A 2004 study into advertising effectiveness by Deutsche Bank in the US consumer packaged goods sector found that only 18% of television advertising campaigns generate a positive return on annual investment, whilst the *Harvard Business Review* reports that for every dollar invested into traditional advertising for consumer packaged goods, the short-term return on investment is just 54 cents.[9] Marketing in the business-to-business (B2B) sector fares no better; an astonishing 84% of B2B marketing campaigns actually result in a fall in market share and brand equity.[10]

Throwing money at the problem doesn't seem to help – doubling campaign budgets for established products can lead to an increase in sales of just 1–2%.[11] And quality over quantity doesn't seem to impact on efficacy either – great ads don't mean great sales, as illustrated by the award-winning Budweiser 'Whassup?' campaign in 2000: Budweiser's US market share actually dropped 1.5–2.5 percentage points during the campaign, with sales in barrels falling by 8.3% – the largest revenue drop the company had experienced since 1994.[12]

Of course, there are exceptions to the rule; sometimes traditional marketing campaigns built around mass media advertising can still work. But the problem is that they are few and far between. A century ago the founder of consumer packaged goods giant Unilever, Lord Leverhulme, lamented: 'Half the money I spend on advertising is wasted; I just don't know which half.'[13] Today, if a company manages to waste only half of its advertising budget then quite likely it is doing better than many of its competitors.

Crisis in mass marketing? The numbers

- 18%: Proportion of TV advertising campaigns generating positive ROI
- 54 cents: Average return in sales for every $1 spent on advertising
- 256%: The increase in TV advertising costs (CPM) in the past decade
- 84%: Proportion of B2B marketing campaigns resulting in falling sales
- 100%: The increase needed in advertising spend to add 1–2% in sales
- 14%: Proportion of people who trust advertising information
- 90%: Proportion of people who can skip TV ads who do skip TV ads
- 80%: Market share of video recorders with ad skipping technology in 2008
- 95%: The failure rate for new product introductions
- 117: The number of prime time TV spots in 2002 needed to reach 80% of adult population – up from just 3 in 1965
- 3000: Number of advertising messages people are exposed to per day
- 56%: Proportion of people who avoid buying products from companies who they think advertise too much
- 65%: Proportion of people who believe that they are constantly bombarded with too much advertising
- 69%: Proportion of people interested in technology or devices that would enable them to skip or block advertising

Source: US figures, various sources[14]

Given falling returns and increasing costs associated with traditional marketing campaigns, it was only a matter of time before shareholders and boardroom directors would begin to look at marketing as the next cost bucket to rationalize. That time is now. The challenge is crystal clear: show that every marketing dollar spent can deliver a healthy, predictable and measurable return on investment.

It is said that necessity is the mother of invention, and over the past few years marketers have been remarkably inventive in coming up with innovative marketing campaigns that demonstrate an enhanced return on investment. One solution has been to reduce media costs by using free media - word of mouth connections to propagate marketing messages. As well as being free media, word of mouth connections are influential media: a 2004 UK survey by consultants CIA:MediaEdge of 10 000 consumers found that 76% cite word of mouth as the main influence on the purchasing decisions, compared to traditional advertising which flounders at 15%. In the US, NOP research shows that 92% of Americans cite word of mouth as their preferred source of product information, whilst advertising company Euro RSCG has found that when it comes to generating excitement about products, word of mouth is 10 times more effective than TV or print advertising.[15]

In addition to being free and influential media, word of mouth connections are becoming increasingly more important in influencing purchasing patterns: a 2002 study by US research company Goodmind found that one third of Americans believe word of mouth is now more important in influencing their purchases than it was three years ago,[16] whilst over the past 30 years NOP reports that the influence of word of mouth has grown 50%.[17] Why should this be? Why should the oldest media available to marketers - word of mouth connections - become even more important in influencing buying behaviour in an age when media formats and channels are proliferating? The answer is five-fold:

1. New **personal communications technology** such as blogs, instant messaging, mobile telephones, email, online review sites and personal websites are increasing the speed, reach and utility of word of mouth.
2. Increased **marketing literacy** among buyers means people increasingly dismiss traditional marketing campaigns as biased propAdganda. Instead, they turn to trusted word of mouth sources for advice.
3. Acute **advertising clutter** is making it increasingly difficult for traditional marketing campaigns to 'break through' and capture people's attention. To avoid the advertising cacophony buyers turn to their friends for word of mouth recommendations.

4. Accelerating **media fragmentation** is shrinking media audiences; more channels, more media are making it harder for advertisers to reach their target markets through traditional marketing campaigns.
5. New **ad blocking technology** is empowering people to skip, stop, or avoid unwanted advertising messages and interruptive marketing campaigns.

Of course, it is one thing for marketers to realize that word of mouth connections are powerful media, but it is quite another to know how to harness word of mouth to stimulate demand. Viral, buzz and word of mouth marketing techniques are all solutions to harnessing word of mouth connections as marketing communications media. *Connected Marketing* illustrates these techniques in detail, but, to introduce them, 15 high profile campaigns stand out as representing the range of solutions open to marketers for exploiting word of mouth connections as media for driving demand.

High-profile viral, buzz and word of mouth marketing campaigns

1. **Hotmail** (1996): The campaign that put *viral marketing* on the map and that helped Hotmail become the leading personal web-based email service provider. Deceptively simple, the campaign involved turning users of the service into brand advocates, by appending all outgoing emails with a small 'P.S.' message as if it was from the sender: 'Get your free email at Hotmail'. By therefore turning Hotmail users into a promotional sales force, the email service recruited 12 million subscribers in 18 months with a marketing budget of only $500 000.[18]
2. **Unilever Dove 'Share a Secret'** (1998): One of the first examples of a *viral sampling* initiative, this campaign turned loyal users of Dove soap into active brand advocates by enabling them to order free Dove samples for their friends. An advertisement invited Dove users to mail in the name and address of a friend with whom they would like to share the Dove secret. The friend would then receive a Dove gift certificate in the post entitling them to a free Dove pack of soap and discount vouchers. By supplying their own name and address, participants in the promotion would also receive a Dove certificate themselves. 90% of people who participated in the offer supplied the contact details of a friend, some 80% higher than Unilever had expected, and during the promotion Dove's market share rose 10%.[19]

3. **The Blair Witch Project website** (1999): The first high-profile online campaign that combined advertising with entertainment (aka *advertainment*) and that was used to successfully promote the word of mouth hit movie. Costing only US$15 000 to produce, the Blair Witch Project website used mystery and intrigue to stimulate word of mouth anticipation for the movie. The site attracted 75 million visits on the first week of the film's release.[20]
4. **Virgin.net 'viral email'** (2000): Popular *viral email promotion* that demonstrated the speed of viral marketing. In just three hours, the UK ISP (Internet Service Provider) signed up 20 000 people to its marketing database after sending out an email offer to only 25 people and asking them to forward it to a friend. By offering recipients and those to whom they forwarded the email a compelling incentive for signing up – free cinema tickets – the email spread like an epidemic, swiftly creating a CRM (customer relationship management) database.[21]
5. **Agent Provocateur 'Proof'** (2001): The most popular online *viral campaign* to date – that happened by accident and happenstance rather than by design. A sexy cinema advertisement for the lingerie brand, featuring an orgasmic Kylie Minogue, was posted online. It was then copied, downloaded, or forwarded an estimated 100 million times. The risqué viral clip, called 'Proof', massively increased awareness for the small luxury boutique brand, putting Agent Provocateur onto conversational agendas around the globe.[22]
6. **Hasbro POX 'Alpha Pups'** (2001): A high-profile experiment in *viral seeding* designed to ignite word of mouth for the toy manufacturer's new portable game console 'POX'. Notable for the ingenious solution employed to identify opinion-leading boys aged 8–13 (aka 'Alpha Pups'). Hasbro sent market researchers to city playgrounds, skate parks and video arcades to ask a simple question: 'Who's the coolest kid you know?' When they got a name, the researchers went in search of this cool kid to ask him the same question, to climb the local hierarchy of kid-cool until someone finally answered 'Me!' Alpha Pups, once identified, were invited to trial the game, in return for which they would be given 10 new pre-release POX units to share with friends – to kick-start word of mouth. The innovative seeding campaign created media buzz, getting reported in the *New York Times* and the *Harvard Business Review*. It has also been included as a case study in the latest edition of the marketing classic *The Diffusion of Innovations*.[23]
7. **Procter & Gamble 'Tremor'** (2001–ongoing): The most comprehensive and extensive *seed marketing* initiative to date: a

national US sampling panel of over 250 000 teen opinion leaders used to optimize product launches. Using 'get it first' sampling (pre-launch freebies), 'inside scoops' (buzzworthy information) and 'VIP votes' (market research) to turn opinion leaders into word of mouth advocates, Tremor seeding trials have been able to produce sales uplifts of up to 30% for P&G's new products and those of its partners - including Coca-Cola, Sony and Toyota. In 2005, Procter & Gamble launched 'Tremor Moms', a second launch optimization seeding panel of 500 000 young mothers.[24]

8. **Sony Ericsson T68i 'Fake Tourists'** (2002): The ingenious, if controversial, *live buzz* campaign to kick-start word of mouth for their new T68i camera phone. The campaign involved hiring teams of undercover buzz agents to pose as tourists in ten US cities. Their mission: to go up to people in the street, hand them the phone and say 'Excuse me, would you mind taking a picture of us?' With the photo taken, the buzz agents would start a conversation 'Thanks a lot, man. It's cool, right?' - then proceed to talk up the new phone. Branded variously as stealth marketing, covert marketing, or undercover marketing, no data on the success of this live buzz campaign has been published, but awareness of the phone was certainly augmented by the extensive media coverage of the innovative, if controversial, campaign.[25]
9. **adidas 'Fevercards'** (2002): Popular *brand advocacy* campaign designed to help adidas brand fans evangelize about their favourite brand. The campaign involved offering sets of free personal contact cards to visitors to its website, featuring adidas brand artwork on one side and customized with personal contact details on the other. Ordered and personalized online, printed using on-demand digital printed technology then sent out by postal mail, one million adidas 'Fevercards' were ordered in 46 days and sent out to fans in 180 countries. Follow-up research showed that 78% of Fevercards sent out were handed out to friends, stimulating brand conversations 65% of the time. This brand advocacy campaign also enabled adidas to generate a high-quality CRM database of 50 000 adidas brand advocates.[26]
10. **Burger King 'Subservient Chicken'** (2004): Highly successful and hilarious online branded game (aka '*advergame*') to promote awareness for Burger King's new chicken sandwich. The game involved directing the actions of a man dressed up as a chicken in underwear through what appeared to be a webcam window. Through online word of mouth ('word of mouse'), the game clocked up 286 million visits, creating double digit growth of awareness for Burger King's chicken sandwich.[27]

11. **Nike 'Armstrong Bands'** (2004): High-profile *cause marketing* campaign to enhance the word of mouth appeal of the Nike brand. In a joint venture with Lance Armstrong Foundation for cancer research and education, Nike began selling US$1 yellow rubber wristbands for the Foundation in May 2004. The bands were engraved with the words 'Live Strong' - the mantra of Lance Armstrong, cancer survivor and seven-time winner of the Tour de France. With no marketing promotion, demand for the bands spread uniquely by word of mouth: in just six months, 20 million bands had been sold in the US and more than 60 other countries worldwide, with proceeds going to charity. (By spring 2005 that number had doubled to 40 million). In 2005, Nike - continuing its brand-enhancing good corporate citizenship drive - followed up with a second word of mouth cause campaign called 'Stand Up, Speak Up', selling interlinked black and white wristbands to promote an anti-racism message.[28]
12. **General Motors 'Fastlane Blog'** (2004): One of the first high-profile *business blog* campaigns to stimulate online word of mouth. The campaign involved getting senior executives from the car manufacturer to publish online diaries of their personal thoughts, opinions and predictions relevant to their industry. Written in a candid, informal and honest fashion, the business blog avoided gimmicks, marketing speak, or sales talk and enabled readers to comment and engage the diarists in dialogue, attracting a large daily readership.[29]
13. **General Motors 'Oprah's Great Pontiac Giveaway'** (2004): Much-cited *buzz stunt* to get General Motors' new Pontiac G6 car onto conversational agendas. The event involved giving away a new Pontiac G6 to each member of a live audience of the popular US Oprah Winfrey television talk show. The 276-car giveaway cost an alleged US$7 million, and certainly succeeded in getting people and the media talking, chalking up 624 news reports, including the cover of *People Magazine*, and driving half a million people to the Pontiac website.[30] The campaign was also noteworthy because it later was the subject of negative buzz - as a campaign that sold itself rather than the car it was supposed to sell, and because the recipients of the free car had each received a nasty surprise in the form of a $7000 tax bill.[31]
14. **Microsoft Xbox *Halo 2* 'ilovebees'** (2004): The first *promotional 'ARG'* (Alternate Reality Game) to create mainstream buzz. An ARG is a cross-media game that blurs the distinction between fiction and reality using fake websites, real-world puzzles, telephone messages and cryptic clues in the media to create intrigue

and excitement. The promotional ARG for the new Microsoft Xbox console game *Halo 2* began with an enigmatic reference to a website, www.ilovebees.com, in a pre-launch advertisement for *Halo 2*. Those curious enough to investigate found only an amateurish website of a bee lover in California - but that would then become unstable and start spewing strange messages, including GPS coordinates of public telephone boxes across the US. If they visited the telephone boxes, participants would receive a message and become drawn into an alternate reality where they would become players in the creation and distribution of a sinister and mysterious plot related to the forthcoming game. Through word of mouth, the 'ilovebees' ARG attracted over a million visitors prior to the release of *Halo 2*, created scores of active online communities, was picked up and widely reported in the mainstream mass media, and introduced the motto of ARG gamers, 'TINAG', into popular language (This Is Not A Game).[32]

15. **Orange 'Orange Wednesdays'** (2004): One of the first high-profile *viral mobile campaigns* - used to generate loyalty, capture customer data and acquire new customers for Orange, the UK mobile telephone network. The 'Orange Wednesdays' promotion, set to run for three years, enables Orange customers to invite a friend for free to any cinema in the UK every Wednesday using vouchers delivered directly to their handsets. The Orange customer simply sends an SMS message to Orange in order to be sent an immediate SMS reply with a buy-one-get-one-free voucher redeemable at over 80% of UK cinemas. If the friend then signs up with Orange, they too can invite a friend for free every Wednesday. Although the promotion doesn't use a viral 'forward a friend' SMS voucher, its virality lies in the fact that it transforms existing customers into promotional agents for the mobile network - by giving them promotional offers to share with friends. Based on 'pull' rather than 'push' technology (people request the mobile voucher, rather than receiving uninvited advertising from the provider or friends), Orange Wednesdays represents a new breed of viral marketing that uses mobile handsets to connect with customers and stimulate word of mouth advocacy.[33]

On the importance of being remarkable

The above viral, buzz and word of mouth campaigns all have one thing in common; they help make the products they promote *remarkable*. Through innovative campaigns that use surprise, humour, intrigue, or delight, these campaigns get the products they are promoting onto conversational agendas. By getting onto conversational agendas they raise the product's salience in the minds of their target buyers and create conversational contexts conducive to sharing opinions. By creating conversational contexts conducive to sharing opinions, more opinions are shared, and if those opinions are positive, sales are boosted.[34]

This simple economic rationale behind viral, buzz and word of mouth marketing shows how campaign buzz is a means to an end and not an end in itself. Campaigns create conversations, and conversations stimulate opinion sharing. As the General Motors buzz campaign for the Pontiac G6 illustrates, big buzz alone does not make big bucks. Campaigns that only generate buzz about themselves rather than the products sponsoring the buzz are no more useful than ads that sell themselves rather than the products they are supposed to sell. The key point to remember here is that it is product advocacy, not campaign buzz, that is proven to drive sales growth.[35]

So when should viral, buzz or word of mouth techniques be employed? The answer is deceptively simple: when your product delivers an experience that exceeds expectations.[36] Products that deliver experiences that exceed expectations are more likely to get advocated when they get onto conversational agendas. Of course, many products and services are not exciting enough, on their own, to get talked about. What viral, buzz and word of mouth campaigns do is add the excitement necessary to get them talked about. When this happens, opinions get shared and superior products benefit, but bad products suffer: viral, buzz and word of mouth campaigns don't create word of mouth in a vacuum; they unlock, stimulate and accelerate the natural word of mouth potential of your product.

One simple way to decide whether your product sales are likely to benefit from a viral, buzz, or word of mouth campaign is to ask your clients, customers or consumers how likely it is that they would, if prompted, recommend your product or service to your friends, family or colleagues?[37] The more likely it is that people would recommend your product, the more likely it is your sales will benefit from viral, buzz and word of mouth campaigns. On the other hand, if people are unlikely to recommend your product, even when prompted or elicited for an opinion, then it is unlikely that viral, buzz or word of mouth techniques will boost your sales. Like advertising that works when you have something worth advertising, viral, buzz and word of mouth campaigns work when you have something worth recommending.

Is your product suited to a viral, buzz or word of mouth campaign?

- To assess whether your product is likely to benefit from a viral, buzz or word of mouth campaign, simply ask yourself, your colleagues or, better still, your customers this simple question:

  On a scale of 0–10 in likelihood, how likely is it that you would recommend this product or service to your friends, family or colleagues?

- Products that consistently score highly (8 and above) are those that are most likely to benefit from viral, buzz or word of mouth campaigns because the campaigns will unlock, stimulate and accelerate the natural word of mouth potential of your product. Remember, viral, buzz and word of mouth campaigns are most likely to drive sales when you have a product that delivers an experience that exceeds expectations.

Throughout the chapters of *Connected Marketing* you will see time and time again that the not-so-secret 'secret' to harnessing word of mouth connections as marketing media is to do something worth talking about - that is, to do something remarkable. In today's cluttered markets, unless you are remarkable, you are invisible. This means that you need to create remarkable experiences for your clients, customers or consumers, both in terms of delivering a product experience that exceeds expectations and by creating campaign experiences that exceed expectations. By being remarkable and creating experiences that exceed expectations, you create word of mouth, and when you create word of mouth, you drive sales - it's that simple. *Connected Marketing* will show you how viral, buzz and word of mouth marketing are helping make marketing remarkable again by delivering exciting and entertaining experiences that exceed expectations. In doing so, they are helping to put marketing back into its rightful place as the beating heart of successful business.

Overview of *Connected Marketing*

In Part One: Connected Marketing Practice, Paul Marsden, a market researcher at the London School of Economics, kicks off with a chapter on seed marketing - pre-launch sampling initiatives with opinion leaders, conducted in the name of market research. The chapter explains how to set up and run effective seeding trials, and uses case studies from Procter & Gamble, Pepsi, 3M, Google and Microsoft to show how seeding trials with opinion leading consumers can boost sales by up to 30%. By triggering a powerful psychological mechanism called the Hawthorne Effect that transforms trial participants into loyal adopters and vocal word of mouth advocates, Marsden argues that seeding trials are a scientific, proven and scalable solution for launch optimization.

In *Chapter 2: Live buzz marketing*, Justin Foxton, founding partner and CEO of live buzz marketing agency CommentUK, describes how to use street theatre and live performance to create word of mouth. Using a number of UK case studies from GlaxoSmithKline, Unilever and London Transport, Foxton also addresses the issues of stealth and transparency in using live performance to create word of mouth: can actors create intrigue and excitement and go undercover, passing their performances off as real life, or should live buzz campaigns always disclose themselves as marketing?

Sven Rusticus, CEO of European marketing communications agency Icemedia, introduces brand advocacy programmes in *Chapter 3: Creating brand advocates*. after discussing a number of introduce-a-friend referral programmes designed to generate brand advocacy, Rusticus shows how brands such as adidas, L'Oreal and O'Neill are turning brand fans into word of mouth advocates by providing them with their own branded contact cards. Featuring the latest campaign artwork on one side and personal contact details on the other, branded contact cards are sent out by mail, and get handed out at social occasions, stimulating brand advocacy.

Chapter 4: Brewing buzz by Liam Mulhall, founder of Australian beer manufacturer Brewtopia, provides an in-depth case study of Blowfly Beer to show how a creative combination of buzz marketing initiatives can be integrated into a comprehensive launch strategy. Based on learning from Blowfly's success, the chapter provides a number of recommendations for planning and implementing buzz marketing campaigns.

In *Chapter 5: Buzzworthy PR*, Graham Goodkind, founder and chairman of UK-based Frank Public Relations, argues that public relations and buzz marketing are two sides of the same coin. Good PR campaigns not only generate media coverage, they stimulate word of mouth by getting people talking. Goodkind uses case studies from Mattel Fisher-Price toys, *New Scientist* magazine, Condomi condoms and Slendertone to illustrate PR campaigns that took word of mouth to a new level. He concludes by revealing his formula for producing buzzworthy PR activity.

Chapter 6: Viral marketing by Justin Kirby, founder and managing director of UK- and Australia-based online viral and buzz marketing consultancy Digital Media Communications, introduces the field of viral marketing. After a brief look at definitions and the history of viral marketing, Kirby lays out why businesses are increasingly including viral marketing in their brand marketing activities, what the risks and issues are, and how to plan a successful viral marketing campaign (focusing specifically on how to use entertaining video-based advertisements to spread awareness virally online). He finishes by presenting and reviewing some key viral marketing case studies from brands including Bacardi, Trojan, Ford, Virgin Mobile UK, Mazda UK and Burger King.

Idil Cakim, director of knowledge development at the global PR firm Burson–Marsteller provides in *Chapter 7: Online opinion leaders* a blueprint for identifying online opinion leaders and demonstrates their influence in the propagation of a viral marketing campaign for software company SAP. Cakim shows that people are almost twice as likely to open an advertising clip when it is forwarded from someone with an opinion-leading profile.

In *Chapter 8: Buzz monitoring*, Pete Snyder, CEO and founder of US-based New Media Strategies, shows how online buzz monitoring – measuring and tracking online word of mouth – can be a powerful business tool. Using case studies from the likes of Burger King and Royal Ahold, Snyder shows how sensitivity to what people are saying about a product or service enables businesses to respond quickly to needs, wants and desires.

Chapter 9: Changing the game turns to the potential of online branded games as a buzzworthy alternative to passive or 'interruptive' advertising. Steve Curran, president of US digital design agency Pod Digital, makes the case that online branded games are not only entertaining; they also provide an opportunity for interactive involvement with a brand. Curran shows how Coca-Cola, Nike, Mitsubishi, Nabisco, Procter & Gamble and many other brands are making branded games an increasingly important feature of their advertising mix.

Chapter 10: Blog marketing has been written by a collaboration of marketing experts all running their own blogs (short for weblogs – frequently updated personal or collaborative websites in the form of diarized journals containing opinions, information and weblinks that reflect the interests and personalities of their authors). The authors suggest that blogs can be effective promotional tools for products, services, brands and companies. Andrew Corcoran (senior marketing lecturer at Lincoln Business School, University of Lincoln), Paul Marsden (London School of Economics), Thomas Zorbach (CEO of German marketing agency vm-people) and business book author Bernd Röthlingshöfer provide evidence that blogs are powerful vehicles of online word of mouth. Using promotional blogs from McDonalds, Nike, Nokia and Microsoft as examples, the authors outline the different ways blog marketing can be used to ignite word of mouth and stimulate sales.

Part Two: Connected Marketing Principles begins with *Chapter 11: Word of mouth: what we really know - and what we don't*, an overview of 50 years of academic research into word of mouth by Greg Nyilasy from Henry W. Grady College of Journalism and Mass Communication at the University of Georgia. Nyilasy discusses the prevailing definition of word of mouth in academia as well as the major theoretical explanations for its power. He reviews the results of over 100 studies about the cause and effect of word of mouth, and concludes with 13 recommendations for marketing practitioners about how to manage and evaluate word of mouth programmes.

Chapter 12: Black buzz and red ink by Brad Ferguson, founder of US marketing consultancy Intrinzyk and founder emeriti of US marketing agency Informative, looks at the financial impact of negative word of mouth. Using an in-depth case study focusing on the US airline industry, Ferguson shows how just a 1% reduction in negative word of mouth for an airline is likely to add US$4 million in operating profit, while an increase of 1% in positive word of mouth is worth just under US$2 million.

Stéphane Allard, associate director at French word of mouth marketing agency Spheeris, puts right some of the common misperceptions about buzz marketing in *Chapter 13: Myths and promises of buzz marketing*. Allard debunks the myths that you can only create buzz around exciting, groundbreaking products, that you don't need to manage buzz, that managing buzz is cheap, and that buzz can't be measured. Allard also provides some working definitions that differentiate viral, buzz and word of mouth marketing from other approaches such as street marketing and stealth marketing.

In *Chapter 14: Buzz marketing: the next chapter*, Schuyler Brown, co-creative director at US-based Buzz@Euro RSCG, looks at innovations and emerging trends in buzz marketing - initiatives designed to stimulate word of mouth among a target market. Identifying five key new realities for buzz marketers and making six forecasts for the future, Brown uses buzz campaigns from Polaroid, Reebok and Burger King to illustrate how to create buzz successfully in change-obsessed consumer culture.

Chapter 15: How to manage connected marketing by Martin Oetting, head of marketing and PR at the Berlin campus of ESCP–EAP European School of Management, places the different areas of connected marketing into an integrated framework. Oetting suggests that, jargon aside, connected marketing boils down to two essential activities: connecting with clients, customers, or consumers via buzzworthy campaigns; and connecting via opinion-leading market influencers who have an effect on the mass market through partnerships and privileged 1-2-1 dialogue.

In the conclusion, Justin Kirby reiterates the essence of what connected marketing is, providing an overview of how best to use it and touching on ethics and measurement issues. He then offers 10 predictions about the future of connected marketing.

Together, the chapters in *Connected Marketing* show how viral, buzz and word of mouth campaigns are leading the charge in a marketing revolution that makes marketing remarkable again and puts clients, customers and consumers, rather than marketers, at the centre of the marketing process. Shunning the old 'command and control' mentality of mass marketing, viral, buzz and word of mouth marketing champions a new 'connect and collaborate' vision, where marketing is done *with* people rather than *at* them. We call it the connected marketing revolution.

Notes and references

1 From Dash, M. (1999) *Tulipomania: The Story of the World's Most Coveted Flower and the Extraordinary Passions It Aroused.* New York, NY: Crown Publishers. See also Mackay, C. (1841) *Extraordinary Popular Delusions and the Madness of Crowds*. London: Wordsworth Reference; and Tarses, M. (1999) 'Tulipmania.com' online article archived at www.sunwayco.com/tulipmania.html.

2 From Dye, R. (2000) 'The buzz on buzz', *Harvard Business Review*, 78 (6): 139–146.

3 Reichheld, F. (2003) 'The one number you need to grow', *Harvard Business Review*, 81 (Nov.–Dec.): 1–11.

4 Known as 'Kevin Bacon Effect', or 'Small World', the interconnectedness of people means that everybody is connected to everybody by no more than six degrees: see Barabasi, Albert-Laszlo (2002) *Linked: The New Science of Networks.* Cambridge: Perseus Publishing, and Watts, D. (2003) *Six Degrees: The Science of a Connected Age*. New York: W. W. Norton.

5 Interestingly, the term 'viral marketing' was originally coined to denote product seeding designed to kick-start copycat effect (so people 'catch' the idea and adopt it by seeing it adopted by others), and not for online word of mouth campaigns – see Carrigan, T. (1989) 'New Apples tempt business', *PC User*, 27 September: 'At Ernst & Whinney, when Macgregor initially put Macintosh SEs up against a set of

Compaqs, the staff almost unanimously voted with their feet as long waiting lists developed for use of the Macintoshes. The Compaqs were all but idle. John Bownes of City Bank confirmed this. "It's viral marketing. You get one or two in and they spread throughout the company."'

6 Stengel, J. (2004) 'The future of marketing', Conference speech presented at the AAAA Media Conference Thursday, 12 February 2004, archived at http://www.pg.com/content/pdf/04_news/stengel_feb_12_2004.pdf.
7 Light, L. (2004) 'The end of brand positioning as we know it', Conference speech presented at the ANA Annual Conference, cited in *The Ramsey Report: The State of the Online Advertising Industry*, 8 November 2004, archived at http://www.sempo.org/research/ramsey_white_nov04.pdf.
8 Quoted in Markillie, P. (2005) 'Crowned at last', *The Economist*, 31 March.
9 Deutsche Bank study findings reported in *The Economist* (2004) 'The future of advertising: the harder hard sell', 24 June, archived at http://www.economist.com/printedition/displayStory.cfm?Story_ID=2787854. ROI figures from Copernicus Marketing Consulting study reported in Clancy, K. and Stone, R. (2005) 'Don't blame the metrics', *Harvard Business Review*, June 83(6).
10 Clancy and Stone, ibid.
11 Ibid.
12 Figures from Zyman, S. (2002) *The End of Advertising as We Know It*. Hoboken, NJ: Wiley, p. 18.
13 Cited in *The Economist*, 'Future of advertising'. The quote has also been attributed to America's first discount-retailer Frank Woolworth and John Wanamaker, father of the department store.
14 '18%: Proportion of TV advertising campaigns generating positive ROI . . .', Deutsche Bank study reported in *The Economist*, 'Future of advertising'; '54 cents: Average return in sales for every $1 spent on advertising . . .' Clancy and Stone, 'Don't blame the metrics'; '84%: Proportion of marketing campaigns resulting in falling sales . . .', Clancy and Stone, 'Don't blame the metrics'; '256%: The increase in TV advertising costs (CPM) in the past decade . . .', figures reported in Garfield, R. (2005) 'Chaos scenario', *Ad Age*, 13 April, archived at http://www.adage.com/news.cms?newsId=44782; '100%: The increase needed in advertising spend to add 1–2% in sales . . .', UCLA study findings reported in Clancy and Stone, 'Don't blame the metrics'; '14%: Proportion of people who trust advertising information . . .' Annual Gallup poll on professional honesty and trustworthiness, figure cited in Ries, A. and Ries, L. (2002) *The Fall of Advertising and the Rise of PR*. New York: HarperCollins; '90%:

Proportion of people who can skip TV ads who do skip TV ads', MPG Study 2004 reported in Carton, S. (2004) 'Advertising is more than just TV commercials', *ClickZ*, 18 October, archived at http://www.clickz.com/experts/ad/lead_edge/article.php/3422331; '80%: Market share of video recorders with ad skipping technology in 2008 . . .', Jupiter Research (2004) forecasts reported in Carton, 'More than just TV commercials'; '95%: The failure rate for new product introductions . . .', ACNielsen BASES and Ernst & Young study reported in Clancy & Stone, 'Don't blame the metrics'; '117: The number of prime time TV spots in 2002 needed to reach 80% of adult population . . .' figure given in Stengel, J. (2003) 'Consumers are reinventing marketing: Will you be in touch or left behind?', Conference speech at World Federation of Advertisers Conference, Brussels, archived at http://www.eu.pg.com/news/speeches/20031028wfastengel.html; '3000: Number advertising messages people are exposed to per day', figure from Shenk, D. (1997) *Data Smog: Surviving the Information Glut*. San Francisco: Harper Edge; '56%: Proportion of people who avoid buying products from companies who they think advertise too much', Yankelovich Partners (2004) research on advertising effectiveness, reported in Elliott, S. (2004) 'New survey on ad effectiveness', *New York Times*, 14 April, section C, page 8, column 3, archived at http://www.nytimes.com/2004/04/14/business/media/14adco.html; more figures from Yankelovich Partners research reported in IAPA (Institute for Advanced Practices in Advertising) archived at http://iapia.org/problem.html; '65%: Proportion of people believe that they are constantly bombarded with too much advertising . . .', Yankelovich Partners (2004) in Elliott, 'Survey on ad effectiveness'; '69%: Proportion of people interested in technology or devices that would enable them to skip or block advertising', Yankelovich Partners (2004) in Elliott, 'Survey on ad effectiveness'.

15 Euro RSCG (2001) 'Wired and wireless: Key findings', report summary archived at http://www.eurorscg.com/starview/doc/ww_summary.pdf.

16 Goodmind (2002) 'Findings show the influence of word of mouth on purchase decisions is rising dramatically', press release archived at http://www.goodmind.net/release_WOM.asp.

17 Berry, J. (2005) 'Identifying, reaching, and motivating key influencers', Conference paper presented at the Word of Mouth Marketing Association Summit Chicago, 29–30 March.

18 Jurvetson, S. and Draper, T. (1997) 'Viral marketing', original version published in the *Netscape M-Files*, 1 May, edited version published in *Business 2.0*, November 1998, archived at http://www.dfj.com/cgi-bin/artman/publish/steve_tim_may97.shtml.

19 See Chapter 3: Creating Brand Advocates by Sven Rusticus for more information on the Dove 'Share a Secret' campaign. See also *Promo-magazine* online article archived at http://promomagazine.com/mag/marketing_health_beauty_care/.

20 Figures from Burson-Marsteller (2000) 'The Power of word of mouth', archived at http://www.efluentials.com/pages/news/061500_2.S.

21 For campaign details see Woffenden, C. (2000) 'Viral marketing leads to healthy e-business', *IT Week*, 5 January, archived at http://www.itweek.co.uk/analysis/104892.

22 Figure reported in Saugar, M. (undated) 'They don't want advertising', online article in IdeasFactory http://www.ideasfactory.com/new_media/features/newm_feature_59.htm. The viral clip can be viewed at the website of the campaign creator cdp-travissully http://www.cdp-travissully.com.

23 See Tierney, J. (2001) 'Here come the Alpha Pups', *New York Times*, 5 August, Late Edition – Final Section 6: 38: col. 1, archived at http://tinyurl.com/7vo64; Rogers, E. (2003) *Diffusion of Innovations*. New York: Free Press; Godes, D. and Ofek, E. (2004) 'Hasbro games', *Harvard Business Online*, 20 December, archived at http://harvardbusinessonline.hbsp.harvard.edu/b01/en/common/item_detail.jhtml?id=505046.

24 See Chapter 1: Seed to Spread by Paul Marsden for more information on the Tremor panel.

25 For more on this and other 'stealth' campaigns see Kaikati, A. and Kaikati, J. (2004) 'Stealth marketing: How to reach consumers surreptitiously', *California Management Review*, 46 (4) (Summer): 6–22. See also CBS 60 Minutes report (2004) 'Undercover marketing uncovered' (July) – online version archived at http://www.cbsnews.com/stories/2003/10/23/60minutes/main579657.shtml.

26 See Chapter 3: Creating Brand Advocates by Sven Rusticus for more information on this and other similar campaigns.

27 Figures from Haas, C. (2005) 'Creating viral advertising campaigns', Conference paper presented at the Word of Mouth Marketing Association Summit, Chicago 29–30 March. Campaign website is www.subservientchicken.com.

28 See www.laf.org for sales of Armstrong bands, for background, facts and figures, see Leskin, S. (2004) '20 million "Live Strong" wristbands sold to date', *PNN Online*, 27 October, archived at http://www.pnnonline.org/article.php?sid=5548.

29 From Wiley, M. (2005) 'Lessons from the Blogosphere: GM's blogging experiment', Conference paper presented at the Word of Mouth Marketing Association Summit, Chicago, 29–30 March, and Baker, S. and Green, H. (2005) 'Blogs will change your business', *Business*

Week, 2 May, Cover story, archived at http://www.businessweek.com/magazine/content/05_18/b3931001_mz001.htm.

30 See de Moraes, L (2004) 'Oprah's fully loaded giveaway', *Washington Post*, 14 September C01, archived at http://www.washingtonpost.com/wp-dyn/articles/A19203-2004Sep13.html. Although the campaign created huge buzz, sales of the Pontiac G6 appear to have remained sluggish. For an analysis see Webster, S. (2005) '"Oprah" buzz works no magic for Pontiac G6', *Detroit Free Press*, 22 March, archived at http://www.freep.com/money/autonews/oprahx22e_20050322.htm.

31 For details on the languishing sales of the Pontiac G6 see Webster, '"Oprah" buzz works no magic'; for details on the bad buzz on the tax bill see Munarriz, R. (2004) 'It's Oprah Winfrey, not win free', *The Motley Fool Online*, 24 September, archived at http://www.fool.com/News/mft/2004/mft04092403.htm.

32 For more information on the ilovebees ARG, see Terdiman, F. (2004) 'I Love Bees game a surprise hit', *Wired Magazine*, 18 October, archived at http://www.wired.com/news/culture/0,1284,65365,00.html. For more on ARGs, see Wikipedia entry http://en.wikipedia.org/wiki/Alternate_Reality_Game.

33 For details, see http://www.orange.co.uk/entertainment/film/cma/overview.html; http://www.flytxt.com/cgi-bin/template.pl?t=cspd&ID=40.

34 Reichheld, 'The one number you need'.

35 Ibid.

36 For more on the psychology of word of mouth see Sundaram, D., Mitra, K. and Webster, C. (1998) 'Word of mouth communications: a motivational analysis', *Advances in Consumer Research*, 25: 527–531. For marketing applications see Rosen, E. (2000) *The Anatomy of Buzz: Creating Word of Mouth Marketing*. London: HarperCollins; Godin, S. (2004) *Purple Cow*. New York: Portfolio.

37 Adapted from Reichheld, 'The one number you need'.

Part One

Connected Marketing Practice

1

Seed to spread: how seeding trials ignite epidemics of demand

Paul Marsden

London School of Economics/Associate Director, Spheeris

Picture this: a marketing research department operating as a profit centre, not only generating intellectual capital but also driving sales through word of mouth outreach programmes with opinion leaders. Sounds like a fantasy? Well, think again. Big brands such as Procter & Gamble, 3M, DreamWorks SKG, Microsoft and Google are all harnessing the power of research to optimize product launches. How? Through seeding trials - sampling conducted in the name of research, designed to transform opinion leaders into loyal adopters and word of mouth advocates.

Seeding trials: 'It's research, Jim, but not as we know it'

When most people think of marketing research, they tend to think of research rather than marketing. Indeed, marketing research is formally defined as identifying and measuring marketing opportunities and problems, evaluating marketing actions, or monitoring marketing performance.[1]

But there is a new breed of research that is putting the marketing back into marketing research: seeding trials. Seeding trials involve targeted sampling with opinion leaders, conducted in the name of research.

Rather than simply offer free samples, previews, test-drives, etc. to opinion leaders, the idea behind seeding trials is to create goodwill, loyalty and advocacy by putting the product or service in their hands *and* giving them a say in how it is marketed. By involving opinion leaders in this way, by effectively inviting them to become part of your marketing department, you create a powerful sense of ownership among the 10% of your target clients, customers, or consumers who have the power to ignite word of mouth demand. By transforming these opinion leaders into word of mouth advocates through seeding trials, companies are using marketing research to ignite word of mouth networks and accelerate sales. As Star Trek's Dr McCoy might have said: 'It's research, Jim, but not as we know it.'

Seeding trials in action: Post-it Notes

The power of seeding trials in transforming the fortunes of a brand is no better illustrated than through the intriguing history of Post-it Notes, the little yellow stickies from the office supplies company 3M. The story started in 1968, when 3M asked one of its researchers, Dr Spence Silver, to develop a new super-sticky adhesive. Unfortunately Dr Silver failed, and did so quite spectacularly. What he came up with was super-weak glue that wouldn't stay stuck. Consigned to the back shelves of 3M's R&D lab for six years, the fruits of the failed innovation project were virtually forgotten.

Then on one Sunday in 1974, Art Fry, a new product development researcher for 3M, had a 'Eureka' moment while cursing scrap paper bookmarks that kept falling out of his church choir hymnal. Perhaps the un-sticky glue could be used to make bookmarks? The idea of Post-it Notes was born. Unfortunately, when this concept of temporary sticky paper bookmarks was tested in research, it bombed. Nobody could see a use for them. However, and despite 'kill the programme' calls from management, Fry convinced 3M to run a limited test launch of Post-it Notes. Unfortunately, that failed too. Post-it Notes were doomed.

Before pulling the plug on the whole sad affair, 3M decided to run a seeding trial with opinion leaders in its target market – a sampling initiative conducted in the name of research. The company identified secretaries to CEOs in large companies all across America as opinion leaders for office supply products, and sent them boxes of Post-it Notes, inviting them to come up with ideas for how the little yellow stickies could be used. The seeding trial generated goodwill and advocacy from these opinion-leading secretaries who – flattered by the invitation to be involved in the development and commercialization of a new product – were transformed into

Post-it Notes brand champions. The 'useless' Post-it Notes soon started appearing on memos, desks, diaries, drafts, reports and correspondence, and spread like an infectious rash through and between companies. The rest is, as they say, history. Post-it Notes had been saved by a seeding trial, transformed from failure to a multi-million dollar and highly profitable brand by a group of opinion-leading secretaries.[2]

The science bit: why seeding trials drive demand

Why did a seeding trial, targeted sampling conducted in the name of research, transform the fortunes of Post-it Notes? To answer this question, we need to understand two things: first, a peculiar psychological phenomenon known as the Hawthorne Effect; and second, the critical role of opinion leaders in driving sales.

The Hawthorne Effect

Back in the 1930s, a team of researchers from the Harvard Business School were commissioned to run some employee research for the telecom giant Western Electric (now Lucent Technologies). Conducted at the company's production plant in Hawthorne, near Chicago, the research programme involved inviting small groups of employees to trial various new working conditions before rolling them out to the general workforce. To the researchers' amazement, the participants seemed to like whatever was trialled, to such an extent that their productivity increased! For example, when researchers invited participants to trial working in brighter lighting conditions, productivity increased. But then when they trialled dimmer lighting conditions, productivity also increased. In fact, productivity kept increasing in successive trials of working under progressively dimmer lights, until the lighting was no stronger than moonlight! In another trial, the research participants were invited to test working shorter hours, and sure enough their productivity increased again. Indeed, subsequent trials showed that the more breaks the research participants were given and the less time they worked, the greater their productivity. But then, when the researchers asked them to trial longer hours, productivity went up again – to an all-time high.[3]

When taken together, the results of the various Hawthorne studies showed that whatever the researchers asked participants to discuss and trial resulted in an increase in productivity. The team of Harvard researchers, led by Elton Mayo, realized that their results had nothing to do with what was being trialled and everything to do with running research trials. By singling out a small group of employees to participate in an exclusive trial, participants felt valued, special and important. The special attention they received gratified their ego and created a positive emotional bond with what they were trialling. The practical upshot was that the research trials effectively transformed the research participants into advocates for whatever it was they were trialling. A series of further trials found this phenomenon to be more or less systematic, and the research team coined the term 'The Hawthorne Effect' to describe the goodwill and advocacy that research trials generate among research participants.

The Hawthorne Effect: how to win friends and influence people

If the psychology of the Hawthorne Effect all seems a bit abstract, try it for yourself and see how powerful it is. The next time you want something from someone (a salary increase, a date, or whatever), first do some 'research' with them by asking them for their advice on some matter. It doesn't actually matter what it is that you ask them their advice on; the important thing is to be seen to be listening to what they have to say, and then to tell them that you appreciate their opinion.

Then, simply ask them for whatever it is you want from them. The chances are that your 'research' will have triggered the Hawthorne Effect and you will get what you want. By asking them for their opinion you will have not only created goodwill but also flattered their ego. At a subconscious level, they will feel indebted to you. This psychological indebtedness makes them significantly more likely to agree to whatever it is you are asking of them.

By seeing the Hawthorne Effect in action, you'll realize that it is a very powerful influence technique; you'll also know to watch out the next time someone asks you for your advice and then asks you for something!

It is this Hawthorne Effect, harnessed by seeding trials, that transforms opinion leaders into loyal adopters and powerful word of mouth advocates. By turning opinion-leading target buyers into product or service evangelists using the Hawthorne Effect, a brand can create a powerful volunteer sales force.

The truth about opinion leaders

> Simply by finding and reaching those few special people who hold so much social power, we can shape the course of social epidemics . . . Look at the world around you . . . With the slightest push - in the right place - it can be tipped . . . (Malcolm Gladwell, *The Tipping Point*).[4]

With the possible exception of Tom Peters's *Thriving on Chaos*,[5] *The Tipping Point* by Malcolm Gladwell is perhaps the most influential and widely read book to date on the power of word of mouth. Voted by Forbes as one of the most influential business books of the past two decades, this international bestseller uses the science of social epidemics (runaway word of mouth) to outline a simple three-point formula for how word of mouth hits happen: 'the Law of the Few', 'the Stickiness Factor' and 'the Power of Context'. While the Stickiness Factor and the Power of Context deal with the 'what' and the 'where' of word of mouth (having something intrinsically worth talking about, in an environment conducive to word of mouth spread), the Law of the Few addresses the 'who', reminding us that the opinions of 10% of any target market will drive the buying behaviour of the other 90%.

Although Gladwell uses the language and jargon of epidemiology to unpack the concept of opinion leadership, the idea behind the Law of the Few is an established business truth dating back to the 1940s. Indeed, evidence for the Law of the Few was first produced in a 1940 landmark study on media influence conducted by Columbia University.[6] The research found, contrary to what might be expected, that mass-media messages do not directly influence the mass market but instead influence a small minority of individuals who then influence their peers through word of mouth. The researchers coined a new term for these hubs of word of mouth mediating mass-media messages - 'opinion leaders' - proposing a new 'two-step flow' model of media influence to replace the discredited 'magic bullet' or 'hypodermic needle' model of direct media influence.

Since the discovery of opinion leaders, research across just about every product and service category has found that the opinions of an opinion-leading 10% do indeed tend to shape the opinions and purchases of the opinion-following 90%.[7] Opinion leaders, simply defined as target buyers who frequently offer or are elicited for category-related advice by their peers, can include high-profile industry experts, journalists, reviewers and media celebrities. However, the vast majority of opinion leaders in any target market are simply regular clients, customers, or consumers, except for the fact that they have a peculiar 'connected and respected' profile - they are highly connected hubs of word of mouth in their social networks with opinions that are respected by their peers.

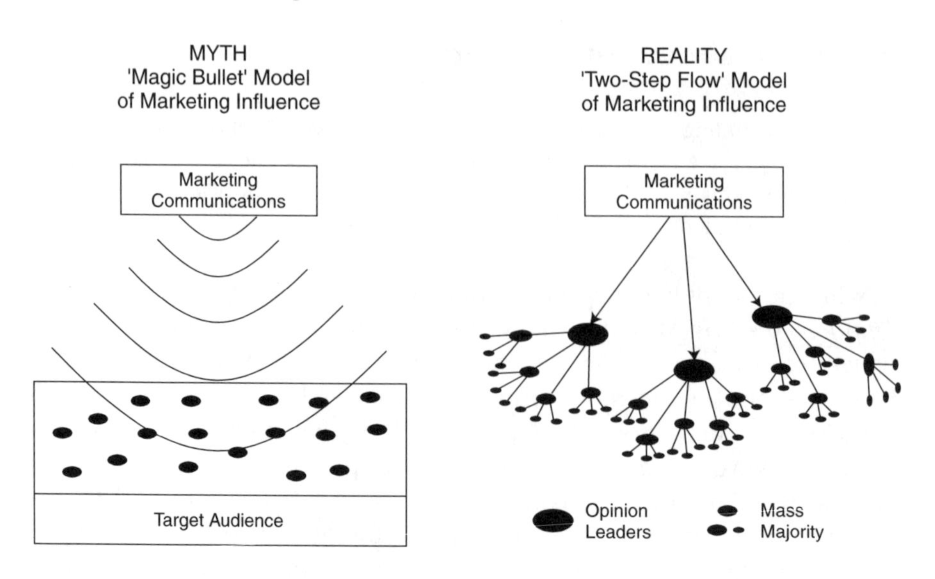

Figure 1.1 Models of media influence

Thus, the influence of opinion leaders derives not from media appearances but from what sociometricians call 'network centrality' - they are word of mouth hubs, who connect everybody to everybody by six degrees of separation, and in doing so connect businesses to their target markets.

Because of the importance of opinion leaders in driving sales, a good deal of time has been invested in (a) finding ways to identify them and (b) among marketing agencies at least, re-branding them with some proprietary label ('Alphas', 'Hubs', 'Connectors', 'Influentials'(SM), 'Sneezers', etc.). While some subtle differences may lie behind the proprietary spin and various trademarked labels, it is worth remembering that the scientifically validated and peer-reviewed scales for identifying this group are called opinion leadership scales - precisely because they screen for opinion leadership (i.e. likelihood of this group's offering or having elicited from them, category-related advice).[8]

What we know about opinion leaders, apart from their connected and respected profile, is that opinion leadership tends to be category-specific - opinion leaders in off-road quad-bikes may or may not be opinion leaders in cosmetic beauty products. We also know that key correlates of opinion leadership are 'category involvement' (interest, knowledge and activity) and 'strength of personality' (persuasiveness and personal charisma). These correlates have enabled reliable opinion leadership scales to be developed and validated in order to be used by businesses for screening existing and target buyers.

The opinion leader screener

Opinion leaders are simply those target buyers in your market who are likely to frequently offer or be elicited for category-related advice. The self-designation technique for identifying opinion leaders involves asking existing and potential buyers to fill out a short opinion leadership screening questionnaire:

How much do you agree or disagree with the following statements?

(1 = Strongly Disagree to 5 = Strongly Agree)

1. My friends/neighbours consider me a good source of advice about [category]
2. I tend to talk a lot about [category] to friends/neighbours
3. In the past 6 months, I've talked to a lot of people about [category]
4. When asked for advice about [category], I offer a lot of information
5. When discussing [category] products, I usually convince them of my opinion

Source: Adapted from H. Ben Miled and P. Le Louarn (1994) Analyse comparative de deux échelles de mesure du leadership d'opinion : validité et interprétation. *Recherches et Applications en Marketing*, 9(4): 23–51

How to find opinion leaders in your target market

To identify the opinion-leading 10% who drive sales in your target market, a number of practical solutions have been developed: 'self-designation', 'professional activity', 'digital trace', 'key informants' and 'sociometry'.

1. **Self-designation:** Asking existing or prospective buyers to fill out a short self-completion questionnaire that screens them for opinion leadership status. For example, Procter & Gamble use a self-designation questionnaire on their website at www.tremor.com to recruit opinion-leading teens into seeding trials. Although this technique is open to self-reporting bias (people tend to overestimate their opinion leadership), it has been validated and found to be reliable.
2. **Professional activity:** Using the job title of target clients, customers, or consumers as an indicator of opinion leadership status. Jobs that

suggest category involvement, a capacity to spread the word and to influence peers by word of mouth are predictive of opinion leader status. For example, just as 3M identified secretaries to CEOs as opinion leaders in office stationery products, Ford identified opinion leaders for its new Focus model as PAs to celebrities.[9] Although this approach may be less scientific than the self-designation approach, it is a quick, easy and cost-effective solution to identifying opinion leaders.

3. **Digital trace:** Identifying opinion leaders through an online search on category-relevant blogs, websites, discussion lists, newsgroups and web forums. For example, Siemens mobile identified opinion leaders to participate in a seeding trial of one of its new phones using an Internet search of popular online user review forums.[10] In an era where opinions are increasingly shared online, the digital trace left by that activity provides businesses with a fast and smart method for opinion leader identification.
4. **Key informants:** Asking a limited number of people assumed to be knowledgeable about the patterns of word of mouth influence who they would designate as opinion leaders. Although this technique is ideally suited to identifying opinion leaders in small markets or in individual organizations, it was used by games manufacturer Hasbro to identify young opinion leaders for its handheld electronic game POX, in 2001. Market researchers headed off to video arcades, skate parks and playgrounds and went up to young boys aged 8–13 asking 'Who's the coolest kid you know?' When they got a name from the young 'informant', the researchers went in search of this cool kid to ask him the same question, in order to continue up the local hierarchy of kid-cool until someone finally answered 'Me!' Once they had identified an opinion leader, the researchers invited the 'Alpha Pup' (as they called them) to participate in an exclusive seeding trial for which they would be rewarded with 10 new pre-release POX units to share with friends.[11]
5. **Sociometry:** Actually mapping the patterns of word of mouth influence in a target market in order to identify hubs of influence. Costly and time-intensive, the use of sociometric techniques is mostly limited to mapping influence networks in organizations for change management purposes.

From principles to practice: learning from drug dealers

Seeding trials – targeted sampling with opinion leaders conducted in the name of research – drive sales because they trigger the Hawthorne Effect amongst the 10% of a target market whose opinions drive word of mouth demand. As a launch optimization tool, seeding trials have been extensively used in the drug industry to transform opinion-leading physicians into loyal adopters and powerful word of mouth advocates of new prescription medicines. So established are seeding trials in the healthcare sector that they have their own industry standard codename: Phase IV trials.

Phase IV trials get their name from the way new drugs are researched in the pharmaceutical industry. Research begins with Phase I trials, which involve testing the new product for safety, usually in a small number (10–100) of healthy people. If the drug is found to be safe, then Phase II trials begin, which involve testing how effective the drug is in doing what it is supposed to do in a slightly larger number of people (100–300) who are actually suffering from the condition the drug is designed to treat. If all goes according to plan, then the product goes into large-scale Phase III trials with many sufferers (300–3000), in order to measure the comparative efficacy of the new drug against other treatments, its side-effect profile and its relative financial value over alternative therapies. Only when these three phases of research are satisfactorily completed can the new drug be cleared for launch by market authorities. But it is at this point, when the drug has been finally cleared for launch, that a fourth phase of non-regulatory research often takes place: Phase IV trials.

Phase IV trials are targeted sampling initiatives with opinion leaders conducted in the name of research. They involve inviting a group of opinion-leading physicians to participate in a trial of a new drug by prescribing it to a certain number of patients and feeding back on their experience. In return for their participation, the doctors are often promised free access to the drug for their patients, as well as additional exclusive information and services to help them use the drug effectively. It's a win–win situation for both the physician and the drug manufacturer. The physicians get their status as opinion leaders reinforced through privileged access to the new products and special VIP services, and they may often receive some financial remuneration for taking part in the trial. For the drug company commissioning the trial, valuable information is captured from influential lead-prescribers, and the powerful Hawthorne Effect is triggered, transforming trial participants into opinion-leading, word of mouth advocates.

Going Google over seeding trials

In the software industry, seeding trials with opinion leaders go under a different name: 'beta testing'. The practice, however, is identical: targeted sampling with opinion leaders conducted in the name of research. The goal of beta testing is to get opinion leaders to trial a pre-release version of software (a 'beta version') in order to (a) capture feedback on any glitches that need to be ironed out and (b) trigger the Hawthorne Effect and transform participating opinion leaders into loyal word of mouth advocates for the new software. This combination of offering opinion leaders a sneak preview, removing the cost barrier to trial, and engaging them in research dialogue is a powerful combination that can drive sales.

For example, to optimize the Windows 95 launch, Microsoft ran a massive seeding trial with 450 000 opinion-leading PC users in the US. A total of 5% of Microsoft's entire target market in the US participated in the trial, each receiving a pre-release sample copy of the software. By connecting with opinion-leading target buyers through research dialogue instead of advertising monologue, the seeding trial generated goodwill and an army of product advocates. When Windows 95 was launched on the stroke of midnight on 24 August 1995, the seeding trial paid off handsomely: one million copies were sold in the first four days, making it the fastest-selling software of all time, trouncing the previous record of 40 days to sell a million copies.[12]

As in the healthcare sector, seeding trials have become a widely used solution for optimizing product launches in the software industry. A recent high-profile example was the seeding trial used to launch Internet company Google's new email service Gmail. In March 2004, 1000 online opinion leaders were invited to sample a beta version of the new service in the name of research. To enhance the word of mouth potential of this seeding trial, Google enabled participants to invite their friends into the trial, who could also invite their friends if they signed up and so on. This 'snowball' or 'viral' recruitment enhanced the Hawthorne Effect, stimulating the transformation of goodwill and ownership of participants into active word of mouth advocacy.

The result was an exponentially increasing number of Gmail evangelists, each recruiting new users with the fervour of religious converts. Driven by the cachet of being invited as an opinion leader to have an exclusive sneak preview of a yet-to-be-released product, and by having a say in how that product was to be commercialized, the Gmail seeding trial generated a reported three million Gmail adopters and advocates in just three months with no advertising spend.[13] Indeed, the seeding trial created so much word of mouth demand that people were prepared to pay to become participants: an online black market emerged on the Internet with invitations being sold for up to US$200.[14] By playing the scarcity card – people value things more when availability is restricted – the invitation-only policy for participation in the Gmail seeding trial resulted in a word of mouth frenzy. Dozens of blogs were set up by participants to share their experiences as Gmail 'insiders' with wannabe research participants, and the mainstream mass media, including *The New York Times*, ran stories on the seeding trial.[15] Through a seeding research trial alone, Gmail became one of the most high-profile and well-known email services in the world, getting as many mentions on the Web as its far more established (and heavily advertised) competitor, Yahoo! Mail.[16]

Teen trials – Tremor style

Although seeding trials are extensively used in the IT and healthcare industries, the most audacious use of sampling opinion leaders in the name of research has come from the consumer packaged goods sector. In 2001, brand giant Procter & Gamble (P&G) (owner of Crest, Clairol, Pringles, Pampers, Tide, CoverGirl, Max Factor, Olay, Hugo Boss fragrances and others) began recruiting teen opinion leaders into a nationwide online seeding trial panel, codenamed Tremor, which now has over 250 000 members - a full 1% of the US teen population.[17] Recruited by word of mouth and banner advertising on popular teen websites, potential Tremor panel members are promised exclusive prelaunch samples and previews of new products from P&G, and to have a say in how these products are marketed. Screened for opinion-leading status with a simple online screener (only one in ten applicants are invited to become part of the 'Tremor crew'),[18] panel members participate in sampling initiatives conducted in the name of research for a wide variety of innovations including beauty products, music, movies, videos and gadgets.

By giving opinion-leading teens a voice in how new products are commercialized, the Tremor panel creates a sense of ownership and involvement, triggers the Hawthorne Effect and transforms panel members into loyal adopters and vocal word of mouth advocates. Tremor seeding trials have included inviting panel members to:

- Help develop Vanilla Coke's 'Nothing Else Like It' billboard campaign and come up with intriguing messages to appear on promotional heat-sensitive cans[19]
- Vote on launching Snoop Dogg's new line of shoes[20]
- Advise on the trailer for the movie *Biker Boyz*[21]
- Choose which Herbal Essence commercial to air for promoting Fruit Fusions Tropical Showers range[22]
- Recommend which fashion model to use in a Pantene commercial[23]
- Select backing music for a Pringles advertisement[24]

- Pick models for a body-spray calendar[25]
- Help design the new Crest Spinbrush[26]
- Vote on a t-shirt design for Vans 'Warped Tour' concert[27]
- Name the DreamWorks SKG movie *Eurotrip*[28]
- Choose the logo for the teen movie Win a Date with Tad Hamilton! 29

The key to success in these seeding trials has been to combine 'Get it first' targeted sampling with an 'Inside scoop' of exclusive information about the product for participants to share with their friends, and what could be called a 'VIP vote' (very *influential* person) that enables the participant to influence how the product will be promoted.[30] The impact of *involving* opinion-leading teens through 'VIP votes', 'Get it first sampling' and 'Inside scoops' is illustrated by the way Caitlin Jones, a Tremor panel participant, reacted when she saw a trailer for a movie she had been consulted on with her friends: 'Oh, my God, I voted for that logo!' she exclaimed. 'So they do listen. It does matter.' The opinion-leading teen was instantly transformed into an active evangelist and set about organizing group outings to see the movie.[31]

As a launch optimization tool, Tremor seeding trials can reportedly generate a 10–30% increase in sales or audiences, measured against a control region where the panel is not used.[32] For example, when the panel was used to optimize the launch of a new line of CoverGirl Outlast Lipcolor lipstick, sales were on average 14% than in a matched control region. Each Tremor participant evangelized about the lipstick to on average nine friends, six of whom said they intended to buy the product. In the words of Tremor chief executive Ted Woehrle on panel participants: 'We offer them the inside scoop and influence [i.e. a say in how the product is promoted]. If you get the right 1%, you have the critical mass required to make a difference.'[33]

In another Tremor initiative designed to measure the effectiveness of the seeding panel, 2100 Tremor opinion leaders from Phoenix were invited to get involved with the launch of a new malt-flavoured milk product from Shamrock Farms, a dairy foods producer. As a result, sales in Phoenix outperformed those in a matched control city, Tucson, and 23 weeks later sales were still 18% higher in Phoenix.[34] In a similar test, Tremor panellists were sent a partial script of an upcoming TV show, and this resulted in a jump in viewing ratings of 171%.[35]

The Tremor seeding initiatives' effectiveness in boosting sales has had third-party brands – including Sony, Toyota, AOL, Warner Brothers, Verizon, and Kraft – queuing up to harness the Tremor seeding trial panel as a launch optimization tool.[36] For instance, the music label EMI Group has retained Tremor and intends to sample panel members with new albums – in the name of research – by asking the opinion-leading teens to vote on which tracks should be promoted on video channels and radio programmes.[37] This is a simple but psychologically smart way of using the panel to harness the Hawthorne Effect and create advocates. In an era where teens are turning away or turned off by interruptive mass-media marketing, Tremor offers businesses a turnkey solution for harnessing the oldest and most powerful media of all: people media. Although P&G does not publish revenue data for its opinion-leading Tremor panel, the predicted income from third-party brands using the panel in 2004 was US$12 million, with the number of campaigns up 30% over 2003.[38] Whatever the numbers behind Tremor, P&G believes the panel to be so effective that it warrants being replicated; in 2005 the company began recruiting a second Tremor seeding panel in the US, twice the size of the original one, made up of 500 000 mothers ('Tremor Moms').[39]

Seeding trials unlimited

While P&G's Tremor seeding panels may represent one of the most systematic uses of targeted sampling in the name of research, P&G is not alone in pioneering this technique. For example, New Line Cinema invited fans of JRR Tolkien's *Lord of the Rings* epic fantasy novel to advise on the film of the book. By giving fans a say in the film production, the Hawthorne Effect was triggered and an army of word of mouth advocates was created.[40] Similarly, the marketing company BrandPort uses its panel of advertising aficionados to seed new advertising campaigns. Panel members are sent previews of new advertising campaigns and asked to comment on them.[41] Again, the involvement generated by research dialogue cues goodwill and ownership – increasing the likelihood the panel member will adopt and advocate the advertised brand.

More generally, when sportswear brands such as Nike and Reebok offer opinion-leading trendsetters free pairs of their latest sneakers, it is not just to capture feedback from cool kids but also to kick-start word of mouth advocacy.[42] By putting the new product in their hands, or rather on their feet, in the spirit of partnership and in the name of research, the Hawthorne Effect is cued and vocal evangelists are created. Likewise, when Pepsi ran an opinion leader outreach trial with 4000 American teenagers in 2001, it wasn't just to find out what they thought of their new soda, Code Red, but also to amplify and accelerate word of mouth advocacy.[43] The Pepsi opinion leader outreach programme was so successful at driving sales for Code Red that Pepsi decided to pull its planned TV advertising campaign. Similarly, when Unilever asked 250 fashion-forward urban girls to participate in a pre-launch seeding trial of the Max Azria BCBGirl fragrance in Canada,[44] it wasn't just to find out what they thought of the perfume; it was to harness the Hawthorne Effect and drive demand. By giving each of the participants not only a full bottle to trial themselves, but also 100 samples each to hand out to friends, Unilever created a volunteer sales force that ensured the fragrance was a best-seller in cities where the research was conducted.

Driving sales with seeding trials: 10-point checklist

There is no single formula for optimizing launches and boosting sales using seeding trials that transform opinion leaders into word of mouth advocates. However, the following checklist covers off the key questions that you need to ask yourself when planning a seeding trial.

The right product

1. *Are we offering something new?* Opinion leaders like to lead with their opinions, so seeding trials work best, that is, they drive sales, when the product or service you are bringing to market is genuinely new and different. Old, generic, or commodity products and services with nothing new to say are unlikely to benefit from seeding trials because people only tend to talk about the new and surprising.
2. *Are we offering something better?* Seeding trials will accelerate your sales if the product or service seeded delivers a superior experience. While the Hawthorne Effect will initially transform opinion leaders into advocates, the advocacy will be short-lived if your product or service is substandard or disappointing. This doesn't mean your product or service has to be groundbreaking, outrageous, or revolutionary for seeding trials to optimize a product launch – some of the most successful seeding trials have involved frozen pizza, canned soda and bath soap. What you do need however is a unique selling point that clients, customers and consumers can articulate to each other.
3. *Are we offering something that can be sampled?* Seeding trials are targeted sampling initiatives conducted in the name of research, so they necessarily involve sampling (credible advocacy can only come from first-hand experience). For many products and services, offering opinion leaders some kind of free limited trial, sample, download, or preview may be relatively straightforward, especially if sampling is a widespread practice in your sector. However, you may have to be more creative if your product or service doesn't lend itself so easily to trial – as can be the case for certain high-value, low-margin

products such as some technology products or perishable goods such as fresh or frozen foods. If this is the case, try and think of novel ways in which you can set up your trial to nevertheless get your product into the hands of opinion leaders (in the name of research): in-store or at-office trials, trials at hotels, conferences or trade shows, trials using redeemable gift certificates or vouchers, special loans or screenings, etc. If you really cannot enable opinion leaders to trial your product or service, you could still enable them to trial the product 'virtually' by viewing it and seeing it in action online.

The right people

4 *Have we identified our opinion-leading target buyers?* Seeding trials are all about influencing the influencers in your target market, i.e. influencing the opinion-leading target buyers who frequently offer and are elicited for category-related advice. So run a brainstorming session to generate a list of people in your target market who would make good 'connected and respected' advocates for your product or service. Think in terms of individual profile - jobs, place of work, leisure activities and club membership - and in terms of networks - what are the big and visible organizational and social networks in your target market, such as employers, associations, interest groups, etc.? The important thing is to be creative, and not stop at the usual suspects when looking for opinion-leading target buyers: experts, celebrities, journalists, bloggers and reviewers. Instead, think of who could be effective word of mouth hubs in your market. For example, PAs, club secretaries, beauty therapists, health and fitness instructors, bar staff and hairdressers are popular choices for seeding consumer goods. Don't forget that investors, employees and satisfied clients, customers or consumers can also make for opinion-leading advocates because they have product experience or a stake in your success.
5 *Are we seeding to enough opinion leaders?* Opinion leaders make up 10% of your target market, and successful seeding trials will seed up to 10% of these opinion leaders, i.e. up to 1% of your entire target market. Seeding trials on such a scale may be prohibitively expensive, especially in the consumer packaged goods markets, but you do need a minimum critical mass of Hawthorne-Effect-enhanced opinion leaders advocating your product or service for an appreciable sales uplift. While this number will vary according to the size of your market, a useful rule of thumb is to seed to a minimum of 250 opinion leaders per major urban centre.

6 *How are we going to deliver the trial experience?* Once you have identified opinion leaders to invite into the seeding trial, you have to solve the logistics challenge of how to get the product or service into their hands as cost-effectively as possible. Downloads via the Internet, by post, by courier, by hand, or by pick-up from a convenient location such as a store, hotel, or mall? Contact a handful of sales promotion agencies or specialist sampling companies for advice on what they would recommend as the most cost-effective and logistically simple approach for you. The advice will cost you nothing and may include some creative solutions you might have missed.

The right action

7 *Does our seeding trial involve exclusive 'Get it first' sampling?* Seeding trials, targeted sampling initiatives with opinion leaders, work best when they enable participants to get their hands on new products and services first - before everyone else. If your seeding trial offers participants an exclusive sneak preview, the sales-driving effect will be optimized because you increase the word of mouth potential. If pre-launch seeding is not possible, then think of ways that you can combine targeted sampling with 'Get it first' access to other services, promotions, or even new advertising. The more trial participants feel like VIPs, with exclusive and priority access to what you have to offer, the more they will advocate. Finally, think of participants not only as clients, customers, or consumers, but as the means to getting more clients, customers, or consumers. What can you seed with the product that will help participants spread the word: discounts, vouchers, promotional gifts, branded merchandise, or special invitations to share with friends, or even further samples to hand out?

8 *'VIP Vote': are we giving seeding trial participants a say in our marketing?* Seeding trials work because they elicit participant advice on how your product or service is marketed. This creates a sense of ownership, loyalty and goodwill that, through the Hawthorne Effect, triggers adoption and advocacy. In practice, giving participants a say in your product or service need only involve a simple online vote on options for a campaign poster, logo, display stand, advertising concept or promotion. Of course, you can go further and involve trial participants in the packaging and design of the product or service itself. For example, the Australian beer Blowfly was built ground-up by research participants through online voting - on all aspects of the

product ranging from bottle design to brand logo (see Chapter 4: Brewing buzz, p. 59). The key is to keep everything as simple as possible in order to minimize the work for the trial participants. The goal is to make opinion leaders feel they have contributed to your innovation or how it is marketed - without them having to do anything more than a couple of mouse clicks, which is why simple voting between options works better than lengthy discussions, questionnaires, or surveys.

9 *Does our seeding trial offer participants an 'Inside scoop'?* Seeding trials work because they make participating opinion leaders feel like 'insiders', that they have the 'inside story' on your product or service. What information can you share with them to reinforce the impression that they have a special relationship with you? For example, can you give them a 'behind the scenes' experience, provide them with insider guides, gossip or stories, or give them privileged access to company discussions, blogs, or marketing materials? Some companies, such as Unilever, go so far as to print personal branded contact cards for seeding trial participants to reinforce the impression that they are indeed insiders[45] (see Chapter 3: Creating brand advocates, p. 47).

Measurement

10 *Have we put in place a mechanism for measuring the effectiveness of our seeding trial?* Ultimately, seeding trials are a sales acceleration tool, and they stand or fall on the sales uplift they produce. To measure your trial's impact on sales, you can do as P&G do and use a 'control' region or group, where the trial is not run, and measure differential sales performance. This may be fine for measuring offline sales, but for measuring the effect on online sales you might need to provide trial participants with a pass-it-on promotional discount code to forward to friends, who forward it to friends, etc., enabling you to track the number of online sales the seeding trial generates. Of course, there are other softer measures you can use, such as the effect of the trial in increasing awareness levels. To do this, it suffices to include an online dimension to your trial, such as a special website, blog, or discussion list and measure the number of visitors it receives. Alternatively or additionally you can measure the number of column inches in online and traditional press, that your trial generates, and calculate the reach of those column inches. Finally, you can also track the effectiveness of the seeding trial using a simple pre- and post-trial poll that measures changes in advocacy rates among trial participants and within your broader target market.

Conclusion: seeding trials as super-charged sampling

When all is said and done, seeding trials are just sampling initiatives on steroids. Not only do seeding trials accelerate sales by removing the cost barrier to trial among the key opinion-leading 10% of your target market, but they also harness the Hawthorne Effect, transforming participants into loyal adopters and vocal word of mouth advocates. While sampling ('tryvertising,' as it has recently been re-branded),[46] offers a first-hand brand experience and is the preferred promotional activity of opinion leaders,[47] seeding trials offer something much more powerful: brand *involvement*. Seeding trials - targeted sampling in the name of research - enable businesses to connect and collaborate with opinion leaders, market *with* them, rather than at them. And by creating a volunteer sales force, seeding trials are a scalable, predictable and measurable solution for driving the one thing known to drive business growth: word of mouth advocacy.[48]

Takeaway points

- Seeding trials with opinion leaders are an effective launch optimization strategy that can enhance sales by 10–30%.
- Seeding trials involve targeted sampling initiatives with opinion leaders conducted in the name of research.
- Seeding trials work by transforming opinion leaders into loyal adopters and vocal word of mouth advocates, and do so by harnessing a powerful psychological phenomenon called the Hawthorne Effect.
- Through 'Get it first' sampling, 'Inside scoops' and 'VIP votes', seeding trials generate goodwill, involvement and advocacy among opinion-leading clients, customers or consumers.
- Companies using seeding trials to optimize product launches and drive sales include Procter & Gamble, Microsoft, Hasbro, Google, Unilever, Pepsi, Coke, 3M, Ford, Dreamworks SKG, EMI, Sony and Siemens.

Notes and references

1 American Marketing Association definition, archived at http://tinyurl.com/dcer6.

2 For more on Post-it Notes story see articles archived at http://tinyurl.com/a8lf and http://tinyurl.com/av6ts.

3 For a description of the Hawthorne trials see Mayo, E. (1933) *The Human Problems of an Industrial Civilization*. New York: Macmillan. For a good online review see http://tinyurl.com/bz8t6.

4 Gladwell, M. (2000) *The Tipping Point: How Little Things Can Make a Big Difference.* Boston: Little, Brown and Company, p. 259.

5 Peters, Tom (1991) *Thriving on Chaos.* New York: HarperCollins.

6 See Lazarsfeld, P., Berelson, B. and Gaudet, H. (1944/1968) *The People's Choice*. New York: Colombia University Press; and Katz, E. and Lazarsfeld, P. (1955) *Personal Influence*. New York: Free Press.

7 For a review, see Weimann, G. (1994) *The Influentials: People who Influence People.* New York: University of New York Press; and Keller, E. and Berry, J. (2003) *The Influentials: One American in Ten Tells the Other Nine How to Vote, Where to Eat and What to Buy*. New York: Simon and Schuster.

8 For widely used and validated scales see Flynn, L., Goldsmith, R. and Eastman, J. (1996) 'Opinion leaders and opinion seekers: two new measurement scales', *Journal of the Academy of Marketing Science*, 24 (2): 137–147; Ben Miled, H. and Le Louarn, P. (1994) 'Analyse comparative de deux échelles de mesure du leadership d'opinion : validité et interprétation', *Recherches et Applications en Marketing*, 9 (4): 23–51; Childers, T. (1986) 'Assessment of the psychometric properties of an opinion leadership scale', *Journal of Marketing Research*, 23: 184–188; and King, C. and Summers, J. (1970) 'Overlap of opinion leadership across consumer product categories', *Journal of Marketing Research*, 7 (1): 43–50. For a good review see Weimann, *The Influentials*.

9 For details on this seeding trial see ACNielsen Report (2003) *Alternative Marketing Vehicles* (June), archived at http://tinyurl.com/azel9.

10 Seeding trial run by author's word of mouth marketing agency Spheeris (www.spheeris.fr).

11 See Tierney, J. (2001) 'Here come the Alpha Pups', *New York Times*, 5 August, Late Edition – Final Section 6: 38: col. 1, archived at http://tinyurl.com/7vo64.

12 Rosen, E. (2000) *The Anatomy of Buzz*. New York: Doubleday, pp. 159–160.

13 Figure from Webmasterworld, archived at http://tinyurl.com/brygh./

14 Chung, J. (2004) 'Beta boosters', *The Globe and Mail*, 7 September, archived at http://tinyurl.com/d73yg.

15 Chung, J. (2004) 'For some beta testers, it's about buzz, not bugs', *New York Times,* 22 July, archived at http://tinyurl.com/b5kco.

16 For example, an MSN search in May 2005 showed Gmail to be as high profile on the web as Yahoo! Mail: 781 675 (Gmail) vs. 870 574 (Yahoo! Mail) citations.
17 Wells, M. (2004) 'Kid nabbing', *Forbes,* 2 February, archived at http://tinyurl.com/cuh3a.
18 Ibid.
19 Ibid.
20 Dunnewind, S. (2004) 'The companies behind teen "viral campaigns"', *Seattle Times*, 20 November, archived at http://tinyurl.com/ac2xg.
21 Press release from 'Tremor' website designers Blue Dingo, archived at http://tinyurl.com/au9qt.
22 Blue Dingo, ibid.
23 Ibid.
24 Levey, R. (2003) 'P&G's buzz on viral marketing', *Promo Magazine*, 19 June, archived at http://tinyurl.com/bv8mx.
25 Dunnewind, 'Teen "viral campaigns"'.
26 Walker, R. (2004) 'The hidden (in plain sight) persuaders', *New York Times*, 5 December, archived at http://tinyurl.com/998we.
27 Ibid.
28 Coolidge, A. (2004) 'Teens virtually perfect for P&G', *Cincinnati Post,* 21 February, archived at http://tinyurl.com/9j9w3.
29 Wells, 'Kid nabbing'.
30 Garrett, F. (2004) 'ADTECH: Word of mouth marketing: tips from Procter & Gamble', *WebProNews*, archived at http://tinyurl.com/8zb3w.
31 Wells, 'Kid nabbing'.
32 Coolidge, 'Teens virtually perfect for P&G'.
33 Ibid.
34 Wells, 'Kid nabbing'.
35 Levey, 'P&G's buzz on viral marketing'. For an early PowerPoint presentation by Tremor, http://tinyurl.com/9hxom, and Mathew, M. (2005) *P&G's Tremor: Reinventing Marketing by Word of Mouth*. ICRAI Business School Case Study Development Centre.
36 Wells, 'Kid nabbing'.
37 Leeds, J. (2004) 'Procter & Gamble: Now promoting music', *New York Times*, 8 November, archived at http://tinyurl.com/azcp7.
38 Wells, 'Kid nabbing'.
39 Baker, P. (2005) 'Word of mouth advocacy: Right people, right message', Paper presented at Alternative Advertising and Marketing Conference Melbourne 24–25 February. Background on Tremor Moms is at http://tinyurl.com/dkt55.
40 Nycz-Conner, J. (2005) 'Look who's talking: word of mouth marketing hits big business big time', *Washington Business Journal*, 6 May, archived at http://tinyurl.com/9ryob.

41 Rodgers, Z. (2004) 'Marketers pay their way to the youth audience', *ClickZNews*, 28 May, archived at http://tinyurl.com/blojt.
42 For background see Gladwell, M. (1997) 'The Coolhunt', *New Yorker*, 17 March, pp. 78-88.
43 From Chura, H. (2001) 'Pepsi-Cola's Code Red is white hot', *Advertising Age*, 27 August, p. 1; Tkacik, M. and McKay, B. (2001) 'Code Red: PepsiCo's guerilla conquest', *Wall Street Journal*, 17 August, p. B5; and Klingbeil, A. (2001) 'The making of a brand', *Gannett News Service*, June, p. 29.
44 Stradiotto, M. (2005) 'Working with agents and activists', Conference paper presented at the Word of Mouth Marketing Association Summit, Chicago, 29-30 March, archived at http://tinyurl.com/7p4y3.
45 In 2004, Unilever ran a clever seeding trial for its new Axe brand body wash that made participants really feel part of the brand: As well as seeding the product, Unilever also provided participants with their own sets of personalized contact cards, with brand artwork on one side and personal contact details on the other. The cards also doubled up as discount vouchers so when participants handed them out, they had a further reason to evangelize about the brand. For more information contact author at paul@viralculture.com.
46 See Trendwatching.com article on tryvertising, archived at http://tinyurl.com/9sdsq.
47 Berry, J. (2005) 'Identifying, reaching, and motivating key influencers', conference paper presented at the Word of Mouth Marketing Association Summit, Chicago, 29-30 March, archived at http://tinyurl.com/9ex2f, and Weiss, M. (2005) 'Buzz sampling', Conference paper presented at the Word of Mouth Marketing Association Summit, Chicago, 29-30 March, archived at http://tinyurl.com/codsr.
48 Reichheld, F. (2003) 'The one number you need to grow', *Harvard Business Review*, 81 (Nov.-Dec.), pp. 1-11.

2

Live buzz marketing

Justin Foxton

Founding Partner and CEO, CommentUK

One-to-one, face-to-face, live buzz, live viral marketing, gossip marketing, word of mouth marketing, stealth marketing, ambush marketing, guerrilla marketing, Tremor marketing, sneeze marketing, theatrical marketing, live point-of-sale, underground marketing, peer-to-peer marketing . . . all strands of connected marketing. The astonishing array of descriptors for what, at its core, amounts to the same marketing phenomenon provides a very good sense of how connected marketing is a young and emergent field. But whatever you call the various activities that go to make up connected marketing, one thing's for sure: it represents marketing's brave new world.

This chapter looks specifically at live buzz marketing. There are parts of the marketing industry – let alone other industries – that are totally in the dark about this activity, what it can do for a product, brand or service, how it is created and, perhaps most importantly, how it can be measured. It courts controversy, debate and column inches, but many are unconvinced that it has any place in the serious business of real marketing. Most marketers quizzed on the subject just put it down to a passing phase: 'it's a fad,' they bluster, 'a leftfield communication alternative that will probably never take root at all!' . . . a little like mobile phones or the Internet . . .

There is, of course, a good reason for the scepticism and confusion that exists around live buzz marketing: it's difficult to define exactly what it is. But if we all knew what live buzz marketing was, it would lose its power.

This chapter unravels and defines this new and extremely powerful technique. In doing so it presents a number of case studies that put live buzz marketing into context and differentiate between the two different strands currently being used by marketers in their connected market strategies: live peer-to-peer marketing and live performer-to-peer marketing.

What is live buzz marketing?

Much like many other approaches to connected marketing, at its most basic level live buzz marketing is simply an attempt to harness, define and formalize what people have done since the beginning of time - talk to one another. Or, perhaps more accurately, it's an attempt to create or manufacture what people have done since the beginning of time. Marketers have now seen and acknowledged the enormous power of word of mouth and have tapped into it, coming up with a host of ways to kick-start it.

In order to understand exactly what live buzz marketing is, it's useful to break the term down and indulge in some dictionary definitions. For example, *Roget's New Millennium Thesaurus* (First Edition V 1.0.5) defines buzz as:

i) Verb 1) to make a low droning or vibrating sound like that of a bee. 2) to talk, often excitedly, in low tones. 3) to call or signal with a buzzer

ii) Noun 1) a vibrating, humming, or droning sound, 2) a low murmur ('a buzz of talk')

Slang: excited interest or rumour ('the latest buzz from Hollywood')

Synonyms offered include the words echo, comment, gossip and word of mouth.

In the modern marketing era, the word 'live' could apply as much to an Internet application as it does to a piece of theatre. So there is no doubt it needs filtering down when it comes to defining the term live buzz marketing. *The Oxford Dictionary Thesaurus and Word Power* guide defines live as:

. . . an actual event or performance

. . . not recorded

('recorded' in the latter case we can assume to imply encoded or programmed, hence distinguishing from the use of the term live in the context of the online environment).

Taking elements of the above, the following could be a succinct definition of live buzz marketing:

> 'A marketing technique that makes use of an actual event or performance to create word of mouth.'

This is easily understood since marketing's new Holy Grail above-, below- and through-the-line is the creation of word of mouth. The following definition, however, is perhaps preferable:

> 'A marketing technique that makes use of an actual event or performance to create an echo.'

This may sound pretentious, but there are valid reasons to use word of mouth and echo interchangeably or even, in this instance, to replace word of mouth altogether. Echo implies something far more complex and explanatory than word of mouth. The use of the word echo also distinguishes this technique from other buzz techniques whose primary intention is the generation of word of mouth. Echo implies that the buzz is begun at a certain point by a live happening or event, which then triggers a sound wave to be sent out.

It begins with a high-impact sound – a clap, a bang, a shout, or even an evangelical testimony – and bounces off one surface, then another, reverberating, repeating and imitating itself until ultimately it fades and falls silent. The clever part though, is creating an echo campaign that takes years, maybe even decades, to fade.

These echoes, and therefore live buzz marketing campaigns, are created by performers. They may include actors and actresses as we understand them to be, but could also include members of the general public who are willing to talk about a product, brand, or service. All it takes for an echo to begin is for somebody – anybody – to be willing to perform a big enough noise up front to send vibrations out into the world. Actors and actresses can be used to do this as their training equips them to project their voices, use their bodies and convince an audience that what they are saying is true. So performers extolling the virtues of a product, brand, service, or message generally have the power to create maximum effect within the target market. The echo, however, can be continued by anyone who hears the message. And, provided the message is loud, clear, appealing and preferably unique, the echo can continue for a very long time indeed, as the case studies in this chapter demonstrate.

The two types of live buzz marketing

There are two main strands of live buzz marketing: live peer-to-peer marketing and live performer-to-peer marketing.

Live peer-to-peer marketing

As in normal peer-to-peer activity, this is when ordinary citizens become brand ambassadors and spread word of mouth about a particular product, brand, or service. Essentially, this is based on people's willingness to talk about something without being paid, in return for which they sometimes receive product samples and loyalty points which can be redeemed for rewards. They also receive the social affirmation of being somebody in the know. Live peer-to-peer marketing is simply when the word of mouth is triggered by ordinary, everyday people.

Live performer-to-peer marketing

This involves the use of trained, qualified, costumed, scripted and rehearsed performers who take on the roll of brand ambassadors in various environments where the correct target demographic is found. It is highly controlled, regulated, targeted and measurable. This chapter predominantly focuses on live performer-to-peer marketing, but you will also find a live peer-to-peer case study later in this chapter on p. 44.

The spectrum of live buzz marketing

There are three main techniques used in live performer-to-peer buzz marketing campaigns: secret, disclosed and overt. All make use of professional, scripted and rehearsed live actors and actresses to trigger buzz. The difference between the three techniques is their potential reach and the degree to which the audience is made aware that they are being involved in a marketing exercise. Techniques two and three (disclosed and overt live buzz marketing) can achieve the broadest possible reach.

The top of the pyramid (which represents the first contact with members of the audience and marks when they are physically exposed to the campaign) is the smallest level of contact in physical numbers but it also represents the highest quality contact. The audience gets to enjoy the experience first-hand and receives the information in an unadulterated form.

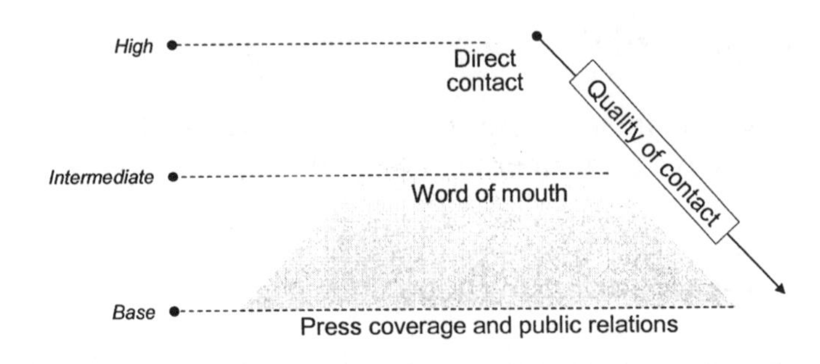

Figure 2.1 Potential reach: three tiers of awareness. Potential reach should be considered when planning and implementing an effective live buzz marketing campaign; it will help to maximize a client's budget

The next tier down sees the reach increase as the first contactors begin to echo the message of the campaign to two, three, four and in some cases up to eight or 10 people. This can increase incrementally as the second contactors tell their two, three, or four friends and so on. This incremental level of awareness is not included in the above diagram (or in any research figures) as it is difficult to measure accurately.

The third and final level of contact has the potential to have the widest reach, but by this stage – as with most press exposure – the marketer has comparatively little control over the message. Brands, however, often benefit from enormous amounts of press coverage that result from live buzz marketing campaigns. This is simply because the uniqueness of the creative applications, combined with a new and different approach, creates a story that's deemed to be worth telling.

The following sections of this chapter describe the three main live buzz marketing techniques, including their strengths and weaknesses, and provide case studies showing how live buzz marketing can apply to everything from a product to a social message.

Technique 1: Secret live buzz marketing

This is a buzz marketing campaign in which the consumer never knows that he or she has been the focus of a marketing exercise. It is rumour, stealth, gossip and viral at its most secretive and undercover. It makes use of product ambassadors who spread information and buzz about a brand, product, service, or message in a marketplace in such a way that the fire is lit but no one ever knows how.

Although the case studies featured here are aimed at consumers, secret live buzz marketing can also be used in business-to-business (B2B)

applications and for the purpose of internal communication. It can be used to great effect to drive consumers to particular stands within trade shows. This is achieved by having performers wander round the show loudly and enthusiastically extolling the benefits of a particular stand.

Case Study 1 (product)

Client: Premium ice cream brand Carte d'Or.

Environment: The ice cream freezers in major multiple grocers around Southern England.

Target consumer: Bridget Jones-type females.

Brief: To raise awareness of the different flavours within the range and to draw attention to the price reduction on offer.

Campaign: Teams of two female performers aged in their mid- to late- 20s were dispatched to 50 stores around the country. Highly charismatic, appealing and fun-loving individuals, they wheeled around a shopping trolley containing nothing but hundreds of tubs of Carte d'Or brand ice cream in every flavour imaginable. One of the girls held a massive shopping list aloft for all to see. On the list, in bold oversized print, were all the flavours of Carte d'Or available in the freezers. The girls spent hours in each store, connecting personally with shoppers who fell into the target market.

The story they told the shoppers was a simple and highly appealing one: they were having a girls' night-in party that evening. None of them could cook so they decided to have an ice cream party. Of course, the only ice cream to eat on such occasions is Carte d'Or. They then went on to tantalize the shoppers by describing out loud the chunks of chocolate and fudge pieces embedded in the delicious ice cream swirls. Shoppers were powerless to resist. They clinched sale after sale by explaining that the reason they could afford to buy the ice cream in such vast quantities was because it was on a special £1.99 price reduction offer.

Results: Over 3500 tubs of Carte d'Or were sold as a direct result of the work of the performers. The campaign also generated enormous amounts of PR on a regional and national basis. It has become entrenched in connected marketing folklore as a result of being featured in a series of television news items and debates.

Case Study 2 (social message)

Client: The Portman Group, Britain's alcohol industry watchdog and campaigning body against the misuse of alcohol.

Environment: Public spaces, including trains, underground stations, bus stops, restaurants and bars.

Target audience: 18- to 25-year-old females.

Brief: The young female market is often overlooked when it comes to responsible drinking messages. Research suggests, however, that it is a key group to target. The brief was to attempt to draw attention to the ill effects of excessive alcohol consumption in such a way that it did not appear top-down or to be preaching in any way. In other words, use target demographic individuals and their horrific experiences to raise awareness and effect potential behaviour change in similar individuals.

Campaign: The project team found appropriate public hangouts for the target audience. A series of short and very hard-hitting scripts were written. Teams of two female performers rehearsed the scripts, which were spoken volubly in public spaces for all around to overhear. They involved one friend cataloguing the amount (and different types) of alcohol she had consumed the night before and the disastrous effect such a quantity had on her. The hung-over-looking youngster then went on to explain that, while walking home alone from the pub, she was stalked by an unknown man. She had to run to get away from him and screamed to alert people of her imminent danger. The other friend was obviously horrified (as were all the bystanders) and reminded the girl of a campaign she saw being run by the Portman Group entitled 'If you do drink don't do drunk'. She suggested that maybe this campaign applied to her. She pointed out that this was a real wake-up call as the next time could end in disaster.

This was just one of a series of scripts that were written and performed over the period of the campaign. Others were more comical, involving exposés of the flirtatious antics of the drunken friend who was convinced that the red-faced old man she was kissing looked just like Tom Cruise.

Results: This kind of activity is difficult to measure, however independent market research conducted on similar campaigns for the Portman Group suggests that of all media utilized (including posters, radio and mobile billboards), live buzz marketing was the most likely to effect behaviour change. The research also demonstrated the high word of mouth potential of such a campaign.

Strengths of the secret live buzz marketing approach

1. The marketing-savvy public - sick to death of being bombarded by messages - is given a welcome break. Marketers can get their messages across without irritating the consumer. Even if the consumer becomes aware that it's a promotion, they tend not to mind because they enjoy the uniqueness of the approach.
2. It enables marketers to create the perfect brand ambassador. By careful scripting, casting and styling, the performers become a living, breathing representation of the target consumer, who can appeal directly to others like themselves.
3. The message can be tailor-made to appeal to precise target individuals. Unlike many other marketing and communication techniques, it isn't a one-size fits all solution.
4. It involves completely interactive, personalized connections with target consumers making it an extremely high-quality contact.

Weaknesses of the secret live buzz marketing approach

1. In some instances, if it's discovered that they are being marketed to, the public can feel duped. This can create a negative backlash both in the consumer base and in the media.
2. By its very nature, the extent of the echo is limited. This is simply because the first interaction (or the first noise or bang which sets off the echo) is not very big or loud. Put simply, if you were to overhear a conversation about somebody who was nearly attacked because they were drunk, you may tell a few people - probably mostly on the day you heard it. If you were a witness to that person being attacked, you would tell hundreds throughout your lifetime.
3. A key factor in the reach of any live buzz marketing campaign is the PR that can be generated. In the case of a secret campaign, this is limited by the fact that it's secret, so any PR usually focuses on the medium rather than the message.

Technique 2: Disclosed live buzz marketing

As opposed to the secret approach, this is a live buzz marketing campaign in which the consumer is not initially aware that he or she is being marketed to, but is informed or becomes aware during the course of the interaction. Its enormous power lies in being able to stage a real life scenario - complete with real life emotions and actions - without running the risk of appearing to dupe or delude the public. It maintains the initial element of surprise before revealing the product, brand, or message and avoids the controversy that can spill out from a completely secretive campaign. In common with secret live buzz marketing, disclosed live buzz marketing can happen wherever the target market is found: cinemas, pubs, restaurants, bars, point of sale, on the street, or just about anywhere.

Case Study I

Client: Transport for London.

Environment: Cinemas across London.

Target audience: Motorcycle riders and motorcar drivers.

Brief: To heighten public awareness of a new road safety TV and cinema advertising campaign aimed at encouraging motorcyclists to ride more defensively and encouraging other road users to be more aware of motorcyclists. The campaign's commercial featured the typical road safety scenario of a motorcyclist not being seen by a car driver and being graphically hit and killed. Realising that the public is largely immune to such social messages (no matter how bloody or graphic), Transport for London wanted to drive the message home in a way that would never be forgotten by the audience and would be talked about for literally years to come.

Campaign: A unique, interactive and dynamic form of communication was designed to create maximum impact across all three tiers of awareness (see Figure 2.1). The project team obtained permission to perform live 30-, 60- and 90-second commercials in six of Carlton Screen Advertising's cinema auditoria around London over a four-weekend period.

A number of two-performer teams were used, each consisting of a male actor dressed as the cinema manager and an actress pre-seated among the audience, masquerading as a member of the cinema-going public. The actor stood at the door greeting patrons as they entered and the actress sat eating popcorn and glancing over her shoulder looking for someone who was obviously running a little late for the movie.

Just before the Transport for London commercial ran on screen, the reel was stopped, the lights came up and, to the surprise of the audience, in walked the cinema manager who had been greeting people at the door. He stood at the front of the auditorium and the following script was performed before the audience's eyes between him and the actress.

ACTOR: Sorry to interrupt your evening ladies and gentlemen, but is there a Sue McNaughton in the house? Mrs McNaughton? Are you here? Mrs McNaughton? (He walks up the aisle, shining a torch into the rows. Sue hesitantly raises her hand.)

ACTRESS: I'm Sue McNaughton. (She smiles, a little sheepishly, at the people around her.)

ACTOR: (As he reaches her.) Would you like to come with me, Mrs McNaughton? (He takes her elbow and guides her into the aisle.)

ACTRESS: What is this about? Where are you taking me? Who are you, anyway?

ACTOR: Sorry, I'm Simon Stevens, Mrs McNaughton. I'm the manager here. Err . . . please come this way.

ACTRESS: (Getting a bit agitated.) Could you please tell me what this is all about? I insist. You can't just come in here and haul me out of the cinema. Tell me!

ACTOR: (She won't move. Simon is at a loss.) Look, Mrs McNaughton, I'm sorry to have to tell you this, but . . . (Simon leans in and says something softly in her ear.)

ACTRESS: (Clearly in denial about the news she has just been given.) No. No, you're wrong.

ACTOR: Mrs McNaughton . . .

ACTRESS: He called me . . . 15 minutes ago . . . from his mobile. He said he was getting on his motorcycle and he'd be here any minute! (Simon takes her elbow, and tries to lead her out.) Leave me! (She jerks away. Then suddenly she turns to him.) Was he wearing his visibility vest? Did they tell you? (Simon shrugs, helplessly.) I bought it for him for his birthday . . . a yellow one. When we spoke a few minutes ago, I asked him to put it on.

ACTOR: Mrs McNaughton, if you'll just come to my office, the police are here to . . . (The sound of a police walkie-talkie crackles outside the door.)

ACTRESS: Oh . . . his lights. He always forgets to switch on the lights. Did he . . . was he . . . is he . . . (Suddenly, she goes limp, and collapses on Simon's shoulder. He leads her out. Sue exits. Simon turns to face the audience.)

ACTOR: Ladies and gentlemen, what you have just witnessed was a fictional scene. However, the story behind it is all too true. Last year 71 motorcyclists were killed on London's roads, and 71 families received the news that their loved ones would not be coming home. As more and more people take to the road on mopeds and motorbikes, we urge you to wear visible clothing, keep your lights on and to ride defensively. If you are not a motorcycle rider, please encourage your loved ones and friends who are, to be careful. Please watch the following screen commercial, and remember, beware of the driver who didn't see you.

(He exits and the Transport for London screen advertisement runs)

Results: A combination of factors ensured this campaign's place in advertising history. Not only is it now the most widely discussed and debated live buzz marketing campaign in the UK, but in its December 2003 List of Lists, *Campaign* magazine voted the live advertisement for Transport for London its third top cinema commercial of all time. Suddenly, live buzz marketing had entered the mainstream.

The press coverage was unprecedented for a campaign of this nature: the campaign attracted prime-time BBC TV news coverage, radio debates and a plethora of newspaper and magazine articles. In all, 387 live advertisements reaching thousands of cinemagoers were performed, establishing a multi-team model for significant first contact reach. The campaign succeeded in maximizing all three levels of awareness: big first contact numbers, even bigger word of mouth, which is still going two years on, and extensive mainstream and marketing industry press coverage.

Case Study 2

Client: GlaxoSmithKline's Lucozade Energy.

Environment: Cinemas across the UK.

Primary target audience: 16- to 24-year-old males.

Brief: Lucozade Energy's proposition of mind and body energy was clearly communicated via an on-pack promotion, which enabled consumers to win the opportunity to learn how to practise a cool skill that required both forms of energy: becoming a DJ, rally driver, or film stuntman. The stuntman promotion was communicated in movie houses around Britain and was appropriately tied in to the release of *Spiderman 2*. Posters explaining the promotion adorned cinema complexes nationwide and a live in-cinema buzz campaign was created that would appeal directly to this target demographic and get them talking to their mates. The buzz campaign ran all over the country during the opening weekend of *Spiderman 2*.

Campaign: In order to achieve maximum awareness from the buzz marketing campaign, three points of intervention between brand and consumer were orchestrated in the cinema complex. On entry into the *Spiderman 2* screening, each member of the audience received a leaflet explaining that the final scene of the new Lucozade Energy commercial was going to be filmed in the auditorium prior to the movie. This was part of the act, grabbing the audience's attention from the start, and building anticipation of and excitement about being part of the live advertisement that was to follow.

Unlike the Transport for London campaign, Lucozade didn't have a big screen advertisement. The live buzz marketing campaign stood alone – a first in the cinema environment. Just before the end of the screen advertisements, the reel was stopped and the following live buzz marketing performance took place.

(A man bursts into the cinema, dressed as everyone's archetypal Hollywood movie director: shades, clip board, peak cap, sleeveless thermal vest – all in black. In a brash American accent, he calls up to the projection box)

DIRECTOR: (TONY)	Okay, Mike, can we have some lights here please! (The cinema house lights come up.) Hey folks, as y'all know, we're here filming the stunt scene for the new Lucozade commercial. What happens is Marty, the stunt guy, gets set on fire and runs through the auditorium, runs down the stairs, runs across the front and then back . . . (By this time Marty, dressed in white heat-resistant overalls, balaclava and gloves has walked slowly down to join the director.)
MARTY: (speaking in strong UK accent)	Ton . . . Tony! 'ang on a minute. My agent said nuffink about running. I don't mind being set on fire, but I'm not running. I've got no energy!
TONY:	Marty . . . buddy . . .

MARTY: No! I don't 'ave the energy, mate. No way!

TONY: (Getting angry.) No way? No way? You call yourself a stuntman? You're not a stuntman, you're a loser! You're fired, buddy! Get off my set.

MARTY: Alright mate, I'm going . . . but I'm not running. (He turns and slowly begins to exit the auditorium. He stops and turns for the final interaction with Tony.)

TONY: (Shouting and gesticulating after Marty out of sheer frustration.) Aahwww!!! I don't need you!! Any of these folks sitting here could do this. Yeah . . . they all stand the chance of making it as a stuntman with Lucozade Energy. (To audience – very excited.) That's right, 10 lucky people get the chance to train as stuntmen at Pinewood Studios and the winner gets £10 000. It's that simple folks! All the details are on the pack. (Calling up to the projection box) Okay, Mike, we're outta here 'til we got us a winner. (To Marty) Ahh, get outta here you loser, get, get . . .

MARTY: Alright, Tone, I'm on my way, but I'm not running, yeah . . .

TONY: Aw, get some Lucozade Energy!! Mike, you can roll the movie!!

(Lights dim and the reel rolls)

Results: Independent market research was commissioned by GlaxoSmithKline and revealed unprecedented results. Two hundred cinemagoers were recruited in the cinema foyer and telephoned two weeks following exposure to the ad. A further 100 recruited in the same way were telephoned one month after exposure. Respondents were not aware they were going to be asked about advertising.

Results after four weeks were: total recall – 89%; propensity to purchase – 59% had bought Lucozade Energy, of those, 42% said the ad had influenced their decision; word of mouth: 66% told someone about the campaign; 25% told five people or more.

The significance of this research was that for the first time it accurately quantified the word of mouth recall and propensity to purchase resulting from a live buzz marketing campaign. While most of the connected marketing industry had until this point used gut feel and educated guesses to determine the success of live buzz marketing, formal, measured results were now available supporting the real power of this approach.

Strengths of the disclosed live buzz marketing approach

1. It causes an initial 'what the heck is going on here?' reaction in the audience. They are put in a position where they simply cannot ignore the message or remain passive to it. It is simply too big, too bold and too arresting to be ignored. What this amounts to is a very big and very loud first bang that sets off an echo that continues for a long time, reaching many people and generating a high level of word of mouth. Recall levels are also extremely high: 100% after 24 hours and down only marginally to 89% after four weeks.
2. If sufficiently unique in terms of its creative application, such a disclosed campaign will also generate significant press interest.

Weaknesses of the disclosed live buzz marketing approach

The only potential downsides to campaigns of this nature are perceptual. They are:

1. The perception that initial first contact numbers will be low. This is not an accurate perception as multiple teams, performing an average of 15 live advertisements per team, per day over a two- or three-day period (as was the case with the Lucozade campaign) will reach hundreds of thousands of people during the period of the campaign.
2. The perception that people will be annoyed at the interruption. Like all communication, provided the creative idea is appropriate to the target audience and the environment in which it is being used, people appreciate the different approach and entertainment value of the pieces. The research conducted as part of the live cinema campaign for Lucozade revealed that audiences are actively seeking new and different experiences in cinemas:

   94% of the audience said that it's good to see different things at the cinema

   85% said the live performance was more memorable than other ads

   89% said it was an original idea.

Technique 3: Overt live buzz marketing

This is where the target audience is never in any doubt about the fact that they are being marketed to. It's highly theatrical, making use of scripts, song lyrics, set and costume elements that feature the brand, product, service, or message prominently. These campaigns combine communication and entertainment and are designed to be high-impact while eliminating the risk of any kind of controversy. If unique and appropriate creativity is applied to overt campaigns, they can be highly effective across all three tiers of awareness (see Figure 2.1), generating maximum first contact take-up, word of mouth and PR.

Like secret and disclosed campaigns, overt live buzz marketing campaigns can happen wherever the target market is found – in cinemas, pubs, restaurants, bars, at the point of sale in retail stores, on the street, in shopping malls and even in internal and B2B environments such as conferences and exhibitions. The echo created by overt live buzz marketing campaigns can reverberate for an extended period of time, touching vast numbers of people.

Overt live buzz marketing campaigns can be extremely effective in terms of creating a fertile seedbed for sampling, selling products, building brands and even collecting data. They can also be used to great effect in communicating difficult social messages. The use of a scripted theatrical production that draws on metaphor and analogy can be a very powerful tool to identify problems and effect change sympathetically.

Over the years, such campaigns have been used in the war against AIDS and other sexually transmitted diseases, in communicating the consequences of global warming, in educating the public on drug and alcohol awareness, and in campaigns promoting road safety and speed reduction (see Case Study 2).

How was overt live buzz marketing born? A surprising strand of history spawned and shaped this technique as we know it today.

South Africa under apartheid rule denied many millions of people access to formal education. This meant that the vast majority of the population could neither read nor write, or at best had very basic literacy skills. The knock-on effect of this was that much of the black population ended up in unskilled employment: mining, factory work, farm and domestic labour. People worked far from their home villages and families, and lived communally at or near their place of work. Low literacy levels, combined with high levels of poverty, meant that the communication mechanisms that some people take for granted were not an effective or viable means by which to reach the vast majority of the population.

As a direct result of this problem, the medium of industrial theatre was born. As its name suggests, the medium was responsible for creating overt live buzz marketing campaigns which toured industrial areas, mines, businesses and factories, delivering government, product, or industrial messages to the workforces. The theatrical pieces were basic but highly entertaining. They employed three or four actors and some simple stage elements, props and costumes, which could be easily transported on the back of a truck and easily set up.

The pieces used song and dance and a basic (but to the audience, highly identifiable) storyline to communicate with and educate people about issues ranging from health and safety and government propaganda on the one hand, to brand and product messages on the other. The 15- or 20-minute self-contained pieces were similar to the first TV soap operas that promoted a particular brand of washing powder.

Since the demise of apartheid in 1994, companies have continued to use industrial theatre to great effect in the corporate workplace - in culture change programmes, conflict resolution, motivation, education, training, induction courses and product launches.

Industrial theatre has now developed to suit the needs of the changing audience. Its name has changed to corporate theatre, the scripting and production of each piece is more sophisticated and different staging environments are used. Campaigns take place in more B2B and internal communications environments and use the boardroom table more as a stage than a flat surface over which to negotiate, argue, or debate. So industrial theatre has transformed the way companies communicate internally and to their customers' customers.

Overt live buzz marketing, this rich and complex mechanism, was essentially born out of the specific circumstances of one socio-political environment. In a sense, South Africa was almost forced to reinvent and use the oldest form of advertising and marketing: live theatre. It drew from ancient civilizations, which gave the people information or government propaganda, and wrapped it all up in a day or even days of entertainment, carousing and eating. Even the Caesars knew that the best way to sell something to the people is to entertain them while you're doing it. Not only will they buy it and keep on buying it - they will tell everyone they know to buy it, too.

Case Study 1

Client: Birds Eye Walls Ice Cream – Magnum Moments (bite-sized).

Environment: The point of sale in major grocery stores.

Target consumer: Female shoppers of all ages.

Brief: Birds Eye Walls launched small, bite-sized Magnum Moment ice creams as a new addition to the Magnum range. They came in boxes containing three flavours: chocolate, hazelnut and caramel. A unique live point-of-sale campaign was required, designed to create awareness of the range via product sampling, to drive sales based on a two-boxes-for-£3.00 special, and to generate both word of mouth and PR. The central above-the-line concept of 'Magnumise your life' needed to be brought to life through an experience that would give consumers, particularly women, a moment of pure unforgettable pleasure and indulgence during their shopping trip. The campaign was to be rolled out to 52 Tesco supermarket stores across southern and central England, over a three-week period.

Campaign: Teams of three handsome male singers, dressed impeccably in dinner suits, were formed. They stood at the supermarkets' ice cream freezers and serenaded passing shoppers with a rendition of the Flying Picket's *a cappella* cover of Yazoo's hit 'Only You'. Lyrics were adapted to reflect the three variants within the Magnum Moments range. The singers held elegant silver trays piled high with Magnum Moments and offered them to passers-by. In between songs, they connected on a one-to-one basis with the consumers, talking knowledgeably about the range and encouraging them to buy the product.

This highly theatrical and unusual sight, combined with beautiful singing, a dose of humour and the taste sensation of the product, gave shoppers a unique experience that many said they would never forget. The experience was heightened by the context in which it was set – a key factor in a successful live buzz marketing campaign. If you can turn a somewhat stressful or humdrum shopping experience into a moment of indulgence and pleasure, you will almost certainly create buzz and affiliation with a brand.

Results: The Magnum campaign was based on a winning recipe: handsome, well-rehearsed men in dinner suits giving away delicious ice creams could not fail to appeal, and the response from consumers and staff was overwhelmingly positive. The campaign took the idea of a brand connecting live with its consumer at the point of sale to new levels, and the word spread rapidly within the target consumer communities and among the Tesco staff and management and their communities. In two follow-up campaigns commissioned by Walls, the performers developed something of a following, regularly being referred to as The Magnum Singers. Extensive PR coverage contributed to this enforcement of the connection between the men, the concept of in-store singing and Magnum ice cream.

Sales increased so much during the period of the campaign that freezers had to be restocked on an hourly basis in all stores. Birds Eye Walls went on to commission a further 52-store roll-out of the Magnum Singers with a Christmas campaign using a specially tailored version of a well-known festive feelgood number – 'Have yourself a little Magnum Moment'!

Case Study 2

Client: London Safety Camera Partnership.

Environment: Shopping malls around London.

Target audience: 16- to 24-year-olds.

Brief: To create a hard-hitting campaign which would drive home the 'Kill your speed' message. To engage the audience in a real-life, hard-hitting way by moving away from conventional shock tactics and/or two-dimensional exhibition displays containing bloody photographs and horrifying statistics. The aim instead was to create a live interactive experience that would be relevant to the target audience, highlighting the dangers of speeding and promoting safety cameras as a deterrent.

Campaign: It's Saturday morning and the usual throng of shoppers heads off to do their weekly shop. Blissfully unaware of what awaits them, they drive – probably a little too quickly – down to their local shopping mall. They enter the car park. As they take their parking ticket, they notice that printed on the reverse is a picture of the Grim Reaper with the tagline: 'Let's not meet by accident!' There are also three shocking statistics about accidents on London's roads. Already, they're thinking about their driving behaviour.

Having parked, they make their way into the mall. In the distance they hear a commotion going on and notice a crowd gathering. Eager to satisfy their curiosity, they move nearer. They encounter a bright, eye-catching stage – again there is the image of the Grim Reaper with bold red lettering: KILL YOUR SPEED. A branded London Safety Camera Partnership representative informs shoppers that a free show is about to begin. They also encourage people in the crowd to participate in a valuable research exercise regarding speeding and speed cameras. Having successfully gauged public perception, the Kill Your Speed rock opera is about to begin.

Atop the heavily branded stage, a high-impact and highly slick 10-minute musical unfolds. It begins with an MC enticing the crowd to draw nearer, as there are 'savings of a lifetime to be had'. Incorporating acting, song and dance, the narrative comprises a group of young friends on their way to a birthday celebration. The driver is torn between his friend telling him to speed up and his girlfriend telling him to slow down. Making the wrong choice results in them being involved in a fatal speed-related collision. It's revealed that this collision took place one year ago and what we have just witnessed is the story being told through the guilt-ridden eyes of the character that caused the accident. The serious messages are delivered through the carefully selected lyrics of songs popular among the target audience, designed to captivate and touch them on an emotional level. Queen's 'Don't Stop Me Now', Pink's 'Get the Party Started', Sting's 'Fragile' and U2's 'One' are but a few of the numbers used to draw the crowds in.

Without even being aware of it, the audience absorbs a number of key facts surrounding speed cameras: cameras are only erected at sites where there have been at least four people killed or seriously injured; the number of collisions is significantly reduced in areas where speed cameras are erected; the cost of road collisions in London approaches £2 billion. As the information is delivered in a highly entertaining manner, the recipients are far more open to accepting the messages.

After the applause, the cast, having taken their bows, step into the crowd and distribute leaflets containing pertinent information. Once again, representatives conduct basic research and gauge public opinion in the light of seeing the show. It is near impossible having undergone this experience not to have a shift in opinion. With leaflets in hand and having being entertained and educated, shoppers move off and continue about their business.

The show unfolds in this manner eight times over the course of a day to ensure maximum exposure to the greatest number of people. Those in the mall for a few hours could encounter it two or three times.

When shoppers make their way back to the car park, they are again reminded of the speed-related messages as they pay for their parking ticket. There is, however, still one final intervention to further solidify the message. At the car park exit boom, dressed in a black hooded cloak with a scythe, stands the Grim Reaper himself. He holds a sign that reads: 'Kill your speed – let's not meet by accident'. This chilling parting shot is communicated exactly at the point where drivers are behind the wheel so it can immediately influence driving behaviour.

Results: A campaign of this nature has an immediate effect on an audience. Its use of music and an emotive storyline enables it to evoke real emotion in a way that's not perceived to be overly didactic or top-down. The fact that it's also highly entertaining balances out the extreme emotions caused by witnessing the horrific death of one of the characters and the crippling of another. Audience members, both young and old, are taken on an emotional journey that often results in laughter and tears. It also results in great appreciation towards the cast and a willingness to take the message they have heard and spread it liberally to people in their social, domestic and work environments.

While no formal market research was conducted on this campaign, a 'before and after' questionnaire revealed increased levels of knowledge and understanding when it came to the key messages. The figures revealed an extremely encouraging shift in perception, especially given the highly emotive subject matter. Before the show, 55.5% of audience members questioned strongly agreed that cameras are there to slow drivers down. This increased to 64.5% after the show. While before the show only 41.3% of the audience members questioned strongly agreed that cameras are there to save lives, this figure increased to 56.8% after the show. Having seen the show, 76.8% now agreed with safety cameras as a means of deterring people from speeding – a huge figure given the public's general aversion towards speed cameras.

The campaign also generated a great deal of press and overall public interest.

Strengths of the overt live buzz marketing approach

1. An overt campaign is unashamed about its objectives to entertain and educate. The audience appreciates the effort to which the brand or organization has gone to add value to their daily lives, rather than just demanding their attention or money. People also appreciate the different and innovative approach, and are willing to tell family, friends and colleagues about what they've experienced.
2. Overt campaigns provide extremely high-quality contact with the consumer. Like disclosed campaigns, they have the ability to cut through the marketing clutter and make a real and lasting impact. Key message take-out is extremely high as the audience can identify strongly with the characters and performers used in the production.
3. Overt campaigns have very high word of mouth and PR potential.

Weaknesses of the overt live buzz marketing approach

1. An overt campaign often relies more heavily on sophisticated props, costumes, stages and even technical requirements such as sound and lighting. This pushes the overall price, so the cost per contact of such a campaign is higher than a secret or disclosed campaign.
2. With an overt campaign you need to ensure that the creative suits the environment. Communicating large amounts of information through the spoken word in environments where people are shopping or relaxing is usually a pointless exercise. These types of content-rich

campaigns are best suited to more controlled environments such as work seminars, presentations, conferences, schools, or university lecture halls, boardrooms, motivational environments and product launches. In public, the message must be short, sharp and entertaining, usually making use of song and/or dance, otherwise you'll lose the audience early on.

Live peer-to-peer buzz marketing

This chapter has focused on live performer-to-peer buzz marketing. But a chapter on live buzz marketing which did not feature a live peer-to-peer case study would be incomplete.

Live peer-to-peer buzz marketing is an enormously controversial area and has sparked much heated debate. Why? It's simply seen by many as being too 'big brother' for their liking, too underground, too insidious – marketers causing their brands and products to seep into every part of our lives, including our personal lives. It involves using your friends, even members of your own family, to market products to you – in some cases without you even knowing that they're being paid or incentivized to do so.

The following example emerged from a US peer-to-peer buzz marketing agency called BzzAgent. The story was uncovered by Rob Walker and exposed in *The New York Times* on 5 December 2004 in an article entitled 'Hidden (in plain sight) persuaders'. Picture the scene . . .

You and your friends are standing around the barbecue on a warm summer's afternoon. Drink in hand, you chat about the events of the week – perhaps your vacation plans or the readiness of the meat on the barbecue. As it's a bring-your-own-meat affair, one of your mates approaches the fire and pulls out a packet of a new brand of chicken sausages. You get chatting and she explains that these are simply the best sausages around. They're absolutely delicious and, on top of it all, they're good for you.

Now, let's pause a moment. Unlike most barbecues, this is not just a friend. It's a friend who has been incentivized to talk up this new brand of chicken sausages to their friends and relatives. Unlike a normal promotion, this person is not an official representative of the sausage company either. She is what's known as a buzz agent, employed by an agency that has thousands of similar ordinary citizens on its database – people eager to receive free product samples and go out and tell their family and friends about the fantastic new product they have 'discovered'. They wear no branded kit and (believe it or not) will fill out a form at the end of the barbecue documenting your and your friends' reactions to the sausages and how many were consumed. The form will note such salient facts as: 'People could not believe they weren't pork!' and 'I told everyone that they were low in fat and so much better than pork sausages.' One very confident agent noted: 'I handed out discount coupons to several people and made sure they knew which grocery stores carried them.' Another said 'My dad will most likely buy the garlic flavour,' before closing with 'I'll keep you posted'.

It's easy to see the potential success of this kind of targeted peer-to-peer campaign, but what of the morality of such an approach?

To play devil's advocate for a moment, if the friend at the barbecue explained to everyone that she'd been sent these healthy new chicken sausages for all her friends to try and share their opinions, would it be quite so creepy? Of course not. If we helped her to fill out her questionnaire, would we feel quite so negative about this form of marketing? We'd probably be rather impressed by a company who was going to such lengths to consider the opinions of its consumers. We are, of course, still being marketed to by one of our own friends, but as long as we know that, we don't seem to mind as much. We just don't like being fooled. It's worth noting that BzzAgent now asks all its buzz agents to disclose the fact that they are buzz agents.

In point of fact, this type of live peer-to-peer buzz marketing and advertising has been going on for ages. It has been practised by organizations that are regularly held up as innovators in the fields of marketing and sales: Amway, Tupperware, Herbalife, Ann Summers. These companies create brand ambassadors, even sales representatives, out of ordinary citizens and get them talking to their friends. The event or performance is usually a party of sorts with some snacks, drinks and a prepared motivational chat by the representative about the wonderful range of products on offer. It's not all that dissimilar to live performer-to-peer buzz marketing when it comes down to it, because the representative will be paid in some way shape or form for their 'performance'. While many people may object to the somewhat forceful style of marketing employed by these companies, far less object on moral or ethical grounds. They know what they're getting in to.

So perhaps all that's left to be said about live peer-to-peer buzz marketing is that marketers should think very carefully about how they wish their brand and brand ambassadors to be perceived in the long term. Is it really necessary (or wise) to use secretive tactics and try to fool consumers in order to generate positive peer-to-peer buzz?

Conclusion

As the live buzz marketing industry continues to develop, and as brands start to recognize its power and enjoy its effectiveness and immediacy, credibility will only be established through accountability. This must include word of mouth reach numbers, recall rates, propensity to purchase rates and (where relevant) actual sales data. Regardless of which live buzz marketing technique is used, solid, reliable, fully accountable research and reporting will differentiate the good from the bad, and ensure that this powerful medium earns its place in the mainstream marketing landscape.

Takeaway points

- If your live buzz marketing message is loud, clear, appealing and preferably unique, its echo can continue for years.
- Secret live buzz marketing enables you to create the perfect brand ambassador, but its covert nature means the echo is limited and the public can feel duped if they find out they have been the focus of a marketing exercise.
- Disclosed live buzz marketing creates an echo that continues for a long time, reaches many people and generates extremely high recall levels.
- Overt live buzz marketing provides high-quality contact and lasting impact with the audience, but its cost per contact is generally more than that of other live buzz marketing techniques.
- Messages delivered by live buzz marketing are very difficult to ignore.
- In any live buzz marketing campaign, the creative idea must be appropriate to the target audience and environment in which the message is delivered.
- Live buzz marketing can generate large amounts of press coverage and PR.

3

Creating brand advocates

Sven Rusticus

CEO, Icemedia

Advocacy and growth

What's the one thing that drives business growth? The answer is brand advocacy. The likelihood that your clients, customers, or consumers will advocate your brand to their friends and acquaintances will be directly correlated to your business growth. This was the finding of an influential study published in the *Harvard Business Review*.[1] Across more than a dozen industries, the research found that companies with high word of mouth advocacy rates (i.e. likelihood of brand recommendation) grow fast, while those that don't have high word of mouth advocacy rates stagnate or shrink. Critically, brand advocacy was found to be far more important than either brand image or brand satisfaction in predicting growth.

This isn't surprising when 91% of people would be likely to use a brand recommended by someone who has used it themselves.[2] Indeed, word of mouth - brand-talk between clients, customers, or consumers - has long been considered the Holy Grail of marketers: it is valued twice as much by people as an information source compared to advertising[3], and 92% of people cite it as their preferred source of information.[4] What's more, word of mouth is increasing in importance; people value it today 50% more than they did in the 1970s.[5] Management consultants McKinsey estimate that over two-thirds of the US economy is influenced by word of mouth.[6]

The power of word of mouth advocacy derives partly from its credibility - while only 14% of people believe what they see, read, or hear in advertisements, 90% believe endorsements from their friends and acquaintances, primarily because they are not seen as having any vested interests.[7] But the power of word of mouth also derives from its exponential reach, if one person was to recommend a product to just two people who each then recommend it to just two others, then a small recommendation chain of just 30 links could reach every single person in combined populations of the US and the EU (2^{30}). This explains how new bands, films, products and services can bubble up from nowhere to become the 'Next Big Thing' driven only by exponential and credible word of mouth.

How to drive brand advocacy

Because word of mouth advocacy is of critical importance in driving business success, a good case can be made for arguing that everything coming out of a marketing department should be focused on increasing advocacy rates. Whether it's advertising campaigns, promotional offers, PR, or any other marketing initiatives, the overall strategic goal of marketing should be to optimize the likelihood of people recommending your brand to each other.

But how do you create campaigns that generate word of mouth advocacy? The majority of word of mouth initiatives involve targeting a special segment of a target market: opinion leaders (variously called 'alphas', 'hubs', 'sneezers', or other proprietary jargon). The logic is that if you can create advocacy among the 10% of your target market who tell the other 90% what to think, say and buy, then you can kick-start an epidemic of demand. If you can convert opinion leaders to your cause, as opposed to the cause of any of the other brands chasing them, then this is no doubt an excellent strategy.

But there's another group of influencers that brands can harness to drive demand through word of mouth: brand advocates. Brand advocates are regular clients, customers, or consumers with no special opinion-leading powers, but who like your brand so much they recommend it to others. Brand advocates, sometimes called brand evangelists, are different from opinion leaders because they derive their influence not from the fact that they frequently offer or are consulted for category-related advice (the definition of an opinion leader), but because, as highly satisfied adopters, they are enthusiastic endorsers of your brand.

One way of looking at brand advocates, is through the brand advocacy pyramid. Any existing client, customer, or consumer base can be segmented by degree of satisfaction with the brand experience. At the bottom of the pyramid, where most people in your buyer base will be

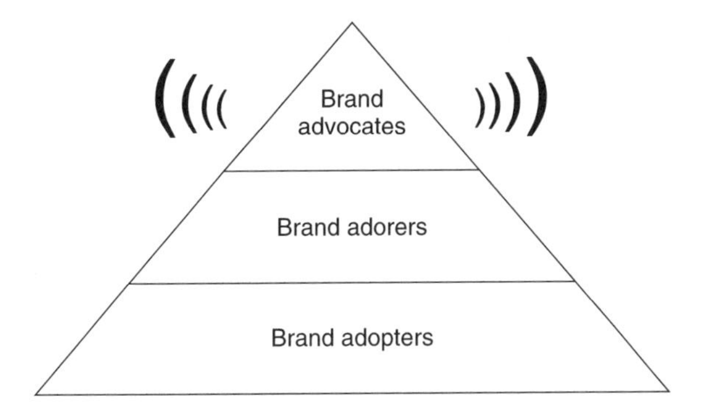

Figure 3.1 Brand advocacy pyramid

found, are your basic brand adopters – people satisfied enough with the brand experience to have become regular users. Some of your current buyers, however, will be particularly satisfied with your brand and display a high degree of affinity and loyalty to it. These are your brand adorers who lie in the middle of the pyramid. Then, at the apex of the pyramid lie your brand advocates: clients, customers, or consumers so satisfied with the brand experience that they are prepared to proselytize about it.

Very few people recruited into any brand franchise will be brand advocates; most will be brand adopters, a few will be brand adorers and fewer still, brand advocates. However, since it is brand advocacy that drives brand growth, it is in every marketer's interest to boost the proportion of advocates in the brand pyramid whilst encouraging existing advocates to advocate more often to more people.[8]

A number of tools and techniques exist to transform brand adorers into active advocates and help existing advocates evangelize about your brand. The most basic is perhaps the classic referral programme (variously known as introduce-a-friend, member-get-member, or customer-get-customer schemes) which rewards brand advocates for their advocacy, but should also reward the people they advocate to, if those people become adopters. For example, eBay, the online personal marketplace, invites members to introduce friends to the service, in return for which both parties receive a US$5 voucher.[9]

The gold standard in referral programmes is perhaps MCI's 1991 'friends and family' scheme. The US telecom company offered its customers a 20% discount on calls to friends and family whom they referred to MCI. If someone who had been referred to MCI by a current customer became an MCI customer themselves, they too would be invited to refer their friends and family and enjoy a 20% discount on calls. This friends and family initiative kick-started a cascading chain of referrals, which generated 10 million new customers for MCI in less than two years.

Although referral programmes are most often used as a cost-effective customer acquisition tool by banks, service providers and clubs, they have also been used to harness product recommendations. For example, in 1999, Unilever ran a brand advocacy referral programme to recruit new users of its Dove soap brand. Called 'Share a Secret', the campaign involved transforming brand adorers into brand advocates and using that advocacy to recruit new adopters into the brand franchise. To do this Unilever invited Dove users to send in proof of purchase, with the contact details of a friend they thought might like to receive a gift certificate for a Dove Gift Pack of two free bars of Dove soap, or US$2.29 off the purchase of Dove Ultra Moisturizing Body Wash (redeemable in store). When sending in their friends' contact details, Dove users were encouraged to include a personal message to their friend, which would be included with the gift certificate when it was sent out. Finally, as a thank you for participating, the Dove user was also sent a gift certificate for the gift pack. The campaign was a success for all concerned: friends received unexpected free gifts with a personal message, Dove users were rewarded for advocating Dove, and Dove boosted product trials, generated a database of names and addresses, and most significantly, boosted market share by 10%.[10]

Another example illustrating how it's possible to turn brand adopters and brand adorers into active brand advocates is the 'Share the Love' initiative from online retailer Amazon. When making a purchase online, Amazon customers could enter email addresses of friends who would be automatically emailed a 10% discount voucher for whatever it was they had just bought. If the emailed friend used the discount voucher within a week, the Amazon customer would be sent a credit to the value of the 10% discount redeemable on future Amazon purchases.[11]

More examples of how to create brand advocates can be found on the website of Ben McConnell and Jackie Huba, authors of *Creating Customer Evangelists: How Loyal Customers Become a Volunteer Sales Force,* where there is a free ebook of case studies called 'Testify' that can be downloaded (http://www.creatingcustomerevangelists.com).

Creating advocacy with Icecards®

Another simple solution to creating brand advocates and amplifying the advocacy of existing advocates is to enable clients, customers, or consumers to order free or low-cost brand merchandise, such as stickers and stationery to share with friends. At Icemedia, this idea has been developed into a turnkey brand advocacy programme that has been adopted by brands such as L'Oreal, O'Neill, adidas, Bacardi and MSN. The idea is simple: enable brand adorers and brand advocates to order free sets of branded contact cards, featuring brand artwork on one side and their own personal details on the other. These Icecards are personalized and ordered online via a simple customized plug-and-play module placed on the brand's website. They are printed using on-demand digital print technology, and then sent out by post to the brand adorer/advocate. Icecards have proved to be hugely popular in the teens to twenties market, with over 40 million cards ordered in the first 36 months since the idea was commercialized.

So what's so great about enabling brand adorers and brand advocates to order sets of free branded contact cards? Well, first the cards enable brands to identify who their brand adorers and brand advocates are: only people with an affinity to a brand will want to put its artwork on their contact cards. Knowing your brand advocates is important, because these are the people who drive your business growth. And because Icecards are ordered online, the ordering module automatically generates a quality CRM (customer relationship management) database of names and

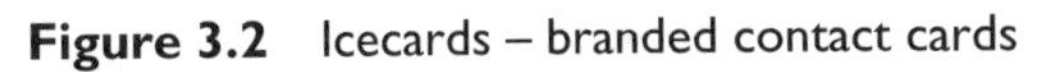
Figure 3.2 Icecards – branded contact cards

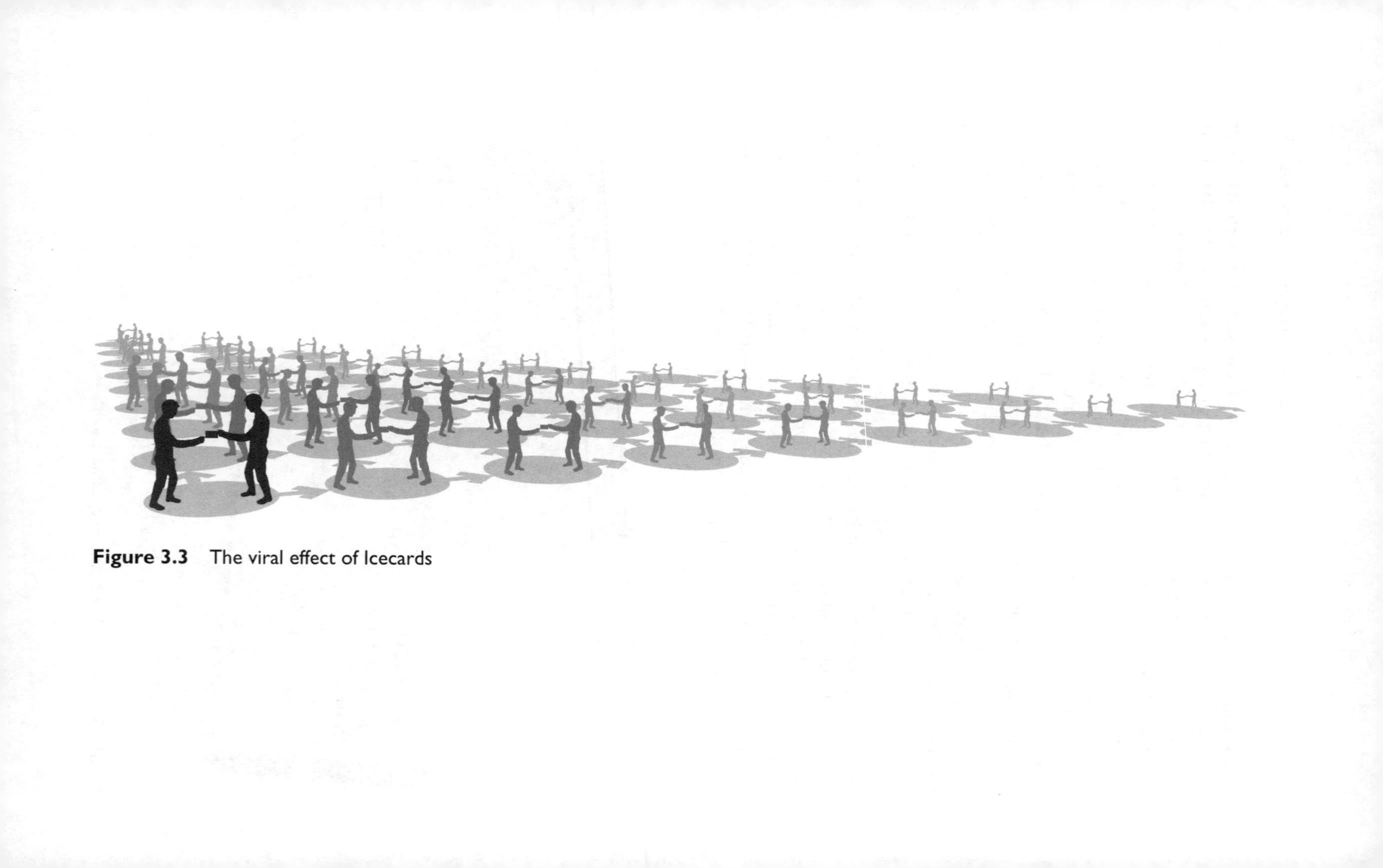

Figure 3.3 The viral effect of Icecards

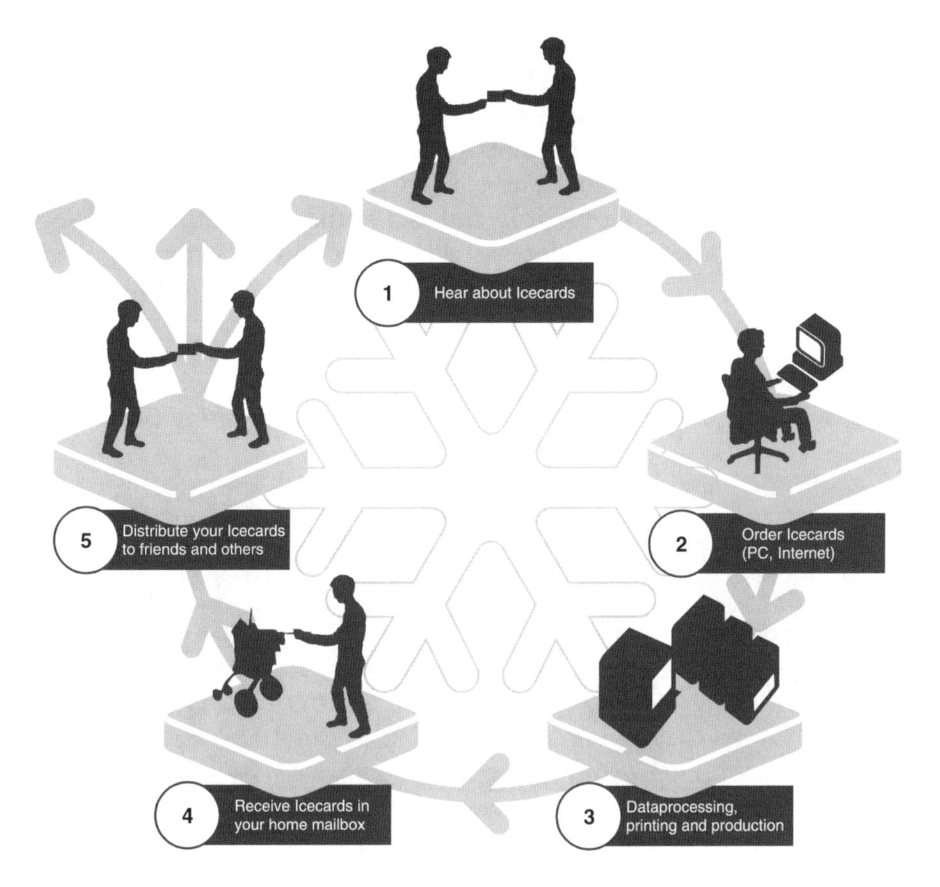

Figure 3.4 Icecards ordering process

contact details of your brand advocates, which can be used for follow-up marketing purposes.

Secondly, Icecards stimulate word of mouth. In research carried out by the Rotterdam Erasmus University on Icecards, it was found that 78% of all Icecards get handed out, and in 65% of cases the handing over of an Icecard triggers a brand conversation.[12] In other words, Icecards provide brand advocates with an excuse to advocate your brand. In addition, a weblink (e.g. get your own free O'Neill cards at www.oneilleurope.com) on every card enables card recipients to order their own cards, adding a viral dimension to the campaign. Thirdly, Icecards strengthen brand loyalty and affinity, making advocates feel part of the brand family. Fifty-nine per cent of people who have ordered Icecards say that the cards improved their perception of the brand. Other advantages of Icecard-type brand advocacy programmes are that they drive traffic to a brand's website, they are adaptable (e.g. can contain promotional codes, feature uploaded photos, be used as VIP member/privilege cards, etc.), they leverage existing brand artwork, they make word of mouth measurable, and most importantly brand advocates think Icecards are cool! They are a fun and handy way to give out contact details when socializing, far cooler than office business cards, and as a status symbol they show off to people that the cardholder has a privileged VIP relationship with the brand.

Case Study 1: Icecards – adidas International Marketing BV

Figure 3.5 Visuals of the adidas 'It's My Mania' Icecards

Campaign: 'It's My Mania'. Icecards campaign for promotion of the adidas Predator football boot.

Objective: To improve branding, develop a direct marketing channel and acquire data on its target market.

Activity: Icecards featured footballers from around the globe wearing the new Predator shoe. The campaign was initiated by sending out a handful of emails to football fans inviting them to sign up for the cards. The email spread exponentially as it was forwarded to friends who forwarded it to friends, reaching 180 countries around the world.

Results: One million Icecards were distributed to 50 000 adidas brand advocates in 46 days. The average age of the people ordering cards was 25 years old; 65% were male. Without any supporting promotional activity, the Icecards sold out in 180 countries around the world.

Case Study 2: Combining Icecards with sampling – Unilever

Figure 3.6 Screenshots of the Axe Snake Peel campaign

Campaign: Launch optimization for Axe Snake Peel, a new scrub shower gel for men.

Objective: To create awareness and buzz around the new shower gel.

Activity: On a special campaign website, accessible from www.axe.nl, consumers could watch Essential Scrub Work Out movies and order Axe Icecards. Two weeks after ordering, consumers received a set of cards together with a product sample. Each Icecard could be used as an in-store discount voucher (worth 1 Euro) for the new product.

Results: 45 000 sets of cards and samples were distributed in 45 days. Card orderers were between 15 and 35 years old, and primarily male.

Case Study 3: Icecards and sales – Cosmo Girl! Magazine

Campaign: A viral communication campaign to launch *Cosmo Girl!*, a new magazine for young women.

Objective: To generate buzz and compel word of mouth around the new magazine, increase brand awareness among the target group of 14- to 18-year-old females, and acquire CRM data.

Activity: The campaign was launched by an email offering Icecards to 14- to 18-year-old females. The link to order Icecards was only communicated on the cards and by using forward-to-a-friend functionality at the end of the online ordering process. This enabled Icemedia to

Figure 3.7 Icecard for *Cosmo Girl!* magazine

measure the spread of the viral effect and demographics of the target group that was reached. Six months after ordering their Icecards, the target group was sent an email questionnaire. Its goal was to measure the effects of branding and purchasing behaviour.

Results: 10 000 sets of cards were distributed in 22 days. The average age of brand advocates was 16 years old and 99.1% were female. Seventy per cent of the Icecards orders used the forward-to-a-friend function – on average, the offer was forwarded to 2.4 friends. The post-order questionnaire revealed the following results:

- 61% of girls who ordered *Cosmo Girl!* Icecards completed the survey six months later, indicating a high degree of loyalty and brand involvement
- 91% of respondents bought *Cosmo Girl!* magazine after receiving *Cosmo Girl!* Icecards
- 70% of respondents said they had learned about *Cosmo Girl!* magazine through Icecards
- 65% of respondents said they had visited the *Cosmo Girl!* website in the last month (this was six months after the campaign)

Conclusion

Getting your loyal clients, customers, or consumers to do your marketing for you by transforming them into active and vocal brand advocates is a powerful and cost-effective word of mouth marketing strategy. Brand advocacy programmes can harness the goodwill and energy that already exists in a market, leveraging the sizeable investment that many brands make in getting people to love them. By making brand users feel part of the brand family with personalized brand merchandise such as Icecards to share with friends, not only is the affective bond with the brand deepened, but also brand advocacy is stimulated and made easy to measure.

Takeaway points

- Helping brand advocates to evangelize about your brand is a marketing priority, because brand advocacy drives brand growth.
- Setting up a brand advocacy programme that enables brand fans to have their own personal branded contact cards is a cost-effective way of stimulating advocacy.
- Brands such as L'Oreal, O'Neill and adidas enable fans to order their own personal branded contact cards from their websites.
- Because they are ordered online, orders for personal branded contact cards generate a quality CRM database of brand advocates and drive further traffic to the brand's website.
- Personal branded contact cards are particularly popular with the teen to twenties age group.

Notes and references

1 Reichheld, F. (2003) 'The one number you need to grow', *Harvard Business Review*, 81 (Nov.-Dec.): 1-11.
2 Keller, E. (2005) 'The state of word of mouth, 2005: the consumer perspective', Conference paper presented at the Word of Mouth Marketing Association Summit, Chicago, 29-30 March.
3 Berry, J. (2005) 'Identifying, reaching, and motivating key influencers', Conference paper presented at the Word of Mouth Marketing Association Summit, Chicago, 29-30 March.
4 Keller, 'State of word of mouth, 2005'.
5 Berry, J. (2005) 'Identifying, reaching, and motivating key influencers'.

6 Dye, R. (2000) 'The buzz on buzz', *Harvard Business Review*, 78 (6): 139–146.
7 Grad Conn, managing director for Grey Direct in Toronto, presenting at iMedia Brand Summit, Florida, 9 February 2004, cited in Watters, L. (2004) 'CRM: Get with it or get left behind', *iMedia Connection* (10 February), archived at http://www.imediaconnection.com/content/2783.asp.
8 For a general discussion of brand advocacy or brand evangelism as it is sometimes known, see McConnell, B. and Huba, J. (2003) *Creating Customer Evangelists: How Loyal Customers Become a Volunteer Sales Force*. Chicago: Dearborn; and McConnell, B. and Huba, J. (2004) Testify e-book, http://www.creatingcustomerevangelists.com/testify/testify.pdf.
9 eBay's refer a friend scheme can be found at http://pages.ebay.com/referafriend/.
10 See *Promomagazine* article archived at http://promomagazine.com/mag/marketing_health_beauty_care/ for more details on this promotion.
11 Details on Amazon's 'Share the Love' referral programme archived at http://www.ennex.com/~Marshall/viral/ref/site/Amazon-ShareLove-order.asp.
12 For details on this study, please contact Icemedia via www.icemedia.nl.

4

Brewing buzz

Liam Mulhall

Founder, Brewtopia

So, how do you get consumers to demand a product that doesn't exist, from a company they've never heard of in the high-volume commodity beverage industry, without spending a dime on traditional marketing and without a production facility, staff, or any money?

Answer: start with a large dash of ignorance, forget industry best practice and do everything 'wrong'!

This may sound reckless but such an approach has enabled a boot-strapped, start-up beer company called Brewtopia to compete with the largest duopoly in Australia in just two years. At the time of writing, Brewtopia is licensing its product in the UK, New Zealand and Singapore, and is finalizing talks to export beer to the US. This is all despite the fact that it has never paid for advertising or any traditional marketing and has never employed more than four people.

Brewtopia's success is due to a highly unconventional approach to business that embraces out-of-the ordinary approaches in every aspect of its operation. This in turn has meant that it has been able to extract huge benefit from a connected marketing strategy that puts the customer centre stage as a vehicle to not only sell the beer by word of mouth but also to shape the company.

How it all began

Unlike the United States and Europe, Australia does not have many microbreweries. The marketplace is dominated by beer from two national companies which have around 94% of the market. So, there is no distribution chain or precedent for smaller brewers to place bottles in pubs, restaurants and retail outlets.

Brewtopia had to crack into a giant market with no apparent footholds or entry routes for new suppliers. But it couldn't compete with the millions traditional brewers spent on advertising. No one at the company had a marketing background and it couldn't afford an agency. Plus, it didn't have a product yet. It had a name – Blowfly Beer – but the beer itself was still in R&D. To get noticed, Brewtopia knew it had to create buzz around the product and do it in a way that flew in the face of how traditional beer companies operated.

Brewtopia's philosophy was: 'If we're going to compete with these guys, we need to re-write the rules of engagement.' This meant every time they had to make a decision they would ask themselves: 'What would our competition do?' and then do the opposite.

Inspiration came from two disparate fields: sports coaching and IT software.

Brewtopia's founder had read a story[1] about the soccer team PK-35 in Helsinki, which hadn't won a game for months and was languishing at the bottom of the minor league. It had a small base of hardcore fans. The coach decided to get the fans involved. Every couple of days he sent them an SMS asking them to vote on the players. The tactic was so popular that the team's wireless messaging list grew from 300 fans to more than 30 000. The team ended up in the top leagues, winning games and involving fans as owners and members in every aspect of the club.

Brewtopia's founder's career had been in IT sales, where he sold software and hardware but mostly vaporware – software concepts that didn't exist in reality but were in demand from customers; enough demand forced R&D to develop the software into a feasible, saleable product. Vaporware is effectively selling something you don't really have.

The idea was to cross-pollinate these two concepts and hopefully develop a model to crack open a vein of gold in the antiquated Australian beer market.

Step 1. Think upside down

Brewtopia would sell the concept of the beer before anything else and see if it could generate a demand for a custom-built boutique beer with a difference. If successful at achieving this, then it would be able to calculate the demand and demographics, and identify the channels required to get the beer to the drinkers. From there, volumes could be calculated and brew runs specific to demand would be made without wastage or unnecessary inventory.

This runs in reverse to popular manufacturing and retail theory, which is to build a brewery, brew a style of beer, get it into licensed venues, then advertise heavily to sell it.

Step 2. Sell the experience

People buy on emotion and justify with logic. All great salespeople understand this. A phenomenon that was occurring at the time Brewtopia was considering its means to market was *Big Brother* and the rest of the reality television craze. It was entertaining and put the customer in the driving and decision-making seat.

Most businesses don't entertain - they're too busy trying to hit next quarter's target. Most businesses don't empower their customers at all - most don't even talk to them after a sale. Brewtopia decided to empower its customers to make their own choices, build their own product and decide the outcome of the company - essentially making the customer the 'experience'.

Step 3. Think like a buyer, not a seller

People love to buy, but hate to be sold to. Brewtopia would refuse to blow its own trumpet, which meant never paying for advertising or doing any traditional marketing. People buy from people and while the brewery certainly understood word of mouth, connected or viral marketing were terms it had never heard and nor were they as yet actively accepted as a bona fide communication route.

Brewtopia's concept was to have friends of friends of friends being the only drinkers of the beer, and the only way to find out about the beer was from a trusted source.

Step 4. Give customers a vested interest

The next stage was to give customers a simple yet emotional reason to buy in initially, and to motivate them to let others know and pass on the message. Without Brewtopia knowing it, this was the key to the success of what became a viral campaign and in turn the key to the company's success.

Brewtopia gave all those who signed up the opportunity to drink free beer at the launch, and offered them the potential to own part of the company that made the beer. While most breweries will only let you buy their shares on the Australian Stock Exchange, Brewtopia would give theirs away.

Actioning steps 1–4: the pre-launch 13-week viral campaign

Brewtopia decided to get potential customers involved from day one and launched a 13-week viral campaign on 5 August 2002.

First it sent an email to close friends and family (about 140 people in all) asking them to register as members on a new beer website. In exchange they would get the unusual opportunity to *vote* on every aspect of the beer development and its marketing. Plus, if the company floated, they would get a single share of stock for each vote they cast, carton they purchased and registering friend they referred. Although the share might not be worth anything in terms of cash-value for a long time, it might help cement a fan's emotional tie to the company.

Length of campaign

Brewtopia chose a 13-week campaign because it felt that was long enough to get people really involved, but not so long that they would lose interest, get bored and drop out. The aim was to build to a crescendo of excitement in time for the launch.

The tough part was coming up with 13 different things that consumers might like to vote on. The votes included:

- name of the beer
- style and taste of the beer
- logo
- type of bottle and packaging
- location of the launch party
- merchandise
- pricing
- where it would be sold.

Email auto responders

People feel more involved when they get a fast reaction from a company, so Brewtopia set up two auto responders: one sent to new members and one sent whenever anyone voted. Emotionally, new members are the most likely to rush out and tell more people about the fun list they just joined. So the new member auto responder featured a PS offer telling them they could get an invite to drink free beer at the launch party in Sydney by inviting at least four mates to join the voting list.

Votes feature only two choices

It's critical that you have only two choices. More than two and the majority won't rule. You want people to think that at least half the time their choice is getting implemented.

Enact decisions immediately

If you take too long people think it's not worth it. They assume 'They're not listening to me'. Brewtopia announced what won and showed a picture of it.

On two occasions the company discovered they couldn't honour the voter's decisions due to unforeseen expenses. It quickly emailed out a personal apology and honest explanation. Voters liked Brewtopia's hand-on-its-heart approach because it gave them faith that their votes were taken seriously. Customers also gained an appreciation of the trials and tribulations of starting from scratch and making a mark.

Post-launch actions: sales growth campaigns

At launch all members were invited to the party in Sydney and to purchase the beer directly from Brewtopia via mail order. The goal was to build enough sales (and enthusiasm) from direct response customers to bankroll the company's expansion into pub and retail distribution.

It focused on four tactics to keep consumer excitement strong after the beer launched in January 2003:

Reach and focus on influencers

Brewtopia started with the people who already knew its product and cheeky nature: its customers. All were asked why they signed up and why they bought the product. It still gets over 80% response rates with customers appreciative that the company is interested in them and what they think. Brewtopia then uses that feedback to ask them to help further.

Employ third parties to accelerate the buzz

Brewtopia was proactive and opinionated in its bid to generate buzz. It got articles written and tried to position itself as an expert in the field of beer. It then realized that its field wasn't beer but rather marketing.

It responded to any relevant article in the press with genuine feedback and backed up what it said with results and tests. This resulted in an article being published about Brewtopia, which led to coverage in *Marketing Sherpa, Australian Financial Review, Melbourne Age, Sydney Morning Herald, Fastcompany Magazine, B&T Marketing, Anthill Magazine, Sunday Life* and *Adnews Today*, and even a national television spot. Every time it gets coverage, sales rise sharply, especially from new customers.

Be outrageous and do the unexpected

Congruent conflict – making people think twice about accepted expectations – gets people talking. (The name of the beer for a start – who would call a beer Blowfly?)

Keeping in line with the 'Do it wrong' credo, Brewtopia looked at its beer deliveries to licensed premises. The system used by other breweries sees the beer delivered first thing in the morning when no one is about. Brewtopia decided from the word go that its deliveries would be done at peak times (lunch and after work) to attract maximum attention, and they would be carried out in a Blowfly beer truck done up as an ambulance with sirens and flashing lights. Two lab-coated delivery boys carry in a stretcher full of beer through the front door and lob it over the bar for all to see.

Use technology and non-industry channels to spread the word

Breweries buy ads on television and newspapers. Period. They sponsor sporting events. Period. That leaves everything else to the brave.

Brewtopia realized quickly that the big companies had to go through multiple levels of sign-off before they even got close to doing something different with their marketing strategy. They were still stuck in the 1950s with just TV and radio, which no longer offered a captive audience.

Brewtopia, however, would have to use everything but TV and radio – so it exploited the potential of the Internet, weblogs, pay TV, SMS and instant messaging. It studied and evaluated every new medium used by its target market.

Continued incentives to spread the word to hot leads

The company had to continue with word of mouth marketing, since it wasn't merely a campaign or project for Brewtopia but its entire marketing strategy. It had to find and harvest new propositions and techniques within this connected approach, and always had in mind the premise: What's In It For Me (WIIFM).

Brewtopia's initial campaign was built on brand impact and maximum coverage at all costs. It got the word out, but that alone rarely guarantees sales. It had to design a programme that brought in business and increased turnover, but in a way that didn't appear contrived or oversold. The idea was to underpromise and overdeliver.

If expectations are low, you can only come in above that. It is almost a reverse sell. For instance, the first page of the Blowfly website still says, 'If you don't want to wait two days for your beer, or you are a dickhead, we don't want you to buy from us. You're a pain in the bum and we don't want customers like you.'

People don't expect this type of talk in a sales brochure. No one wants to be a dickhead, so they buy the beer and they all send it to their friends because it's out-of-the-ordinary.

Ask for negative feedback

Forget testimonials and good feedback, Brewtopia only ever sends bad feedback to sales and administration staff. This is the only way to improve in its book. If someone has gone out of their way to complain, the company sees it as the perfect opportunity to win them back, get referral business and improve how the business works. Good feedback rarely gets any of that. Brewtopia put an email link (dirtyemail@blowfly.com.au) on the first page of the site with the line 'Tell us if you're pissed off, what we stuffed up, didn't do, or buggered up'.

Give recognition

Giving recognition to customers is one of the most powerful tools Brewtopia has found. Right back to Maslow's Hierarchy of Needs,[2] people need to be needed, to be part of a social community and to be recognized. And it doesn't matter where that recognition comes from. Take the unknown backroom boy thanked by the lead star for his part in helping secure the best actor award. Chances are no one in the room knew the person being thanked, what they looked like, or what they really did. They'll be forgotten in a gigasecond, but you can bet your last dollar that the backroom boy was beaming from ear to ear and that one mention will be recorded in memory to the grave.

Brewtopia devised a member of the month scheme. Members had to do something to earn the accolade so they didn't feel they were simply picked out of a barrel. For example, they could introduce Brewtopia to a retail outlet, be the first member in a new country, or get the company some press coverage. Following the award they would typically tell everyone they knew to head for the Blowfly website to see their name up in lights. Such recognition then secures that customer's loyalty for good.

A reason to keep loyal to the product/company

Too often a viral campaign is just that: a one-off campaign that does what it sets out to do and afterwards the company retreats back to its usual routine. This is generally contrary to the viral campaign's core intent.

Brewtopia was at times guilty of this and it had to keep looking to past victories for new inspiration. All of its developments, innovation and creativity have roots in customers' suggestions or feedback. Nothing it does has been contrived or put out in the marketplace to see merely how it goes.

And when customers see that they are not only being listened to but also their suggestions are actually helping to design your business, they feel empowered and it inspires loyalty. Business is like a marriage or partnership: a two-way street built on trust and good outcomes that satisfy both parties.

Lessons learned

Brewtopia also learned some lessons worth noting and that, while not fitting into a specific word of mouth strategy, were necessary to the overall success of the business. So many times organizations get locked into a rigid business plan and miss opportunities, the best of which come from the customer.

Throw out the business plan, it's limiting

Brewtopia uses it as a rough guide only, specifically when it needs to look at why it went into business. Everything else in such a plan is outdated as soon as the first sale comes in. A business plan, if followed day-to-day, is predicating what should happen and doesn't answer the right questions when things don't happen.

Customers determine the business and everything else follows. The biggest and most profitable part of Brewtopia's business today wasn't even mentioned in the business plan. The company founder had to fight to get the Board to change direction when he decided they would do what no other brewer in the country would do: bastardize their brand and recipes to customize, personalize and privatize beer, and contract to other players.

No sales targets. Again, it's limiting

Brewtopia noticed how every year it would just get its sales target, or that sales would come in when overachievement bonuses or accelerators were up for grabs. Yet year-on-year targets would increase by 10–20 per cent and sales guys would still just come in over the line.

Sales targets are predicating again. Brewtopia has a minimum it needs to sell every month to break even and everyone knows that figure. Anything over that and everyone gets to share in the profits. This prevents sandbagging and incentivizes all staff within the organization.

Be humble

Most companies have some systems that don't work effectively – they keep customers on hold too long, they take too long to deliver, products never work like in the demo . . . Brewtopia realizes it's no better. But its idea is to realize this, embrace it, and try to foul up a little bit less than the competition in every area.

In large numbers, people are predictable

Brewtopia understands that people have thought and acted in the same way for 5000 years. The same things trigger greed, hate, love and other emotions. The trick is to understand the process the brain goes through in making a buying decision, and identify what elements must be present before it insists the hand pulls out a wallet. And then you must understand what turns people into walking, talking sales machines for your product.

Delayed gratification, patience

Brewtopia's theory was that if it could sell one carton, it could sell 10,000 – just through multiplication. Having sufficient patience is the most difficult thing about starting a business. If you expect to reach critical mass quickly, chances are you won't. It is important to keep doing the things that are working and eliminate those that aren't – whether it be a pure viral campaign or, as in Brewtopia's case, its whole marketing strategy. Brewtopia's founder had an accounting background, knew the power of compound interest, and that, at some point, the critical mass and multiplication would be exponential.

Conclusion

Brewtopia and its product, Blowfly Beer, represent one of the best demonstrations of the power of connected marketing. Metrics are still developing to measure the effectiveness of such strategies but, arguably, Brewtopia doesn't need them to demonstrate what viral and word of mouth campaigns can do. Bearing in mind that it has never paid for advertising or any traditional marketing, the facts below make inspiring reading.

In two years since launching, Brewtopia:

- has *never* paid for advertising or any traditional marketing – ever
- doesn't sell its beer in retail outlets, other than a handful of bars
- doesn't own a brewery
- has never had more than four staff
- competes with the largest duopoly in Australia, which has 94 per cent of the market
- has sold to 20 000 individuals
- has members in more than 28 countries
- turns over AU$50 000 to AU$100 000 every month in beer sales alone
- conducts 90 per cent of sales direct over the Web
- has never made a proactive sales call, let alone a cold call

- 54 out of 100 referrals register as members (up from 32 out of 100 in December 2003)
- 1 in 2.25 members buys the beer at least once
- has sold beer to Fortune 500 companies and branded beer for companies such as Cisco Systems, Telstra, Foxtel, television channels 7 and 9, Commonwealth Bank of Australia, Electrolux, Yahoo, and the Australian Parliament
- is currently licensing its product in the UK, New Zealand and Singapore
- is finalizing talks to export the beer to the US.

What Brewtopia learned

The people least likely to buy via mail order were that initial group of 140 close friends who got the first email. They all registered but 20% of them never bought any beer.

When it came to voting, the results were reversed. Initially, around 50% of the list voted each week. In the third week Brewtopia had just over 1000 members and ended up getting about 500 votes back with very quick turn-around.

But, as the list grew to friends of friends of friends of friends, the results decreased accordingly. When it got over 5000 people, voting dropped to about 33% and when it hit over 10 000 it dropped to 27%.

The ambulance-style delivery has resulted in orders. When making deliveries, the team has found notes tucked into their windshield from other pubs owners who want to chat with them.

The ultimate lesson: following other people's best practices is the surest route to mediocrity. Rewrite the rules of engagement and build the company around the customer, not the product.

Takeaway points

- It is possible to sell a concept before you have a product.
- Empowered customers can sell your product and shape your business. Give them a vested interest.
- Focus your marketing strategy on the influencers first.
- Being up front with customers when you have a problem can earn you points.
- Businesses can entertain.
- Negative feedback from customers is more useful than praise.
- Someone else's best practice isn't necessarily yours.

Notes and references

1 http://www.fastcompany.com/magazine/57/finland.html.
2 Maslow, Abraham (1970) *Motivation and Personality*, 2nd edn. New York: Harper & Row.

5

Buzzworthy PR

Graham Goodkind

Founder and Chairman, Frank PR

How many times have you heard the phrase 'and then we can go and get some good PR on that'? It has normally been spoken by someone who believes that the 30-second television spot is the only thing there is when it comes to consumer communication, and that their most humdrum, run-of-the-mill, seen-it-a-hundred-times-before ad campaign can make it onto the front page of the tabloids. Unfortunately, these people still remain, probably in the majority, in the marketing industry.

Alas, there would be no story behind the ad. Nothing to differentiate it from the previous campaign or any others in its category for the last 50 years, no one remotely interesting in the ad and nothing ironic, witty, funny, or controversial about it.

Public relations is the art of getting publicity for free. The PR industry has never enjoyed the budgets or creative kudos that advertising has, but increasingly it is seen as a far more powerful tool when it comes to building brand. This notion was at the centre of the book *The Fall of Advertising and Rise of PR*,[1] co-authored by Al Ries. When interviewed in *B2B* magazine,[2] Ries says that advertising is good for maintaining and reinforcing brand but that PR should replace it 'as the major communications vehicle for launching or repositioning a brand'.

The role that PR has to play in generating buzz is vastly under-rated. In this chapter, we look at the relationship between PR and buzz marketing, and illustrate the power of their combined effect in connecting with consumers today.

PR: that was then, this is now

In 1992, the father of PR, Edward Bernays, defined a PR consultant as 'an applied social scientist who advises a client on the social attitudes and actions he or she must take in order to appeal to the public on which it is dependent.'

In the same talk entitled 'The Future of Public Relations',[3] Bernays, the nephew of Sigmund Freud, aligned public relations to Jefferson's principle that 'in a truly democratic society, everything depends upon the consent of the public'. PR, said Bernays, 'embraces the engineering of consent'.

This may seem a lofty interpretation of what is often referred to today as 'spin', but it helps to underline the link between PR and connected marketing. Increasingly, activities such as viral, buzz and word of mouth marketing are what help to engineer that consent.

For a more modern definition of PR, the Institute of Public Relations defines it as follows:

> PR practice is the discipline which looks after reputation, with the aim of earning understanding and support and influencing opinion and behaviour. It is the planned and sustained effort to establish and maintain goodwill and mutual understanding between an organization and its public.

It's a perfectly adequate description, but to understand how PR works and the role that buzz or word of mouth marketing can play in it, an examination of the dynamics of crisis PR is a good starting point.

The media has the power to turn something relatively small into a monster. Many companies have all sorts of bad things happen to them. Scarily, a lot more than you think: consumers getting ill, untoward things found in the product, electrical goods causing house fires, kids in danger from badly designed appliances, the list is a long one. However, we only get to hear about a fraction of these things. Why?

Well, if the media doesn't pick up on the scent of the story it just remains an issue that the company needs to deal with. In the vast majority of potential crises, a call is taken directly by the company, either via its customer services department, a friendly tip-off, or from the emergency services dealing with an incident. An unhappy, hurt, or injured customer is then very well looked after. If relevant, a tidy sum is paid to make things better, free (presumably non-faulty) products are showered on them and their family, special privileges are granted. And the public at large are none the wiser.

But, if the company doesn't handle the situation very well, which happens a lot less than you think, and the media get its teeth into the story, that issue becomes a fully-fledged crisis. With it comes the potential for panic, product withdrawal and severe financial damage, or even litigation and further bad news.

From an evolution of the crisis perspective, the media won't mean the national media in the first instance. Very often, if you look at how media coverage of such incidents develops, it's the local free sheet or sometimes the trade magazine that gets wind of the story first. The nugget of a story is picked up by the daily regional, the local radio station and then a local television news crew visit the scene. The regional desk of a national newspaper twigs this and the news editor in the big city wants to know more and splashes a big story. The other nationals pick up on their story, websites publish their versions, the story goes global. Within 48 hours the crisis has penetrated everyday conversation.

Looking at these dynamics in reverse makes for an interesting approach to PR. In a nutshell: what can you do as part of your marketing communications that will have at its core something that will spread through the media, get them excited and in turn have the power to break through into the subject matter of the daily chitchat in the pub or around the water-cooler? Doing this also forces you to go through some of the thought processes needed to create a positive buzz dynamically and proactively for a brand, product or service.

And this is why a PR take on creating buzz and connected marketing strategies in general is so important. A good PR mind can sniff out the ideas that the media will like and therefore write about and/or broadcast. Most importantly perhaps, someone looking at the world through PR-tinted spectacles should be in a better position to generate the ideas in the first place. He or she should know instinctively if that idea will grab the media's attention.

So given the fact that PR people are in a good position to generate ideas with the best buzz potential, why don't they?

Without doing the profession a disservice, it's perhaps because PR people have always thought a little one-dimensionally about life and the way they go about their work. The end product of a PR campaign, in other words the media coverage in itself, has been a barrier for most practitioners. PR people tend to talk of coverage as if it's the only result that matters. Nobody ever asks: 'What does the media coverage that we've generated actually do?'

The stories that get into the media should go further than just being read, seen or heard. They should be the sort of stories, headlines, pieces of information and sound bites that provide the content for conversations that readers have in the course of their everyday lives. Going back to the crisis analogy, it's similar to how people would talk about a product that was killing everyone that ate it, was really dangerous to get on your skin, or was hurting small children if they got too close to it.

This is where there has always been a missing link. PR generates something that gets written about in the papers, then what? An individual may casually read an article that is, in fact, a piece of great PR. They flick to the next page and read another article about, say, the political situation in Kazakhstan. Assuming that person is perhaps not Kazakhstani, so what? The missing link is when that article actually gets him or her talking.

When he or she gets to the office and makes a hot drink at the communal coffee station, it needs to be the subject that they'll bring up in conversation with a colleague. At a dinner party that evening, they should share their take on that story with fellow guests. Truly breakthrough PR thinking will come up with ideas that have inherent buzzworthiness.

The way that PR has traditionally been evaluated has gone some way to propagating the blinkered attitude towards merely getting something in the paper, as opposed to creating coverage that gets people talking about what they've read in the paper.

The prime method of measurement of the value of PR has been based on the equivalent advertising cost of that space, or its total reach. Occasionally, evaluation will look at what key messages have been communicated. But who has seen a technique proposed or developed yet which tries to measure how well that particular piece of coverage or campaign has prompted people to talk to others about it and therefore spread the message even further? This is what PR today should be all about.

The following case studies show that a really good idea, which is deliberately conceived to be media-friendly, can also be used to kick-start word of mouth, creating a buzz and connecting with consumers.

Case Study 1: New Scientist *magazine – 'Live forever with the* New Scientist'

Background: The magazine was being re-launched and the publishers wanted to do something that would get people talking about science and the *New Scientist* again. Outside of its core of loyal readers and subscribers, many people didn't fully appreciate the magazine and what it did. The re-launch was also hardly the stuff of which headlines are made: a change of typeface, some new editorial sections and thought-provoking columns contributed by professors.

Campaign: Considerable thought went into deciding what was science's lowest common denominator: what scientific concept would be mass market enough for as many people as possible to understand and talk about it freely? At the same time it had to be scientifically worthy of merit and had to be a good story for the press. The answer: in a bold front cover headline face, the magazine told every reader they could 'Win the chance to live forever!' It was, quite simply, the weirdest competition ever.

The competition worked as a token-collect across five weeks with the eventual winner, drawn at random from all the entries, 'winning' something quite unusual. Upon death, he or she would be immediately packed in a crate of ice and flown over to the Cryonics Institute in Michigan, USA. The winner would then be frozen in liquid nitrogen to prevent any further physical decay and maintained indefinitely in cryonic suspension at –196 degrees celsius. At such time in the future as medical science has developed enough, the winner would then be defrosted, revived and cured of what had caused their death in the first place.

Cryonics ticked the box of being scientific, albeit extremely fringe, and easy enough for everyone to understand, whether they were a scientist or not. Most importantly, the fact that the competition was so unique, unusual, and in a way a bit silly, meant that the media was interested – big time.

News of the *New Scientist* promotion was featured in all the UK broadsheets and tabloids on 19 September 2002 – from *The Times*'s 'Science prize to last an eternity' to the *Daily Mail*'s 'The Immortals' through to the *Sun*'s 'Life after death is magazine prize'. Television news also got the bug and ran in-depth stories, using footage of movies such as *Vanilla Sky* and Woody Allen's *Sleepers* to ask the question of whether life after death was indeed possible. The story was a major segment of all BBC TV News broadcasts the day it was announced.

A cryonics volunteer from the UK was also lined up to talk to the media. The volunteer was paying a life insurance policy that paid out, on her death, the US$28 000 to the Cryonics Institute to ensure her suspension costs were met. She provided interesting anecdotes for the broadcast news stories that were proactively set up. Her point of view was that even if there was a 0.1% chance of cryonics working, then that was 0.1% greater than the alternative. The *New Scientist* website got in on the act, running a countdown clock that predicted when you might keel over and die, and therefore need the prize.

The media coverage soon went global with publications as far afield as India, Australia and Chile all covering the story.

Perhaps the most powerful aspect of the whole campaign was how it got people talking and generated buzz on a number of different levels. Would you put your name down to be frozen? Would you really want to live forever? Would it be nice to come back in a world without your family and loved ones? Imagine what a freak you'd be considered when you were brought back from the dead? Can it actually work? And perhaps most importantly, what other stuff is in the *New Scientist*?

The issue of the magazine featuring the promotional concept was the best-selling issue to that date, and the publishers calculated that at least £2million worth of positive PR was generated for the title. And the winner still hasn't used his prize.

Case Study 2: Condomi – Size Him Up

Background: Condomi is a German company which, from 2002 to 2003, was trying to get established in the UK. It was way behind market leaders Durex and number two brand Mates, and didn't have the marketing bucks of Trojan, which was also trying to grab a slice of the market at the time. Prior to the start of the PR campaign, a third of condoms were bought by women, a fact that gave the PR company a directional steer.

Campaign: The campaign revolved around 'size', and central to it was an online tool that would 'guesstimate' the size of a person's manhood. In the case of a female, it might well be her prospective partner's, and in the case of a man, more often than not, it would be to see how accurate it was in predicting the size of his own.

A fun website called Size Him Up (www.frankpr.it/sizehimup) was built, which asked the visitor to input hand, feet and nose size. It then took a few seconds to calculate an answer and came back with a likely size. Alongside the guesstimate was a suggestion as to the most appropriate condom from the Condomi range.

Links were seeded to the campaign website on a range of online viral sources, which attracted significant initial site traffic. But the real breakthrough came when news of the online tool reached the media. Instead of merely publicizing the website on a press release, the tool was used to guesstimate the size of a number of celebrities and a fun league table was formed.

The pick-up was tremendous: each story about Size Him Up created a large influx of people to the site with obvious spikes in unique visitor numbers following every article that was published.

In turn, the new traffic brought its own viral effect with it as new visitors recommended the site to their contacts (a recommend-a-friend link was built into the site). The story was then adapted and it succeeded in penetrating what were traditionally out-of-bound areas for a condom brand. Most memorably, it was picked up by Atticus, the political diary section of *The Sunday Times*, which used the site to guesstimate the size of politicians' manhoods.

Within the first month, over one million individuals used the online tool, all of them interacting with and talking about the brand at a much deeper level than they had with any other condom product range.

Case Study 3: HP Sauce – 'Official sauce of Great Britain'

Background: The HP Sauce condiment brand, ultimately owned by Danone, wanted to get people talking about it. The brand was about to embark on a new ad campaign that set about positioning it as the 'Official sauce of Great Britain'. HP Sauce is one of the few iconic British brands that could pull off this invented status, and the creative material for the advertising was looking to bring that positioning to life in typical situations associated with everyday life.

Campaign: A PR campaign was formulated in order to stimulate the media coverage that would make people aware of the desired step-change in perception and get people talking about the brand in the language of today. True to its principles, the brand allowed the campaign to be a bit cheeky, slightly naughty and lots of fun.

When looking at things that were happening at or around the time of the mainstream campaign launch in February 2005, snooker cropped up on the radar screen. It was a sport that was typically British, everyone knew about it and it got a lot of television broadcast hours. The next trick was to think disconnectedly about the sport.

'Balls', was shouted out in frustration during one of the PR think tanks – and the nugget of an idea was born.

With a bit of persuasion, a deal was struck with the organizers of the Masters snooker tournament, taking place during relaunch, whereby HP Sauce became the official sponsor of the brown ball at the event. These days in sport usually everything has been sponsored. No one had ever sponsored the balls in snooker, let alone just one ball.

The second part of the deal was even trickier, but it was the trigger that brought the whole campaign to life. Jimmy White, one of snooker's most well-known faces, a people's favourite and – a bit like HP – an icon in his own right, joined in the fun. He agreed to change his name, officially and legally via deed poll, to Jimmy Brown to commemorate the HP sponsorship of the brown ball. He actually preferred the new moniker of James Brown, so the new strapline of 'the godfather of snooker' was fittingly used for the campaign in keeping with his soul music namesake!

At the official announcement of the Masters tournament, the 'double deal' of the HP Sauce sponsorship of the brown ball and Jimmy White's new persona was announced. The media went wild. The press release fell on the day when the Palestinians and Israelis sat down for the first peace talks in more than a decade, but all that was trumped by the colour brown.

Over the next few days radio talk shows, columnists and feature writers, even the cartoonists, all developed the story further, on and off the sports pages and shows. The media practically door-stepped the new Mr Brown as the tournament approached, and he, to his credit, enjoyed the fun and milked the story further.

There were lots of levels for buzz to occur on, because of the fun nature of the whole campaign. The UK Press headline writers had a field day, which helped the word of mouth factor. 'Jimmy's a sauce pot' exclaimed the *Daily Star*. 'Brown is the new White in saucy promotion' reported the *Guardian*. 'Change of name may be sauce of inspiration for White' said *The Times* pithily. 'White adds sauce to a spicy image' echoed the *Evening Standard*, with Metro getting attention with 'Jimmy drops White and pockets Brown.' The *Observer* provided a rather succinct analysis of the events of the week under the headline (borrowed from another coloured brand) of 'The future's Brown'.

There was also a more serious side to the whole initiative that played out in the media. Snooker as a sport has been struggling over the past few years as the tobacco companies, the traditional sponsors of the tournaments, had been forced to withdraw their involvement. HP's involvement, in its own unique way, signalled a way back from the brink for the game. Maybe other brands would follow suit? The equally legendary UK snooker player Steve Davis in particular went on the record saying this deal was a great fillip for the game. It was more words for mouths to talk about.

Stage two of the campaign was to stir up a bit of a hornets' nest with the BBC, the broadcaster of the Masters tournament. All the producers and commentators working on the broadcast of the tournament were sent a copy of the signed and sealed deed poll certificate, along with a personal letter from Jimmy Brown asking them to please refer to him by his new name in all coverage of the tournament.

This letter was then 'mysteriously leaked' to the wider media whose next step was to call the BBC to see whether it was to be Brown or White in their coverage. The BBC's response was, on the back foot, that they weren't really sure what they were going to do, but they weren't happy with giving exposure to commercial sponsors such as HP. 'Browned off' cried the *Sun* newspaper the next day, with Jimmy claiming that if they didn't call him by his new HP-inspired name then he wouldn't do any pre- or post-match interviews, and wouldn't come to the table if his former name was used. Cue more radio show debate, news stories, columnist comment . . . and buzz.

The campaign achieved millions of pounds worth of free publicity for HP Sauce, and it all came out as the new TV advertising campaign was airing, working alongside it to communicate the new positioning. A one-off piece of market research showed that 49% of all adults were aware of the story, and of those 49%, 47% had discussed the whole thing in a conversation with a friend, colleague, or member of their family. A concrete measurement of how this story spread through word of mouth.

The campaign also proved a success for Jimmy in terms of image and performance. A 66/1 outsider at the start of the Masters tournament, he made it all the way to the semi-final. In the final frame that sealed his fate, he missed a fairly straightforward pot and his opponent went on to clinch victory. How ironic that the ball he missed was the brown.

Case Study 4: Slendertone – 'Does my bum look big on this?'

Background: Slendertone was launching its latest innovation, the Bottom and Thigh Toner. Using technology that had driven sales of tens of thousands of Abdominal Training Systems, the Toner was a natural product extension.

But the problem with Slendertone had always been two-fold: credibility and public apathy. The most common objection that Slendertone got from customers was 'too good to be true'. Perhaps because the Abdominal Training System had been positioned for years as a magical solution, though the technology behind it works. This lack of belief was coupled with a feeling that people had heard it all before, with the brand having been around for over 25 years.

The new product needed someone who could vouch for it and who could be believed. Getting a buzz about that belief was everything. Slendertone had tried unsuccessfully to use celebrities to front their marketing campaigns in the past. This wasn't necessarily a problem, however, because celebrities can taint a product with a lack of credibility since people know that they are being paid for what they say. More effective is to front a campaign with someone who is persuasive and believable, and who has an ability to get people talking about the brand. This person could then become a celebrity.

Campaign: Anita Hart, 'the world's leading bum double', was signed up to endorse and be the face of the new Bottom and Thigh Toner. Her list of credits included being body (and stunt) double for Cindy Crawford, Liz Hurley and Pamela Anderson. More often than not she was either their bottom or thighs. Indeed, for Hurley she had been both in one of the Austin Powers movies. She wasn't really a bottom double but a stunt-woman, but as a piece of spin that was the *pièce de résistance*.

Anita was flown over to the UK to shoot a one-off ad. A mega poster site had been booked next to the Dominion Theatre at the junction of Oxford Street and Tottenham Court Road (this location reportedly has the highest amount of footfall anywhere in London). The aim was to create the world's largest poster of a bottom. Anita's bottom was body painted with an evening dress, and the tagline used on the poster was 'Does my bum look big on this?'

The next step was to make Anita a star. To coincide with the mega billboard being erected, the *Sunday Times Style Magazine* did a cover story on 24 June 2001 about Anita and her career as bottom double to the stars. 'How does she keep her bum in such good shape?' was the question they just had to ask, and the readers would want to know (and in turn tell their friends). The answer was, of course, by using the new Slendertone Bottom and Thigh Toner. The product was given instant credibility.

It didn't stop there: the unveiling of the poster site became national news headlines and Anita came back to the UK to spend time on chat shows, revealing her rear and demonstrating the product. A celebrity had been created and in the end media requests had to be fended off. Every time she appeared in a newspaper or magazine, she was accompanied by the giant poster, along with her advice on keeping the perfect bum in shape using Slendertone.

Through the *Sunday Times Style Magazine* feature, the US media picked up on the fact that this previously unknown person had suddenly become flavour of the month in the UK. Interview and appearance requests from her native country then started to come in.

Slendertone still hadn't launched its product range in the US and were awaiting FDA approval. The US media were told to hang fire and as soon as the approval had come through, Anita was unleashed on media channels across the US – media channels that previously hadn't been the slightest bit interested in her 12-year career. She was billed as 'the butt that is sending the Brits crazy'. Again, the billboard ad visual was used and the product rationale communicated with every appearance. Slendertone USA's sales activities were off to a flying start, and in the UK the new product started rivalling sales of the Abdominal Training System in a very short space of time.

Case Study 5: EMAP – 'Get your kit off'

Background: EMAP's *Zoo Weekly* lads' magazine was locking horns with IPC's *Nuts* in a fight for young male readers. Getting young men talking about your brand is not the easiest task. They are considered more marketing-savvy and marketing-cynical than the average mainstream audience. They are also difficult to engage and interact with. EMAP wanted a PR campaign to generate ideas that stirred up interest outside of the magazine's weekly content. If you looked at the two magazines as an outsider it was sometimes difficult to tell the difference. The news and features, typeface and layout were so similar. Brand saliency was going to be a clincher for a reader.

Campaign: Getting into the hearts and minds of potential *Zoo* readers meant getting into their everyday conversations too. What would they talk about over a pint? The answer was to connect with the audience by getting behind, quite literally, the most outrageous soccer goal celebration ever.

In January 2004, Roman Abramovich's high-flying Chelsea Football Club was drawn away to play the minnows of Scarborough Town in the third round of the FA Cup, traditionally a stage of the competition that throws up David versus Goliath tussles. As it turned out, Chelsea against Scarborough proved to be a timely draw as far as the battle of the lads' magazines went.

Each Scarborough player was asked to have letters painted on their bottoms that would spell out Z-O-O W-E-E-K-L-Y in a line (the remaining outfield player would be the space between Zoo and Weekly). The match was to be covered live on TV (and transmitted around the world) and if Scarborough managed to score, they were to run towards the cameras, turn around, bend over and do the ultimate moonie, revealing the name of the magazine to millions watching.

The hope was that this unique deal was going to cause a furore at the Football Association (FA) headquarters in Soho Square, London, and be the trigger for the level of publicity that was needed to get into the lads' consciousness and conversations. And that's exactly what happened.

The legendary football agent, Eric Hall, was chosen to be the man to broker the deal between *Zoo* and Scarborough. He would add credibility, deflecting any notion that it was nothing more than a blatant PR stunt. In effect, the magazine was prepared to pay reasonably big money to the players' pool at Scarborough. Having said that, it probably amounted to no more than the daily earnings of a medium-ranking Chelsea player.

After the announcement of the deal and the planned goal celebration 'brought to you by *Zoo Weekly*', the FA came down on Scarborough like a ton of bricks, as expected. However, it turned out that the provoking of the FA to react strongly wasn't required; the next day plenty of articles appeared in the UK national press about this daring goal celebration plan. Special behind-the-scenes footage for Sky Sports was also arranged in which the players, instead of practicing free kicks, corners and set pieces, were getting the sequencing of their bum cheeks right so that they spelt the name correctly in the event they scored a goal.

The FA, direct and via the media, warned Scarborough that if it went ahead with the celebration, then the referee was entitled to book or even send off the offending players, which would have left them with just the goalie on the pitch. This provoked more column inches and was perfect fodder for all those football chat programmes and phone-ins on radio.

Match day arrived and the media was still debating the 'Will they, won't they?' question. Scarborough had been briefed to be completely tight-lipped on the subject and not to give anything away. *Zoo* spokes-people were equally tight-lipped, although a little more confrontational. The fact that it was only a bit of fun, as well as representing quite a lot of money for these part-timers, got the fans on the campaign's side.

The match commentators were even debating the scene should the underdogs stick one in the back of the net. Alas, the red-hot favourites triumphed and Scarborough didn't manage to score a goal. But the *Zoo* activity filtered down to the lads talking about soccer in pubs and bars around the UK, exactly where *Zoo* wanted to be, paving the way for the new publication to achieve a circulation of over 200 000 within six months, well ahead of initial targets.

Case Study 6: Mattel – IQ test for babies

Background: In October 2004, Mattel's Fisher–Price wanted to generate buzz for its Laugh and Learn range of children's toys.

Kids are massive conversation currency when parents get together. Very often the other parents that you meet at parties and kids' related activities are not 'real' friends but friends you've acquired as a result of your children. You have the relationship through the children and they are your connection point with other mums and dads in conversations.

Campaign: The idea was to get people talking about the new range of Fisher–Price toys by encouraging parents to talk about and compare notes about their kids. The baby IQ test concept was born.

Its official name was The Baby Development Test. The media was likely to dub it an IQ test, but Mattel was conscious not to call it that officially as that could appear to be irresponsible.

An expert child doctor was recruited to develop, in conjunction with Fisher–Price, a 10-point test that a toddler could take. The result wasn't prescriptive, but from a media perspective – and the way the story was shaped – inevitably many people leapt to the conclusion that here was a test for your child which would be able to predict whether he or she was going to be a genius. Research[4] unearthed when developing the concept had revealed that around 75% of parents desire some kind of reassurance that their child is reaching developmental milestones at the expected time.

News of this new test was released and spokespeople from Fisher–Price were on hand for what became a massive news story. The UK's *Daily Mail* newspaper questioned its readers with 'So you think your baby's brainy? Take this test', and reprinted a selection of the activities that had been devised. The *Independent* was more matter-of-fact with 'Baby quiz aims to reassure parents about child's aptitude'. All editions of *Metro* were reporting that 'Baby IQ test is just child's play', whereas the *Sun* challenged its readers with 'Is your baby a genius?'

This tranche of coverage created more widespread buzz, and the story reached the home of Fisher–Price and Mattel in the US and became headline news there, too. Groups of children were brought together for other sections of the media to create more editorial. The UK's *Richard and Judy* TV show even went as far as building a mini-Mastermind set in its studio and having children sit the test on live TV.

The campaign took buzz marketing to another level. Fisher–Price and its Laugh and Learn range came from nowhere to become the most talked-about baby product at a key time of the year sales-wise.

The Talkability® process

In the *New Scientist*, Condomi, HP Sauce, Slendertone, *Zoo* and Fisher-Price case studies, PR is at the core of what became buzz phenomena. But the ideas weren't just arrived at by accident. A deliberate process had been followed.

First, the brief was analysed and the current situation assessed both from the client's own perspective and through market and competitor analysis. Next, opinions were discussed and an informal strategic brief was set. These stages in the process are quite similar to other creative processes.

After this came the idea development phase and this is where the buzz marketing approach kicked in, ensuring that the ideas generated for the campaigns had an ingrained buzz hook or angle that was going to break through the clutter and penetrate peoples' personal deflector shields.

At Frank PR, we call this buzzworthiness 'Talkability'. In order to come up with that golden egg during the creative process, we recommend examining five key parameters:

1. What are the current conventions in the market? Sales channels, design, naming, behaviour, media, anything. How can we disrupt those conventions? What can we do that's different – but for a reason, not just for the sake of doing something different?
2. What's the imagery? What are the visual cues that stick in the mind? What could be done from a visual perspective that would get people thinking? And talking?
3. What's going on in the world? What trends can be identified and with which we could tie in? Importantly, what are the trends going to be going forward? What's going to make up the zeitgeist? What do the future gazers, experts, tea leaves say?
4. Perception is reality. How is our brand, product or service actually perceived now? (Not 'what is the reality?'.) How would we like it to be perceived? How could it be perceived?
5. What are the brand's, product's, or service's unique selling points? What other USPs could we identify? What USPs would we like it to have? What would its USPs be in another market sector?

Sometimes the buzz idea will come from playing with just one or two of those parameters, sometimes from all five. Surrounding yourself with great people during the process is also crucial. You need people who you can bounce ideas off, who you can draw inspiration from and have fun with. Many of the thoughts that have come out of this process have at first been laughed off, only to come back as key elements of the final idea.

Once you've got an idea with inherent Talkability, you just need to shape it and sell it to the media. I say 'just', but this is where experience in the PR industry is vital. Experience gives you a real feel for what makes a story, in which media that story could fit and how to shape the story to make it fit – it's got to work and be relevant for the media you're targeting. And of course it's also got to get across the client's message.

As a PR practitioner you've got to advise how the brand, product, or service message needs to be adapted in order to make the news. You've got to get to some sort of middle ground where your story doesn't come across as a blatant plug for the brand, but at the same time does get across the right messages. Normally, the balance is found by making the messages more implicit in the concept than explicit. The Fisher–Price case study is a good example of this. To make the story too product-led would have taken away from the clever new test that had been developed. What was implicit was that its range of toys was good for a baby's development.

To develop your story so that it makes waves in your chosen media, look through most newspapers. What makes news (that will get people talking)? Is it 'dog bites man' type headlines? No, 'man bites dog' is much more likely to make a good story. Injecting controversy, albeit often quite lame bits of controversy, is a tried and tested shaping technique. Think back to the *Zoo* idea with the team baring their bottoms and the FA up in arms. The idea was not nearly as controversial as some of the things that happen in everyday life without the involvement of PR people, but if it makes for a good story and gets across the client's message, use it.

That old chestnut, sex, also sells and helps to shape buzzworthy ideas. Bottoms and the Slendertone 'bottom double'. Condomi and the measurement of penises. It's not sex in the hardcore sense – that could inhibit make-ability – just sex in a very accessible and everyday way.

These days humour is also becoming more and more important with the media. We're surrounded by such serious events going on in the world that ideas such as asking someone to change their name to James Brown break that reporting cycle of grim news for the media. They like stories that are shaped for them in an entertaining way, just as we as readers like to be entertained. Reality, celebrity, surprise, humanity and unbelievability are also other good shapers that can be used to turn an idea born of the Talkability process into something that's going to get into the papers, and in turn get into that morning's conversation around the watercooler.

Conclusion

PR is no longer the poor cousin of advertising, its achievements simply aiming for, and measured in, dry column inches. Combined with a buzz marketing approach, PR is a powerful tool that can play a large and linchpin role in generating buzz for brands, if brand marketers and PR practitioners take a step back and think 'out of the box'. A good PR campaign born of the Talkability process will then connect with consumers and generate the kind of buzz that can be amplified when integrated with an advertising campaign or other marketing communications.

Takeaway points

- The media has the power to turn something small into a giant (or monster!)
- You can't get publicity if you don't have a compelling story that excites the media. If it also excites consumers it will spread and break through into daily conversation.
- PR combined with buzz, viral, or word of mouth marketing activities will help engineer public consent to connect with and talk about your story.
- Coverage is not the only PR result that matters; what the coverage does (e.g. how far and to whom it spreads, how it changes people's awareness and perceptions of a brand, etc.) is equally important and should also be measured.
- The Talkability process will help generate buzzworthy ideas and shape them ready to 'sell' to the media.
- Brand messages should be implicit, not explicit, in the buzz concept.
- Controversy, sex, humour, reality, celebrity, surprise, humanity and unbelievability can help shape buzzworthy ideas.
- PR is a powerful tool that's moved on from simply generating news coverage; it's now also used to kick-start word of mouth, creating a buzz and connecting with consumers.

Notes and references

1 Ries, Al and Ries, Laura (2002) *The Fall of Advertising and the Rise of PR.* New York: HarperCollins.
2 Callahan, Sean with Ries, Al (14 October 2002) 'Proclaiming the fall of advertising', *B2B*, http://www.btobonline.com/cgi-bin/article.pl?id=9952.
3 Bernays, Edward L. (1992) *The Future of Public Relations.* Talk presented at the 1992 Association for Education in Journalism and Mass Communication (AEJMC) Convention, transcript at http://www.prmuseum.com/bernays/bernays_1990.html.
4 Marsh, Dr Peter (September 2004) *The Developing Baby Report.* Prepared by the Social Issues Research Centre.

6

Viral marketing

Justin Kirby

Managing Director, Digital Media Communications (DMC)

Connected marketing techniques such as word of mouth are firmly back on the marketing agenda. This phenomenon has been formalized during the past decade by specialist books, including Malcolm Gladwell's *The Tipping Point*,[1] Emanuel Rosen's *Anatomy of Buzz*[2] and Seth Godin's *Unleashing the Ideavirus*[3]. The hype about word of mouth marketing's potential for achieving viral-like message spread has also been fuelled by well-known success stories, such as Hotmail and The Blair Witch Project (see the Introduction). However, one of the most significant contributing factors to the current vogue for word of mouth marketing is the rapid uptake of digital media – particularly the Internet and its peer-to-peer technologies such as chatrooms, forums, instant messaging programmes, blogs, and file transfer and social networks – which enable messages to spread faster and more exponentially than ever before.

Digital media's capabilities to turbo-charge the viral spread of information means that well-planned and well-executed connected marketing initiatives – particularly those that integrate more traditional marketing communications techniques in their activities – can help business messages reach the mass market in a way that would require a significant investment if left to more traditional techniques alone.

This capability for mass-market reach is important. Some connected marketing techniques, such as influencer marketing (see Chapters 1 and 7) and live buzz marketing (see Chapter 2) that rely on offline communication, can help improve brand advocacy, which in turn aids business growth.[4] However, they are not necessarily the most effective ways of creating brand awareness.

Viral marketing - especially when used as an integrated rather than isolated approach - can both improve brand advocacy *and* increase mass-market brand awareness. And it can achieve those objectives very cost-effectively, even if your brand, product, or service has no standout, buzzworthy characteristic.

This chapter reveals how and why viral marketing can achieve these aims, examining some of the techniques and the high-profile campaigns that have helped catapult connected marketing into the marketing communications limelight.

A brief word about definitions

There are various debates about how to define viral marketing and viral advertising.

> Viral marketing describes any strategy that encourages individuals to pass on a marketing message to others, creating the potential for exponential growth in the message's exposure and influence. Like viruses, such strategies take advantage of rapid multiplication to explode the message to thousands, to millions.[5]

Exponential growth and the pass-on of a marketing message to others (rather than making a product recommendation) are the key factors in that definition.

Some people maintain that viral marketing is any marketing activity that accelerates and amplifies word of mouth *in the digital domain*. (Arguably, it's digital media's ability to hypercharge the exponential spread of messages that has brought the whole area of connected marketing to the forefront of marketing communications practice.) So some of the connected marketing activities mentioned in other chapters of this book, such as influencer marketing campaigns, could also be called viral marketing if they were distributed via digital media.

What's viral *advertising* then? Viral advertising consists of creating contagious advertising messages or material that get passed from peer to peer in order to increase brand awareness (as opposed to amplifying and accelerating word of mouth advocacy such as product recommendations). Viral advertising is often used when the product itself doesn't have a 'wow' factor that can generate buzz - you make the creative agent or message contagious instead.

Viral advertising could become more strategic viral marketing when - as in the Virgin Mobile UK and Mazda Motors UK examples on pp. 97 and 100 - it not only raises brand awareness cost-effectively, but also generates response such as brochure requests, test drives and eventually sales.

Ultimately, debates about definitions aren't that useful at this stage in the development of the connected marketing field. What's more important is what practitioners and marketers are actually doing and why, particularly as far as the outcome of their activities is concerned. You measure the outcome, not the terminology. So rather than get bogged down in definitions, let's look at why people are doing viral marketing, how they go about it and how success is measured within the context of connected marketing.

A brief history of viral marketing

Before we launch into the hows and whys of viral marketing practice, it's worth putting viral marketing into context by briefly considering its history.

The earliest known use of the term 'viral marketing' is in a 1989 *PC User* magazine article[6] about the adoption of Macintosh SEs versus Compaqs:

> At Ernst & Whinney, when Macgregor initially put Macintosh SEs up against a set of Compaqs, the staff almost unanimously voted with their feet as long waiting lists developed for use of the Macintoshes. The Compaqs were all but idle. John Bownes of City Bank confirmed this. 'It's viral marketing. You get one or two in and they spread throughout the company.'

So viral marketing initially denoted seeding designed to kick-start the copycat effect whereby people 'catch' the idea and adopt it by seeing it adopted by others.

The term cropped up again in a 1996 Fast Company article[7] by Harvard Business School professor Jeffrey Rayport.

Another noted use of the term was made by Steve Jurvetson and Tim Draper from Silicon Valley venture capitalists Draper Fisher Jurvetson who invested in Hotmail. In 1997, Jurvetson and Draper wrote a White Paper[8] describing the high-profile Hotmail phenomenon, hence the frequent attribution of the term to Jurvetson and Draper.

Wherever the term came from, the practice of viral marketing in the digital domain has been around for almost a decade and has registered three major blips on the wider marketing radar screen to date:

1. In the mid-1990s when digital media started coming into its own and Hotmail went from 0 to 12 million users within 18 months.
2. At the turn of the millennium when the dot.com bust squeezed marketing purses and put the onus on accountability.
3. Nowadays, as advertisers try to stand out from the increasing clutter across fragmented media to connect with more cynical, marketing-savvy consumers.

Aside from Hotmail – who simply added a short line of promotional text to the bottom of every email message sent via their service, clocking up 12 million sign-ups within 18 months from a marketing spend of US$500 000 – other high-profile early adopters of viral marketing included Budweiser and John West Salmon. The latter two campaigns consisted of allowing digital video files of cool and funny TV ads to 'escape' onto the Web before they became available via other media. This seemingly unintentional approach made the material exclusive and therefore more desirable to online users, giving them a kind of cachet among their contacts. It caused them to talk about it and pass it on in droves. As a result, the catchphrase 'Whassup' became widely known in the UK even before the planned TV campaign was aired there, while the John West Salmon 'Bear' ad has become a classic of its kind.

Although those early examples achieved significant viral spread and helped generate buzz, they were more 'happy accidents' than carefully planned campaigns with ongoing brand benefits. In fact, Budweiser saw a drop in market share and sales during the 'Whassup' campaign.[9] Nevertheless, marketers started trying to replicate the perceived success of this kind of viral marketing.

Another factor driving the increased use of viral marketing was purely economic: Customer Relationship Management (CRM) projects were expensive and complex to implement and manage, particularly during the dot.bomb era when marketers were more rigorously tasked with delivering return on investment (ROI). Falling online banner ad click-through rates, as users grew more techno-literate and 'immune' to old-format ads, also contributed to marketers turning to the potential 'magic' of viral marketing in the early 2000s.

Initially, viral marketing was used predominantly as a standalone marketing tactic focusing on the creative material which could be a photoshopped image, online game, digital video clip, or even text (jokes, anecdotes, excerpts from reality TV show transcripts, etc.). This tactical use of viral marketing was, and still is, very hit-and-miss; it relies entirely on the creative material or 'agent' alone striking a chord with users. And as the amount of (often mediocre) viral-wannabe material increased, the clutter made it more difficult for campaigns to stand out.

By 2002, brands and practitioners at the forefront of viral marketing realized that the viral agent within a campaign needed to be used as a means to an end, rather than simply as an end in itself. Accordingly, some companies refined their approach to viral marketing, putting the emphasis on strategic use and long-term benefits.

The viral marketing arena has developed in various other ways since 2002. Most notably:

- The development and increasing adoption of digital technologies such as broadband have enabled people to enjoy richer online content, making the Internet not simply a practical medium for activities such as research and shopping, but also a burgeoning entertainment medium. Rising user demand for online entertainment has spawned a parallel rise in the emergence of both mainstream and underground websites dedicated to providing entertainment content.
- Brands have now realized that in order to achieve their objectives they must invest more realistic budgets in the strategic planning and implementation of viral marketing campaigns – Toyota, for example, in 2004 committed a US$10 million budget to viral marketing alone.[10]
- Marketers have learned that to stand out from the growing clutter of wannabe viral campaigns, they must be more groundbreaking and creative in their use of digital media (see Subservient Chicken case study on p. 102).
- Over time, as we shall see from some of the case studies later in this chapter, viral marketing has evolved into a technique that can now be used successfully not only to create a buzz about a brand or product, but also to help generate sales.

Why viral marketing makes sense

Why are marketers increasingly using viral marketing as part of their overall brand marketing activities? It's all about power and money.

Packaging, billboards, branded clothes, signage, trademarked sounds, free food samples, the smell of coffee or baking bread, TV ads, interruptive online ads, unsolicited cell phone clips . . . our every sense is assaulted constantly by both overt and covert marketing messages delivered via a growing plethora of media channels. No wonder people have learned to tune out a lot of marketing communications – and even choose to avoid them altogether thanks to ad-skipping technologies such as personal video recorders and pop-up blockers.

Consumers are also more involved than ever before in controlling communications and message delivery at a global level, thanks to the aforementioned rise of digital media, such as blogs and forums. Consequently, advertisers are finding it more and more difficult to reach marketing-shy, fragmented audiences, let alone engage with them.

In tandem with this situation, many brands are finally realizing that 'the most powerful selling of products and ideas takes place not marketer to consumer but consumer to consumer.'[11] Enter viral marketing.

Unlike traditional 'top-down', marketer-to-consumer techniques, viral marketing focuses on personal experience of the brand and taps into the new power of the consumer. One of the reasons consumers find viral marketing campaigns appealing is because the campaigns tend to be non-interruptive, so they enable consumers to choose to interact proactively with a communication (and the brand behind it), or not, rather than be passively dictated to. This 'bottom-up' approach respects that the consumer is in control; viral marketing campaigns are ultimately driven (or derailed) by consumers themselves.

The result of this user-driven process is ultimately very valuable exponential brand endorsement by influencers and consumers – if the campaign is successful. The difference between one campaign succeeding and another failing is dependent on the campaign's ability to connect with consumers and inspire them to engage and interact with the advertising material, the brand and ultimately the product or service.

Now it's time to show you the money. In 1965, you could reach 80% of a mainstream target audience with three TV ad spots. By 2002, 117 spots were required to achieve the same reach.[12] No business, no matter how rich, can afford to maintain constant mainstream media brand awareness. So less expensive online media routes, and the possibility of peer-to-peer-driven spread whereby the audience effectively becomes another (free!) media channel, are obviously very attractive to brands, who can use viral marketing to maintain a cost-effective level of brand awareness during mainstream media spend 'downtime'.

Continuing the economic benefits of viral marketing, for some time now viral campaigns have been used successfully not only to build widespread brand awareness, but also to help generate sales. For example, Eidos' Hitman 2 online viral marketing campaign[13] helped shift a million units thanks to the buzz it generated, which was picked up in the Top 10 Movers of 2002 on the Yahoo! Buzz Index.[14]

Two other viral marketing benefits to the bottom line are the direct result of using digital media:

1. Viral marketing campaigns can provide accountability when tracked, thereby measuring and proving ROI.
2. Viral campaigns have no fixed cut-off point, so they can provide an ever-increasing ROI. For example, Mazda UK's first online viral and buzz marketing campaign,[15] released in June 2002, is still generating brand awareness and driving significant traffic to their website.

Another string to viral marketing's bow is that it can be used if your brand, product, or service has no compelling 'wow' factor – great news for makers of toothpaste, cars, beer and other generic products. Instead of creating a buzz around your product, you create it around the viral campaign agent. Of course, at the end of the day, no matter how many of the right kind of people talk about and pass on your viral agent, you've still got to have a desirable product that your target audience wants to buy. As General Motors discovered with the Pontiac G6 give-away on the Oprah Winfrey show,[16] you can lead a horse to water and even make it drink, but if it doesn't like the taste . . .

Finally, viral marketing integrates well with traditional marketing activities, giving brands the best of both the 'bottom-up' and 'top-down' approaches. No wonder it's fast becoming an integral ingredient in the overall brand marketing mix.

Viral agent formats: risks and issues

If viral marketing is anything that amplifies and accelerates word of mouth in the digital domain, or even any strategy that creates an exponential spread of a marketing message, then that covers a great many types of marketing communication.

Even the subset of viral advertising alone can include a wide range of execution types, agents, or mechanics – including images, jokes, reality TV show transcripts, quiz promotions, advergames, digital video clips, e-cards, interactive microsites, Alternate Reality Games (ARGs), and more. (Examples of typical viral material can be seen on dedicated viral entertainment websites, such as the Lycos Viral Chart (http://viral.lycos.co.uk), Viralbank (http://www.viralbank.com) and Viralmeister (http://www.viralmeister.com).)

Deciding what kind of creative agent to use in your viral marketing campaign is not only a question of what's most appropriate to convey the creative idea; it's also a question of examining the risks and issues related to using one digital format over another. And given the enormous power consumers have over messages these days, there are obvious risks associated with the use of several types of viral material.

The dangers of using a simple text email, for example, are illustrated by a Carlsberg-branded viral email that was passed from contact to contact during the UEFA Euro 2004 football tournament.[17] The familiar Carlsberg 'probably the best' creative theme impels user appropriation in the same way as Mastercard's 'Priceless' theme. Unsurprisingly, the original Carlsberg email was hijacked at some point in its viral travels – the words 'Shame their lager tastes like p*ss' were added to the punch line.

The same risk is associated with the use of digital image files; they can be doctored easily by users, so that the original messages are lost or take on new meanings. Often this user input makes the agent more entertaining and viral, but not necessarily to the benefit of the brand behind it.

There's no doubt that the most populous viral material format is online games or 'advergames'. Trade marketing news lists don't seem complete these days without the announcement of at least one or two new viral campaigns featuring advergames. Unfortunately, many marketers mistake populous for popular; they believe that advergames are a great, low-cost way to attract flocks of users. They can be. But the sheer volume of (largely mediocre) advergames being offered to users is a major issue. You have to be smarter, more innovative and more creative than anyone else – and (shock, horror!) maybe even spend more money – in order to grab users' attention, make them want to engage with your brand, endorse your message and pass it on to others. Who said viral marketing was a doddle?

As for ARGs, it's difficult to see how they can really be integrated successfully into an overall marketing strategy. They can certainly work in an entertainment context, such as promoting a film or video game as Beta-7 for Sega's *ESPN NFL Football*[18] and 'ilovebees' for Microsoft Xbox's *Halo 2*[19] have done. But they are aimed at people who love mysteries and puzzles and have a lot of time on their hands. As a viral mechanic, ARGs may well be too involved and expensive for the vast majority of mainstream brands looking for mass-market reach. They are reminiscent of the BMW Films project[20] – one of those brand-building initiatives that are high-risk. If pulled off, they create a lot of noise, which you would expect after having so much money thrown at them, but they are unlikely ever to be a mechanic that most brands can integrate into their core marketing strategy.

Chapter 9, *Changing the game* covers advergames and ARGs in greater detail, and shows examples that have succeeded in sticking their head above the crowd and providing benefits to the brands behind them.

Some of the most well-known and successful viral marketing campaigns use video-based creative agents. The benefits of using digital video files include:

- They are small enough to be passed from peer to peer via email after download from multiple distribution websites, encouraging greater user-driven spread.
- They are trackable after download, as they are passed from user to user via email, so they provide brands with greater campaign accountability.

- There is less risk of user interference with the agent.
- Video is a familiar, ad-like/film-like format to users, with the added advantage of interactivity (digital video files can include hotspots that enable users to link through the file to a web page).

Let's take a look at campaign planning before we review some viral marketing case studies.

Planning a successful viral marketing campaign

Viral marketing, as its name implies, aims to reach the widest possible number of people (among which any brand's target market lies). Campaigns kick-start buzz generation and viral spread - then it's up to consumers, who become brand advocates and free media channels as they pass the buzz talk and viral agent on, providing valuable peer-to-peer endorsement.

However, the ultimate point of a viral marketing campaign is not only to 'go viral' (though this is still important while CPM-type (cost per Thousand views) spread remains a major gauge of campaign success), but also to deliver tangible, ongoing benefits to the brand. As Budweiser discovered,[21] getting eyeballs is no guarantee of subsequent market share and sales success. To be a significant part of the marketing communications mix, viral marketing needs to be doing more than just generating low-cost brand awareness; it needs to deliver measurable response.

Viral marketing is being used for two main purposes:

1. To maintain or boost a cost-effective level of brand awareness during mainstream media spend 'downtime', usually by releasing Web-exclusive viral material that retains brand and campaign themes.
2. To kick-start consumer-driven interest in new marketing communications activity - which often means pre-launching a mainstream ad virally (perhaps using a Web-exclusive edit) before it hits TV, in order to create buzz and exploit exclusivity.

Other purposes and benefits specific to each campaign may include:

- Reach well beyond a business's core target market.
- Create buzz around products and brands that have no compelling 'wow' factor (so the agent itself must have the 'wow' factor).
- Accelerate and amplify natural buzz and viral ability for products that do have a 'wow' factor.
- Reinforce existing advertising and branding messages.
- Extend other marketing communications activities.

- Provide accountability when tracked, thereby measuring and proving ROI.
- Help add to the bottom line in terms of response and/or increasing recommendation rates.

The planning stage of a viral marketing campaign should involve setting out feasible objectives, developing the campaign strategy, coming up with a viral idea, story, theme, or angle that can generate buzz, and developing the creative brief. Planning activity may also include helping the brand to integrate and amortize media, PR and creative development activity if the viral campaign is part of a wider marketing initiative.

There are then three core components to any viral marketing campaign:

- Creative material: developing and producing the viral agent that carries the message you want to spread in a digital format.
- Seeding: distributing the buzz story and viral agent online in places and with people that provide the greatest potential influence and spread.
- Tracking: measuring the results of the campaign to provide accountability and prove success.

One important point about seeding: a big mistake that many ad agencies and brands make is thinking that viral marketing is simply about creating entertaining material and finding some websites specializing in viral content on which to promote it. They miss the point that viral marketing needs to create conversations, not simply spread the viral agent. The seeding process for a viral marketing campaign is not the same as the process of online media buying; it's more like a PR process. Seeding is not just about knowing the 'where', i.e. finding appropriate places to locate the viral agent; it's also about knowing the 'how' and 'who' – determining how the campaign can best be advocated and by who, and communicating the buzz story in the most appropriate way to each source route. Just as the results of a viral marketing campaign have to be about more than the number of people who see the viral agent, so the seeding needs to include telling a buzzworthy story that engages the right people with the brand in a way that maximizes the likelihood of meeting the campaign objectives.

In conclusion, three key factors will increase the likelihood of a successful online viral marketing campaign:

1. Specialist strategic planning to ensure that viral marketing is used to deliver tangible, measurable, ongoing brand benefits.

2. Appropriate 'wow'-factor material that users want to seek out, talk about and pass on of their own freewill.
3. Appropriate specialist seeding of the buzz story and the viral agent to places where viral and brand influencers already gather.

Case studies

The following eight case studies have been chosen for review not only because they illustrate some of the pros and cons of viral marketing well, but also because most of them use the technique with long-term, strategic goals in mind.

As James Kydd, brand director for Virgin Mobile UK, states: 'Online viral marketing is best used not as a one-off tactical end in itself, but as an integrated strategic part of the overall marketing mix. It's a means to an end whereby it not only generates buzz, but also provides ongoing, quantifiable brand benefits, such as increased awareness, peer-to-peer endorsement and ultimately more sales.'[22]

Case Study 1: Virgin Mobile

Brand: Virgin Mobile UK, virtual network operator and one of the fastest-growing cell phone companies worldwide.

Campaigns: A series of nine campaigns from 2003 to 2005.

Description: Virgin Mobile UK's viral marketing activity falls within the remit of its brand marketers. The brand first used online viral and buzz marketing in May 2003 as a strategic part of its wider 'Idle Thumbs' marketing initiative, pre-launching online a more risqué (and therefore sought-after) edit of a TV ad.

Eight more viral marketing campaigns featuring video-based creative material have followed to date, including 'Bendy Babe', 'Busta Butt', 'Xtina Spotted' and 'Best Hands'.[23]

The brand has embraced viral marketing as a process not an event. On one hand they have championed the Web-first release of TV ads in order to help kick-start wider marketing campaigns and create a 'watercooler' buzz effect, in the same way that more risqué versions of TV ads are sometimes released in selected cinemas. Obviously, the use of celebrities adds to the buzz. This Web-first approach is very cost-effective because the TV ads are being made anyway. The cost of pre-launching a different edit online is less than the cost of a single insertion in a tabloid newspaper, and the buzz that can be generated is much greater than that from a print ad.

On the other hand, some of Virgin Mobile's viral marketing campaigns have been used as booster activity to widen awareness of specific promotions. This kind of lower-cost, Web-only execution also helps maintain brand awareness when TV advertising isn't running.

James Kydd, Virgin Mobile's brand director, said: 'We have developed the way we use viral marketing strategically as a key part of our overall marketing mix. It has made a big impact on our brand exposure for a fraction of the cost of traditional marketing methods.'

When brands don't have pockets as deep as their competitors' and don't own their own channel, they have to use an array of methods to be seen and heard, and viral marketing is an increasingly important part of that arsenal.

Case Study 2: Trojan

Brand: Trojan, condom manufacturer.

Campaign: Trojan 'Games' (2003).

Description: This online viral marketing campaign featured sex-and-games spoof video clips by The Viral Factory. They were made available on a dedicated microsite and released during the build-up to the 2004 Olympic Games. The site was said to have received over 9 million visits by March 2004. Other brand benefits gleaned from a survey for Carter Products by consumer market research agency QuickWise include:

- 73% of respondents gave a positive rating of the overall impression of the campaign.
- 80% perceived the campaign to be unique.
- 50% indicated they would be more likely to consider purchasing Trojan products in the future.[24]

An important ingredient in the success of this campaign was sex – as always, it gets noticed, it gets talked about and it sells. But in this case, it was also highly appropriate to the brand and product. Executions of this kind would be more difficult and less appropriate for more mainstream brands and products.

Case Study 3: Agent Provocateur

Brand: Agent Provocateur, boutique lingerie manufacturer.

Campaign: 'Proof' (2001).

Description: A digital video clip of a sexy advertisement by cap-travissully, featuring pop icon Kylie Minogue dressed in Agent Provocateur lingerie and riding a bucking bronco machine, was placed online. The ad was also tailored for cinema release, making the viral campaign part of a wider, integrated marketing initiative. The online file was copied, downloaded, or forwarded an estimated 100 million times, a small massively increasing brand awareness and getting Agent Provocateur, a small boutique, suddenly talked about the world over.

The success of this campaign was somewhat of a happy accident, piggybacking off Kylie's celebrity status and her unexpected 'orgasmic' performance. (Again, sex sells!) There was no widespread seeding strategy – a necessity for campaigns executed these days – partly because very few specialist viral entertainment routes existed at the time. Equally, tracking capabilities now exist to provide a more precise picture of quantified campaign results when using standalone video files.

Case Study 4: Ford SportKa

Brand: Ford, automotive manufacturer.

Campaign: Ford SportKa Evil Twin 'Pigeon' (2003).

Description: This viral marketing campaign featured a digital video clip by The Viral Factory showing a pigeon being knocked out (or possibly killed) by the car hood. The campaign raised a large amount of buzz, including many pages of user-driven chat on car forums, complaints from pigeon fanciers and very high-profile spillover into offline media. It peaked on UK TV car show *Top Gear*, one of the BBC's most-watched programmes, which spent 10 minutes showcasing a 100-mile race between a SportKa and a pigeon. (The pigeon won.) You just can't buy that kind of editorial coverage.

Hot on the heels of the planned 'Pigeon' campaign came a 'Cat' execution – literally. The allegedly unauthorized Cat video clip (aka a 'subviral') shows a cat seemingly decapitated by the car's sunroof. This was extremely controversial, causing an even bigger stir than the 'Pigeon' campaign and generating massive buzz online and offline.

Regardless of whether it was officially sanctioned or not, the 'Cat' subviral is a good example – as are the Nokia 'Swinging Cat' and Volkswagen 'Bomber' subvirals – of how viral material can be successful in terms of generating spread and creating buzz, yet be of limited ongoing benefit to the brand. The 'Pigeon' campaign and subsequent 'Cat' clip are credited with repositioning a car that was not a particular standout in its class as a gnarly sports beast. But did this activity increase car sales, and, if so, did that tactical gain during a limited period outweigh the longer-term strategic loss from the potential alienation of Ford's wider audience?

Case Study 5: Mazda

Brand: Mazda Motors UK, automotive manufacturer.

Campaign: 'Parking' (2003).

Description: At the time of this campaign, Mazda's viral marketing activity was driven by and integrated with their CRM activity, response being a core focus.

The campaign, featuring a video clip and buzz story that piggy-backed off the contemporary rash of girl versus boy magazine articles, driving shows on TV, etc., struck a major chord with online users, sparking global debate on blogs and forums about male and female parking capabilities. (For example, 'All I can say is clever, very clever. Now let's see her get out.')[25] The campaign created high-profile, wide-spreading conversation without resorting to overly controversial, dark, or shocking themes. Mazda is seen to be overtly having fun with its brand without alienating consumers, and getting across its brand messages without bashing consumers over the head. The campaign also clocked up a lot of free editorial exposure (including being voted Best Viral Campaign of 2003 in the UK's *Campaign* advertising trade magazine and its German counterpart), and the initially Web-only clip ended up being used as a TV ad in several territories.

The 'Parking' campaign was the second in a series of eight online viral marketing campaigns undertaken by Mazda Motors UK to date. Most of the campaigns have featured Web-exclusive viral agents (whereby the material is made specifically for online use only), and the 'Parking' campaign epitomizes the reasons why. To date, it has generated over 5.5 million calculable clip views and driven over 200 000 visits to the Mazda website – and those figures are increasing daily, even two years after the campaign's launch. It has also helped sell a product that is very similar to many others in its class.

This campaign is another example of using online viral marketing as part of a process, rather than a one-off event – it's a means to an end, delivering awareness *and* driving response. As Steve Jelliss, then-CRM Manager for Mazda Motors (UK), stated: 'Our ongoing series of online viral marketing campaigns have proven their value in providing high brand exposure to a wide-as-possible audience, and ultimately contributing to car sales.'[26]

Case Study 6: Dr Pepper

Brand: Dr Pepper, drinks manufacturer.

Campaign: Raging Cow (2003).

Description: This online viral marketing campaign is noted for its pioneering seeding within the blogging community (see Chapter 10 for in-depth information about blog marketing). Unfortunately, it generated a consumer-driven backlash for what was perceived at best as a clumsy attempted use of viral marketing.

As ClickZ News reported:[27] 'Dr Pepper showered teen bloggers with gifts and indoctrinated them on how to blog its new Raging Cow beverage. The plot backfired, with a well-publicized boycott and global media covering the debacle.'

Ironically, bloggers may have liked the product; they just didn't like their visitors questioning the integrity of their site content. Blogger Carlo Orozco said: 'The day after ragingcow.com went public, I sent out an e-mail to the other five bloggers and told them, "You know, we sold out?". The e-mails I got back were like, "At least I'm enjoying the drink." The funny thing is, I do like it. But my credibility is gone.'[28]

As e-business strategy consultant Rick Bruner said, this campaign has sadly become 'the embodiment in the minds of many of the idea that blogs and marketing don't mix'.[29]

Where it seemingly crossed the line was after briefing interested bloggers, as Bruner reprised: 'As best as I can tell, where it fell afoul of the blogosphere was that it then asked those young bloggers not to mention that they had been briefed about the product, as if their sudden new enthusiasm for flavored milk was purely their own idea.'[30]

The campaign was an innovative, early example of using blogs in viral marketing and its faux pas has subsequently been blown somewhat out of proportion.

Case Study 7: Burger King

Brand: Burger King, fast-food chain.

Campaign: Subservient Chicken (2003).

Description: This online viral marketing campaign created by Crispin Porter + Bogusky used an interactive video of a man dressed as a chicken in a humorous take on webcam activity. The product it aimed to promote was a new chicken sandwich, which in itself does not inspire user-driven word of mouth. Instead, the viral agent delivers this buzz, having been seen by 46 million people according to Burger King.[31]

If the campaign's objective was simply to create a buzz as part of an overall raft of activities trying to breathe life into a brand that was a bit tired and was trying to move away from the kids market and align itself with the slightly older tween demographic, it has been successful. But does the campaign miss a trick or two on the ongoing brand benefit front? How many site visitors realized that the campaign was for a chicken sandwich, or indeed for Burger King rather than Kentucky Fried Chicken or A. N. Other? And how many went on to recommend and buy the new sandwich? As AdAge.com asked: '. . . will it make the flagging burger giant cool again with young men – or even sell sandwiches?'.[32]

The campaign is in danger of having generated a character that outshines the brand, à la Levi's 'Flat Eric'. Again, it illustrates that 'wow'-factor advertainment content and high exposure is not enough; viral campaigns should also deliver strategic, response-related brand benefits such as quantifiable sales, otherwise their value is hit-and-miss, or at best only short-term.

Case Study 8

Brand: Bacardi Global Brands, distillers.

Campaign: Planet Party (2004–2005).

Description: Bacardi's first online viral and buzz marketing campaign was planned and produced by members of the Viral + Buzz Marketing Association (see http://www.vbma.net). The campaign features a Web-exclusive video clip that's spreading around the Web luring partygoers to the Planet Party microsite (see http://www.planet-party.net) where they can download another clip, explore a virtual nightclub, load up

Bacardi DJ, find cocktail recipes and undertake other entertainment-focused activities.

The most significant point about this campaign is that it's the most comprehensive example here of a truly connected marketing initiative. It's part of an integrated wider campaign that includes the profiling and recruitment of influential consumers, so that Bacardi can develop customer relationships as part of an ongoing CRM process. The campaign is helping to kick-start that CRM process, with the viral marketing technique being used both to generate awareness and, more importantly, to follow up end-user interaction with the brand.

This holistic connected marketing strategy epitomizes the use of viral marketing as a process not an event, appreciating that it needs to be integrated with, not isolated from, wider marketing initiatives.

Conclusion

This chapter has revealed how businesses have used viral marketing to date and how it is becoming an important ingredient in the connected marketing brew.

The ultimate goal of any viral marketing activity you do should be to deliver measurable brand benefits - and that means response, not just eyeballs. It doesn't matter how many times your video clip gets seen or your game gets played if nobody remembers, recommends, or buys your product.

When you get the approach right, viral marketing is an extremely valuable addition to any brand's overall marketing communications activity - and to the bottom line.

Takeaway points

- People no longer use the Internet only for practical purposes, such as research and shopping. The development and adoption of technologies enabling easy access to richer content has spawned a growing user demand for online entertainment, which in turn has spawned a rise in the emergence of both mainstream and underground websites dedicated to providing entertainment content.
- Consumers have learned to tune out a lot of marketing communications and they have become more involved than ever before in controlling message delivery globally, thanks to the rise of digital media such as blogs and forums. Advertisers are therefore finding

it increasingly difficult to reach and engage with marketing-shy, fragmented audiences.

- Viral marketing focuses on personal experience of the brand and taps into the new power of consumers and their connections to other consumers.
- Viral marketing – especially when used as an integrated rather than isolated approach – can both improve brand advocacy and increase mass-market brand awareness. It can be used successfully to create a buzz about any brand or product, and to help generate sales.
- Even innovative products need viral marketing to accelerate and amplify their natural buzzworthiness.
- To stand out from the growing clutter of wannabe viral marketing campaigns, brands must be more groundbreaking and creative in their use of digital media.
- Viral marketing campaigns can provide accountability when tracked, thereby measuring and proving ROI. Viral campaigns have no fixed cut-off point, so they can provide an ever-increasing ROI.

Notes and references

1 Gladwell, M. (2000) *The Tipping Point: How Little Things Can Make a Big Difference*. Little, Brown and Co.

2 Rosen, Emanuel (2000) *The Anatomy of Buzz: Creating Word of Mouth Marketing*. New York: HarperCollins Business.

3 Godin, Seth (2001) *Unleashing the Ideavirus*. Chicago: Dearborn.

4 Reichheld, F.F. (2003) 'The one number you need to grow', *Harvard Business Review,* 81 (Dec.): 46–54.

5 Wilson, Dr Ralph F. (2000) 'The six simple principles of viral marketing'. Web *Marketing Today,* Issue 70. 1 February.

6 Carrigan, T. (1989) 'New Apples tempt business', *PC User,* 27 September.

7 Rayport, Jeffrey (1996) *The Virus of Marketing*. Fast company at http://www.fastcompany.com/online/06/virus.html.

8 Jurvetson, Steve and Draper, Tim (May 1997) *Viral Marketing*. Original version published in the *Netscape M-Files,* edited version published in *Business 2.0,* November 1998. Archived at http://www.dfj.com/cgi-bin/artman/publish/steve_tim_may97.shtml.

9 Zyman, S. (2002) *The End of Advertising as We Know It*. Hoboken, NJ: Wiley.
10 Cuneo, Alice Z. (2004) *Toyota Names Ground Zero for Viral Marketing: $10 Million Campaign to Lure Young Consumers*. AdAge.com at http://www.adage.com/news.cms?newsId=40423.
11 Gladwell, *The Tipping Point*.
12 Bianco, Antony *et al*. (2004) 'The vanishing mass market', *Business Week*, 12 July.
13 MarketingSherpa case study (2002), archived at http://www.dmc.co.uk/index.php?bz00MA==.
14 See http://www.dmc.co.uk/index.php?bz0zNg==.
15 Mazda Motors UK's 'Brats' online viral and buzz marketing campaign (June 2002), announced at http://www.dmc.co.uk/mazda6
16 Neil, Dan (2005) 'An American idle', 6 April, LATimes.com archived at http://www.latimes.com/news/custom/showcase/la-hy-neil6apr06.story.
17 17 June 2004, archived at http://news.bbc.co.uk/sport1/hi/funny_old_game/3815633.stm.
18 Lippert, Barbara (2004) 'Wieden's Great Hoax', *AdWeek*, 24 May, archived at http://www.adweek.com/aw/magazine/article_display.jsp?vnu_content_id=1000517454.
19 Terdiman, F. (2004) 'I Love Bees game a surprise hit', *Wired Magazine*, 18 October, archived at http://www.wired.com/news/culture/0,1284,65365,00.html.
20 See http://www.bmwfilms.com.
21 Zyman, *The End of Advertising*.
22 Kirby, Justin (12 July 2004) *Online viral marketing: strategic synthesis in peer-to-peer brand marketing*. Brand Channel, White Paper, 12 July, archived at http://www.brandchannel.com/images/Papers/viral_marketing.pdf.
23 See http://www.virginmobilemissy.com.
24 QuickWise (2004) Research for Carter Products profiled in *QuickWiseNews*, Edition 2, March, archived at http://www.quickwise.co.uk/pdf/QWN_UK2.04.pdf.
25 'evil-c' (posted 1 May 2003). Newgrounds BBS.
26 Kirby, *Online viral marketing*.
27 Mara, Janis (2004) 'Burger King Hen Whets Chicken Yen', *ClickZ News*, 16 April, at http://www.clickz.com/news/article.php/3341301.
28 Werde, Bill (2003) 'The Web diarist as pitchman', *New York Times*, 27 March.
29 Bruner, Rick E. (June 2004) *Raging Cow: The Interview*, archived at http://www.businessblogconsulting.com/2004/06/raging_cow_the_.html.

30 Ibid.
31 Mara, 'Burger King Hen Whets Chicken Yen'.
32 MacArthur, Kate (2004) 'Can chicken-porn gag boost Burger King sales?', *AdAge* Online edition, archived at www.adage.com/news.cms?newsId=40292.

7

Online opinion leaders: a predictive guide for viral marketing campaigns

Idil Cakim

Director of Knowledge Development, Burson-Marsteller

The Internet is often overlooked as a channel for strategically relaying corporate messages to key constituencies. Today's influential audiences actively use the Internet, along with other media, to collect and distill information. Using a unique formula to spot outspoken and driven public opinion leaders, Burson-Marsteller identified the online influencers, or the *e*-fluentials®, in the online communities supported by SAP, a global enterprise software provider. This case illustrates how Web-based technologies and research can be used to generate lists of powerful audience members and connect marketers with special interest groups online.

Who are the e-fluentials?

In 1999, collaborating with RoperASW, Burson-Marsteller conducted a survey among US online adults and did a segmentation analysis to identify online public opinion leaders (i.e. *e*-fluentials) who exert an extraordinary impact on online and offline content and commerce. The study results showed that *e*-fluentials compose 10% of the US online adult population, but they generate most of the buzz about brands, products and companies. On average, *e*-fluentials share an experience they had with a company with 14 people. Their circles of influence expand as their peers pass along *e*-fluentials' messages.

Compared with the average Internet user, *e*-fluentials are far more active users of email, newsgroups, bulletin boards, listservs and other online vehicles when conveying their messages. They are dynamic Web surfers who forward news and website information to people (86%), email companies (64%) and post to bulletin boards (52%) at least several times a month.[1]

While extremely influential online, *e*-fluentials also spread their opinions in the offline world. Civic-minded *e*-fluentials are more likely than average online adults to vote (77% vs 70% respectively), attend public meetings (38% vs 26% respectively), serve on local committees (40% vs 27% respectively) and make speeches (31% vs 17% respectively).[2]

Their families and peers regularly approach them for information, opinions and advice on a wide range of subjects, from business and politics to entertainment and health/lifestyle issues. For example, 44% of *e*-fluentials say they are asked about companies, businesses, or new technologies. More than half (55%) offer their views on hobbies to friends. Forty-one per cent frequently give advice about family issues (e.g. child care, coping with teens, relationships, etc.).[3]

Because *e*-fluentials' opinions have such extensive reach, it is critical that companies establish brand recognition and win these opinion-brokers' approval to expand their customer base. Whether planning a promotion, a crisis response, or a customer-relationship-management (CRM) campaign, companies should consider those among their audience who are experts in collecting and spreading information online.

Since *e*-fluentials highly value one-on-one dialogue and information exchange, companies can better manage their reputations by inviting feedback and providing a forum where *e*-fluentials can chat about their positive and negative experiences, and query others. The majority of *e*-fluentials visit corporate websites before advising on products and companies (85%),[4] and turn back to these areas to confirm hearsay information about companies (73%).[5] Hence, company-sponsored Web areas are critical in managing relations with *e*-fluentials and protecting reputational assets.

A proactive approach to reach e-fluentials

While it is essential to communicate with all Internet users who make the effort to send a comment to a website, companies should also consider profiling target audiences who express interest in getting information about company products and services. The stakeholders in these opt-in lists have already identified themselves as active information seekers. They are therefore more likely to be an *e*-fluential than the average Internet user.

As an initial step in identifying and reaching out to *e*-fluential stakeholders among a company's larger audience, website visitors, online community members and e-newsletter subscribers who have agreed to receive communications from the company can be invited to participate in an online survey. The survey should consist of three main sections:

1. A screening section to identify *e*-fluentials.
2. A psychographic section to enquire about survey respondents' attitudes towards issues concerning the company and/or its industry.
3. A demographic section to learn more about respondents' product ownership, media usage, employment history, purchasing power and background.

After understanding their audiences' communication habits, attitudes and demographic characteristics, companies can engage those who qualify as *e*-fluentials in viral marketing initiatives.

How to screen for e-fluentials

Burson-Marsteller uses a proprietary algorithm to identify *e*-fluentials. The formula is based on the frequency (i.e. almost daily, a few times a week, a few times a month, time to time, never) with which Internet users engage in the following types of activities (www.efluentials.com/quiz):

- Participate in chatrooms
- Post to bulletin boards
- Post to newsgroups
- Post to listservs
- Send emails to companies
- Send emails to politicians
- Make friends online
- Make business contacts online
- Provide feedback to websites
- Forward news and website information to others.

When conducting customized research projects for clients, Burson-Marsteller uses this set of questions to screen for *e*-fluentials. After everyone in the sample completes the survey, the database specialists

the responses to the screening questions and create a new
parating the *e*-fluentials from the non-*e*-fluentials.

Identifying and communicating with e-fluentials: the SAP case study

e-fluentials empower themselves with information they receive from trusted sources. The overwhelming majority (90%) of *e*-fluentials say they open and read emails they receive from known sources, whereas only 60% of them open and read messages from unknown sources.[6] Building a trusting relationship with *e*-fluentials is therefore a prerequisite for successful campaigns in which *e*-fluentials talk about the company to their friends and colleagues.

The global enterprise software provider SAP has the optimal basis for such an *e*-fluentials viral marketing campaign. SAP boasts a global online community of thousands of IT professionals, clients and prospects who frequently participate in online chats, events and message boards sponsored by the company. In May 2003, SAP approached Burson–Marsteller to develop cost-effective marketing strategies to spread messages within and beyond these communities.

The field work

Burson–Marsteller developed a survey to identify the *e*-fluentials among SAP community members and to gain insights about the audience group. The survey started with screener questions on online communication habits and continued with questions regarding the respondents' jobs and companies.

The survey was posted on www.efluentials.com and was hosted by Burson–Marsteller. SAP sent emails to 20 000 randomly selected SAP community members, inviting them to take the survey. During this project, Burson–Marsteller did not handle the actual email addresses of SAP community members. To protect the respondents' privacy, SAP assigned unique ID numbers to each potential respondent. Burson–Marsteller used these ID numbers to tally individual responses and analyse data.

Between the dates of 11 June and 23 June 2003, 1978 SAP community members from over 90 countries completed the survey. The highest number of responses came from the US (19%), Germany (19%), the UK (9%), Canada (5%), Australia (4%) and Holland (3%). Once the data were collected, Burson–Marsteller applied the answers from the screener questions to the algorithm and gave a score to each respondent. Almost

4 in 10 (39%) respondents qualified to be *e*-fluentials. Burson–Marsteller created a list of SAP's *e*-fluential clients and prospects using the unique ID numbers.

Results

The survey results showed that SAP *e*-fluentials were:

- Similar to the typical US *e*-fluentials described in Burson–Marsteller's 2001 benchmark study.
- Significantly different than the non-*e*-fluentials in the SAP communities.

These key trends confirmed that the algorithm worked. A unique group with strong tendencies to generate and participate in word of mouth activities was identified.

Similar to the larger population of US *e*-fluentials, SAP's online influencers forwarded news and Web information (100%), emailed companies (97%), and provided feedback to websites (95%). As a professional IT community, SAP's online influencers (95%) were more likely than the general US *e*-fluential population (81%) to make business contacts online (Figure 7.1).

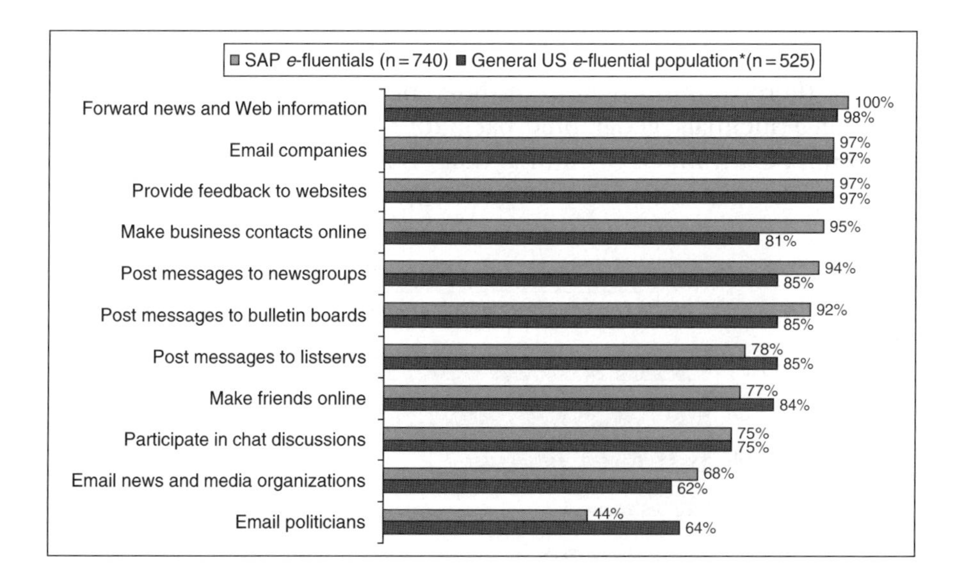

Figure 7.1 e-fluentials' online communication habits
*Source: Burson–Marsteller, e-fluentials Study, 2001

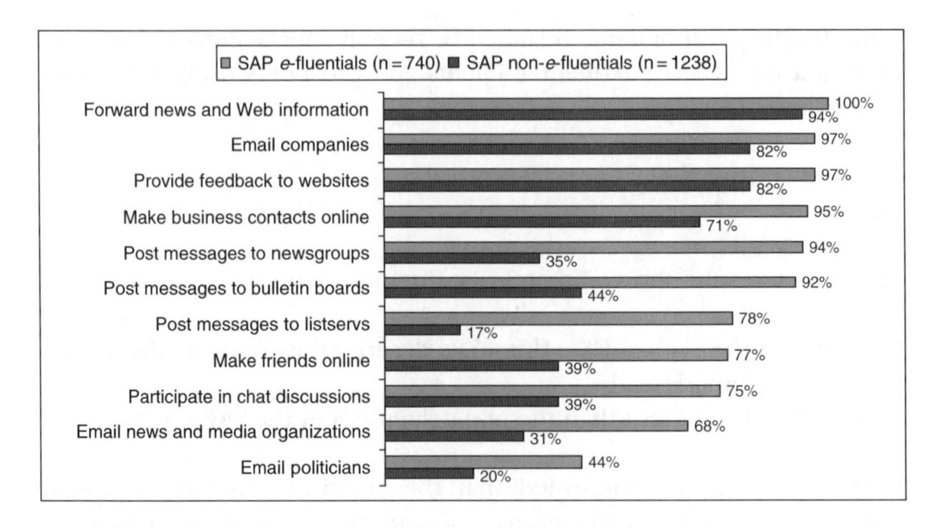

Figure 7.2 SAP e-fluentials vs SAP non-e-fluentials

Burson–Marsteller also compared the *e*-fluentials among the SAP audience to their non-*e*-fluential counterparts. SAP *e*-fluentials were consistently more active users of online communication channels. For example, they were significantly more likely than non-*e*-fluentials to email companies (97% vs 82%, respectively), provide feedback to websites (97% vs 82%, respectively), make business contacts online (95% vs 71%, respectively), and post messages to newsgroups (94% vs 35%, respectively) and online bulletin boards (92% vs 44%, respectively) (see Figure 7.2).

Since there were no significant differences between SAP *e*-fluentials' and non-*e*-fluentials' geographic background, their behavioral variance could not be attributed to cultural traits. Both groups were mainly from Germany, the USA and the UK (Figure 7.3).

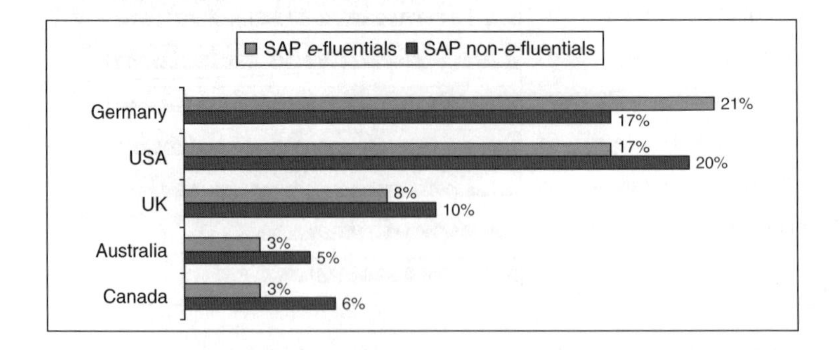

Figure 7.3 Geographic distribution of SAP e-fluentials

The viral marketing campaign

Following this analysis, SAP devised a viral marketing campaign targeting *e*-fluentials. The company developed an interactive webclip informing its community of IT professionals about its new enterprise solutions. SAP emailed the webclip to its *e*-fluentials (n = 776) and to a control group of non-*e*-fluential community members (n = 776). It also tracked the number of emails that were opened, read and passed on.

Seventy-one per cent of SAP *e*-fluentials opened and read the message. Each recipient interacted with the message at least twice – clicking on a link, printing information, or filling out the 'contact us' section. SAP also traced the way its *e*-fluential members spread the message online and measured the extended reach of its online promotion. Among the community members who opened and read the message, SAP's *e*-fluentials were slightly more likely than non-*e*-fluentials to forward the creative piece to their peers and colleagues (8.5% vs 6.6%, respectively).

There were significant differences between how third parties responded to messages received from *e*-fluentials and non-*e*-fluentials. Third parties were more likely to consider *e*-fluentials as trusted sources of information than non-*e*-fluentials. The emails forwarded by SAP's *e*-fluentials were almost twice as likely to be opened than those forwarded by non-*e*-fluentials (27% vs 14%, respectively).

Measuring success

When measuring the results of a targeted viral marketing campaign online, it is important to consider the context in which the messages are sent and received. There are several external factors which affect the communication between a company and its audience, as well as the buzz following a viral marketing campaign (Figure 7.4).

Evaluating the SAP viral campaign

Awareness of SAP

SAP used online marketing research to derive a list of *e*-fluentials among its audience who could spread the word about its products online and offline. The creative email was read thoroughly by most (71%) SAP *e*-fluentials who clicked on and downloaded information

Checklist for measuring the success of viral marketing campaigns

1 The audience's familiarity with the brand, product and company
- Do they know about the company?
- How often have they heard the brand name in the past year?
- Have they used the product?

2 The audience's level of involvement with the brand, product and company
- Do they use the product regularly?
- Do they hear or talk about the company regularly?
- How many times have they received calls, mailings, or emails from the company in the past year?

3 Current nature of attitudes towards the brand, product and company
- Have they heard positive or negative comments about the brand?
- Do they say or write positive or negative comments about the brand online and/or offline?
- What is their overall impression of the brand, product and company?

4 The issue's urgency and relevance for the audience
- Does the message require immediate attention and follow-up?
- Is the information related to the audience's work or demographic background?

5 The perceived benefits for the audience
- Does the promotion help audience members improve their work performance? Will they see the improvement in the short term or long term?
- Does the information have an impact on their personal lives? Will they note the impact in the short term or long term?

6 The creative appeal of the content (i.e. design and writing style)
- Does the email message have a catchy subject line?
- Is the written copy clear?
- Does the design draw the reader's attention to the key messages?

7 Content accessibility
- Is it easy to read and navigate through the main message areas?
- Are there additional links or pop-ups?
- Do readers have to scroll below the main message area to get additional information?

8 Cross-cultural communication
- Is there a uniform interpretation of key messages across social, cultural and work-field boundaries?

Figure 7.4 Checklist for measuring the success of viral marketing campaigns

about SAP solutions. The campaign generated awareness of SAP's new product offerings and planted the seeds for expert discussions on SAP solutions among the *e*-fluentials and their peers.

Involvement with SAP

The SAP *e*-fluentials had been active members of the SAP online community for two years prior to the campaign. They posted opinions on SAP discussion boards, attended industry events and received e-newsletters and other messages from the company on a regular basis. The *e*-fluentials who recognized SAP's brand name in the subject line might have felt inclined to click open the emails they received, as they were familiar with the vendor and they had participated in discussions about its products. Additionally, SAP community members' existing engagements with the company might have bolstered their response rate to the email survey and their interest in the creative email content.

Email content and access

The creative email contained evergreen information that could be used to improve the recipients' work at any time they chose to learn about the products and contact SAP. The SAP products introduced in the email offered long-term benefits to the audience. The pass-on rates could have been even higher if the email had been about an urgent matter that needed immediate attention (e.g. product recall, corporate discount, etc.).

The high click-through rates and the number of activities SAP *e*-fluentials performed within the email creative suggest that the recipients did not have trouble accessing the main message areas. They were engaged in the content.

e-fluential psychographics are universal

The typical *e*-fluentials described in Burson–Marsteller's studies of US online adults exhibited communication patterns similar to the multinational group of SAP *e*-fluentials (see Figure 7.1). The parallel behaviour of the two *e*-fluentials samples suggests that online influencers' Internet communication habits are universal. Within the group of SAP *e*-fluentials, IT community membership and common industry experience may have additionally neutralized cultural differences. While language, custom and culture may affect the way *e*-fluentials perceive and react to creative marketing materials, their online behaviour appears to cross borders and follow the model described in Burson–Marsteller studies.

This project demonstrated the value of implementing quantitative research to identify online influencers and focusing marketing efforts on these powerful stakeholders. As a result of its online outreach, SAP familiarized itself with its most influential community members and further enhanced relations with them. The *e*-fluentials became

SAP's brand evangelists and put their expert-approval seal on the new products. SAP successfully used Internet-based communications to bolster its reputation within the IT community.

Recommendations for identifying online public opinion leaders

The results of viral marketing initiatives can be measured online, providing an accurate account of campaign reach and brand awareness. The *e*-fluentials formula enables marketers to predict which individuals are significantly more likely to pass along and discuss messages online and offline. Email forwards, site-registration entries, messages posted on discussion boards and blogs are some of the metrics that illustrate how small, but powerful target audiences pass along messages to critical masses.

Companies can apply the *e*-fluentials knowledge to their business, identify their own set of online public opinion leaders and turn them into brand advocates. Below are steps for measuring campaign results, demonstrating the return on viral marketing investments and managing relations with opinion leaders:

1. Identify the universe: What audience groups would you like to reach and influence? This group can be identified in terms of demographic characteristics (e.g. women under the age of 45), behaviours (e.g. people who manage their own finances), or attitudes (e.g. people who think technology unites families).
2. Think of opinion-leader criteria and classify the target audience group: In addition to *e*-fluential communication habits, what are some relevant characteristics for opinion leaders in your field? Opinion leaders who discuss automobiles online and attend car shows every year might be a desirable segment for automobile manufacturers. Opinion leaders who watch cooking shows on a weekly basis and exchange recipes online might be a more suitable group of opinion leaders for a supermarket chain or a cooking magazine.
3. Generate list(s): What channels can you use to create an opt-in list of people who might fit the opinion-leader profile? You can screen for opinion leaders among stakeholders who agree to receive communications from your company by registering on the company website, giving their information at a conference booth, or signing up during an industry event.
4. Create a poll for list members to take: Develop a survey including screening questions to identify the opinion leaders among your

stakeholders. Ask questions about behaviours and attitudes that would qualify them as thought leaders in your field. Partner with an online research firm to host the survey online. Invite the list members to take the survey.

5. Identify the opinion leaders based on study results: Analyse the answers to screening questions to define the *e*-fluentials and non-*e*-fluentials in your sample. Study the *e*-fluentials' responses to other questions regarding their attitudes towards industry issues, use of your company's products, and the buzz they generate online and offline about your business.
6. Start communicating with your opinion leaders: Opinion leaders crave knowledge. They like to have insider information from primary sources. Offer to send them e-newsletters and white papers about new products and industry trends. Give them a chance to try your services then take their feedback. In times of crisis, alert them to messages from senior-level executives demystifying rumours about the company.
7. Be prepared to greet them, if they visit again: *e*-fluentials often visit company websites before making up their mind about an issue. Track your *e*-fluential stakeholders' visits to your company website. How frequently do they return to the website? Which areas do they visit? How long do they stay? Be sure to respond to their queries in a timely fashion. They like to check multiple sources to verify the facts. Be transparent and provide them with objective sources of information.

Conclusion

As the SAP case demonstrates, viral marketing campaigns targeting online public opinion leaders can successfully incorporate research and customer relationship management. The *e*-fluentials formula also shows how marketers can prove return on investment by quantifying viral marketing campaigns. Opinion leaders' profiles reveal who receives brand messages, how information disseminates through peer-to-peer networks, how far product- or company-related stories travel, and the overall impact of the campaign in the marketplace.

The Internet is no longer an abyss of chatter. Online research can identify opinion leaders, track their behaviour and illustrate how word of mouth works. Marketing practitioners can extend the lifetime of viral campaigns by getting to know the opinion leaders interested in their brands, earning their trust and communicating with them as business partners on an ongoing basis.

Takeaway points

- Online opinion leaders (e-fluentials) are influential gate-keepers and diffusers of information on the Internet. They should be a high-priority target group in viral marketing campaigns.
- e-fluentials make up 10% of the online population, and are characterized by their frequent and extensive online communication habits.
- On average, e-fluentials will tell their experience with a company to 14 other people.
- e-fluentials are trusted experts. People are more likely to open and read company, brand, or product information when it has been forwarded by an e-fluential.
- Companies should focus CRM initiatives on developing partnerships with e-fluentials, soliciting e-fluentials' feedback on product/service developments and informing them about new initiatives.

Notes and references

1 Burson-Marsteller, *e*-fluentials Study, 2001.
2 Burson-Marsteller, *e*-fluentials Study, 1999.
3 Ibid.
4 Burson-Marsteller, *e*-fluentials Study, 2001.
5 Burson-Marsteller, Corporate Rumors Study, 2003.
6 Burson-Marsteller, *e*-fluentials Study, 2001.

8

Buzz monitoring

Pete Snyder

Founder and CEO, New Media Strategies

When the Internet started to gain some serious traction in the mid 1990s, it was as if the world's largest focus group had suddenly appeared. Granted, the Web is a virtual space filled with a great deal of superfluous information, but it's also home to a wealth of useful consumer, shareholder, brand and competitor intelligence. In many ways, the Internet is like a map of the US. It may be composed of millions upon millions of acres of information, but consumers' population density – their demographic concentration – is relatively finite. What's more, this virtual landscape is readily available for analysis, feedback and two-way communication.

Put simply, the Web provides businesses with an unrivalled opportunity to listen to customers. And listening to target audiences and other stakeholders (such as shareholders and public interest groups) provides businesses – especially those with consumer brands – with exceptional opportunities to learn from, protect and grow their customer base.

Since the late 1990s, the challenge has been to find an effective way to gather that data and then categorize, analyse and transform it into actionable information which positively impacts a company's bottom line. In short, the task at hand was, and remains, to collect and collate 'business intelligence'.

From a technological perspective, there is an entire hardware and software industry built up around data storage, gathering and analysis. But the real challenge, and the ultimate goal, has far less to do with technology and more to do with knowing how to interpret this data and how to use such intelligence hand-in-hand with connected marketing strategies to generate buzzworthy PR for clients. This chapter explores how to achieve this goal and, in doing so, demonstrates how maximum benefit can be extracted from Web users as the ultimate online focus group.

Business intelligence – the basics

Before discussing online monitoring and business intelligence in the context of connected marketing, it's worth going back to basics to define the general term 'business intelligence'. Searchcrm.com offers this contemporary definition:

> A broad category of applications and technologies for gathering, storing, analyzing, and providing access to data to help enterprise users make better business decisions. BI applications include the activities of decision support systems, query and reporting, online analytical processing (OLAP), statistical analyses, forecasting and data mining. . . .

It's a decent definition, but it doesn't really get to the meat of things; it underscores the technological bias and limitations of so many corporate intelligence endeavours. More accurate is a qualitative description such as: business intelligence refers to companies' efforts to *obtain* and closely *assess* knowledge gathered from data about their enterprises - and/or those of competitors - that generates *actionable analysis*, which means it improves efficiencies, enhances strategic planning and increases profitability. If it doesn't benefit the bottom line, there's not much intelligence to it.

Before the emergence of the Web, business intelligence was, primarily, a private effort for corporations whose internal departments, such as customer service, manually tracked incoming information or captured it via technologies (for example, automated call logs). On the consumer side that effort was, in many ways, all about companies talking to themselves. They may have collected a small amount of objective customer data from focus groups, phone interviews, or printed surveys, but most information took the form of in-house statistics or inbound communications such as sales numbers and customer complaints.

A brief historical overview of business intelligence reveals how far this field has developed. From the heyday of punk rock in the 1970s through the British New Wave invasion of the 1980s, if you were a corporate executive and wanted access to company intelligence, you usually had to turn to the IT department, which reigned over information gathering, exchange and storage. The response to such a request often consisted of a spreadsheet of sometimes useful but often useless statistics without a clear methodology for interpretation. As club kids were moving from the glam rock of the late 1980s to electronic house in the 1990s, the business world saw a shift toward client/server systems, data warehousing and data mining. In this era, some companies recognized the value of intelligence gathered on a regular, often daily basis, such as transactional records, customer enquiries and complaints. Data analysis tools emerged, most of which were based on SQL (Structured Query Language), and there was an increased interest in standardizing information-gathering, reporting and dissemination methodologies.

According to data warehousing analyst Tom Burzinski, business intelligence grew out of technological developments that enabled companies 'to use historical data collected over a period of time to predict trends'.[1] That predictive role remains important today, as does the recognition that developing and using technology is only the first step in most corporate data gathering endeavours. Automated trend reports and number crunching are, ultimately, as useless as they sound unless they offer insights into what's working (or what isn't working) for a company to achieve its goals. The true *intelligence* in business intelligence, at least in the advertising industry, is the value delivered by marketing communications professionals. They are trained to assess information and transform it into actionable knowledge that protects corporations, maximizes promotional efforts and increases customer acquisition, satisfaction and loyalty.

Amass intelligence

The 1990s are over and grunge rock is dead. It's a moment in history when many of us who work in online marketing in the new millennium are so savvy about the Internet that we hardly remember a time when it was a vast unknown resource yet to be tapped. We take in stride the fact that the Web has created immense opportunities and significant challenges in polling efforts, and in marketing and communications, including providing brands and corporations with early warning signals from a PR perspective. We all nod in agreement that there's a lot of garbage online, that porn and gambling run rampant. But we also take pride in knowing that the Web is filled with diamonds in the rough and that when you can access the opinions of and conversations among tens of millions of people, you're bound to find some customer relationship management (CRM) gems.

That's why the Internet's affinity for passion interests is such a boon to marketers - whether these interests are topic-related, such as sports and technology message boards, or conventionally demographic, as is true of portals for the young and old. Audience segmentation on the Web allows for more targeted, ongoing and effective two-way communication between corporations and their stakeholders than ever before.

The Internet is a medium in which consumers give voice to opinions in much broader and diverse ways than in letters to the editor or on talk radio. Imagine, for a minute, a TV executive who wants to find out what viewers think about one of her soap operas. She used to read fan mail in the rags and once in a while she would shell out cash for a few focus groups. What does she do now? She hires an online monitoring specialist to tap into what 15 000 of her show's fans are saying on the Web about the series, the actors and the storylines, and obtains consumer intelligence more immediate, detailed and informative than she had ever dreamed possible.

On the Internet, consumers talk at all times of the day or night, even if no one listens. We're all aware of the 24/7 nature of the medium – it's on all day, every day so that news, opinions and controversies can break at any moment. All of which begs the need to monitor what's being said, to listen to and discern what matters most and, when appropriate, to respond as quickly and effectively as possible.

In contrast to the portrait of the Web just outlined, when business intelligence was coming of age in the late 1990s, most companies with an online toehold were focused myopically on counting eyeballs, measuring stickiness and logging clicks. While those firms were hiring designers to build expensive, splashy sites (they believed the hype of the *Field of Dreams* online marketing mentality – 'If you build it consumers will come'), practitioners such as New Media Strategies were founded on the premise that organizations need to listen to and talk with their customers where they live online. This philosophy is as powerful today as it was in the late 1990s: marketers need to take brands, promotions and conversations to online communities instead of waiting for the communities to come to clients.

Online monitoring and business intelligence should be focused on CRM and return on investment (ROI) from the start. Best practices in this industry have to include active listening to and communicating with consumers. They must also include the collection and analysis of actionable data that enables clients to tap into, understand and respond to what customers, shareholders and competitors are saying about them, their brands and their services. The intelligence industry is as simple and as complicated as that.

With the rise of the Internet, business intelligence source material has become more diverse, public and complex. It now includes stories published on legitimate news sites, as well as consumer attitudes and opinions posted on message boards, blogs, in chatrooms and other online venues. Unlike conventional market research, if you know where to look, and what you're seeing on the Internet, you're dealing with intelligence that's unfiltered, honest and, when delivered to clients in a user-friendly and effective format, actionable. But it's important to remember that unfiltered information cuts both ways - on the one hand, it can offer positive feedback on brands and products and prompt enthusiastic consumer-to-consumer (C2C) buzz; on the other, it can showcase the Web's penchant for unfounded rumours, breaches of confidentiality and can generate negative C2C communications. Capturing and understanding the nuances of this unfiltered data are, then, basic components of the online monitoring and business intelligence enterprise, as is knowing what to do with the knowledge once you have it.

The online intelligence game can be as rudimentary as automatically scanning the Web for logos, trademarks and patents, and alerting clients to copyright infringements. It can involve monitoring digital assets (for example, software, music, electronic books and films), tracking piracy cases and issuing Cease and Desist letters.

While brand infringement and related services are helpful - especially for industries that adhere to government regulations and/or demand compliance and consistency among affiliate and partner networks (for example, pharmaceuticals, financial services and insurance) - the most dynamic and useful intelligence applications are far more complex.

When executed well, online monitoring and business intelligence serves multiple purposes:

- It's an early warning PR system essential to brand development and protection - it taps into what information is circulating on the Internet that may not be intended for public knowledge, or may be damaging to a corporate or brand identity.
- It assesses brand awareness - it scans who's talking about a brand and its products and what they're saying.
- It evaluates the effectiveness of marketing campaigns - it tracks promotional themes that resonate with or alienate customers.
- It gathers competitor data - it evaluates aspects of competitor products that excite consumers.

These types of business intelligence provide organizations with diverse benefits, especially when coupled with messaging designed to amplify positive online conversations or intervene in and reduce negative buzz. The following case study demonstrates this point.

Case Study 1: Royal Ahold

Issue: Royal Ahold was overshadowed in the US by its subsidiary brands and was intent on maintaining a strong standing among its investors. In 2003, the company found some internal accounting irregularities and was concerned about the negative impact of this discovery on investor and consumer confidence and brand perception, especially in an era of Enron and World Com scandals.

Background: NMS conducted online research on the perceptions of and responses to Royal Ahold's accounting challenges as soon as the news broke. The research revealed negative buzz focused on Royal Ahold's financial stability, plus concern that the accounting discrepancies reflected larger corporate problems. Analysis of these online conversations generated two primary responses: (1) to assure all stakeholders that the accounting issues were an isolated incident; and (2) to underscore that the mishap did not and would not impact Royal Ahold's long-term viability and valuation. A PR strategy was developed by NMS in keeping with this analysis, and strategic messages were successfully targeted to industry insiders and Royal Ahold's investors and consumers.

Results: While online intelligence initially offered evidence of damage to investor confidence and to the brand, the tenor of online conversations shifted and improved dramatically soon after the PR campaign was launched. This response provided the client with evidence of the importance of listening to and conversing with online stakeholders and industry leaders.

Whether a project is simple or complex, online monitoring and business intelligence depends on understanding the industry sector, data needs, and PR and marketing objectives of the business in question. It also requires you to define the scope of the online universe and categorize the information generated. In other words, online intelligence uses the best of technology to pull knowledge from the Web, and the right methodology to determine what data to collect and how to slice and dice it. The goal is, then, to quantify and qualify. This means business intelligence practitioners have to ask and answer a range of incisive questions:

- What constitutes a client's universe (industry sector, existing consumers, target demographics, competitors), and what are their priorities and objectives (PR crisis management, brand awareness, tracking trade regulations)?

- Where on the Web are consumers and other stakeholders talking about clients, competitors and/or their industry sector?
- Where are consumers and other stakeholders not talking about clients, competitors and/or their industry sector?
- What is being said about clients, brands and products (positive, negative, neutral) in static content (such as newspaper articles online), and in consumer-generated interactive content (news groups, message boards, blogs, chatrooms)?
- Who's talking about competitors and what are they saying?

The art of intelligence

Asking the above questions is an important first step, but the real art of online monitoring and business intelligence is making sure the answers are germane to a client's bottom line and business objectives. In the case of Royal Ahold, that meant not only listening to what was being said on the Web, but also crafting a strategy for how to improve the tone of those conversations and how to quell investor and consumer concerns effectively.

If technology and methodology are the first two key components of data gathering efforts then the third is professional analysis, which means turning the intelligence into business intelligence, and translating analysis into action. Businesses need to know: what's being said about their brands and products? What does it mean? And what can they do about it? Or put another way: how can they influence or magnify conversations online? The goal is to keep an eye on corporate watchtowers and make sure the brand's perspective is focused enough to guard current market share, and broad enough to grow it strategically.

Like copyright infringement and compliance projects, high-level online monitoring and business intelligence analysis involves asking and answering questions. But it's less about information gathering and more about drilling down into and interpreting data, and packaging the results in a way that's informative and actionable. This suggests one of the primary ways in which online intelligence can be part and parcel of connected marketing strategies. If a company's goal is to track brand awareness during a new marketing campaign, then the intelligence gathering process might look like this:

- Determine the scope of the Web universe relevant to the brand and product, and understand the consumer demographics and targets.
- Conduct a baseline audit of brand awareness prior to the launch of the marketing campaign with a view to understanding: who is talking about the brand (and competitor brands); where are those

conversations taking place; and what is being said (positive, negative and neutral)?

- Track the volume of online discussion and perception before the campaign launch and during the campaign in order to monitor buzz peaks and valleys, and to map shifts in sentiment and attitude over time.

While marketing efforts often work hand-in-hand with data gathering, sometimes campaigns precede the need for online intelligence and brand protection, as we see in the following case study:

Case Study 2: Burger King

Background: Phase 1 of the project was to develop and execute an online viral campaign to market one of Burger King's toy promotions. During the course of the campaign, two children allegedly suffered injuries while playing with the toys. This launched Phase 2 of the project, which was designed to evaluate and quell public perception of the allegations and to facilitate Burger King's recall of 25 million toys.

Strategy: During Phase 1, a viral campaign was launched that targeted a broad range of online communities. During the course of this campaign, NMS worked closely with webmasters and online influencers to promote Burger King's toys. For Phase 2, the focus was on brand reputation management, brand protection and consumer intelligence to determine (1) what people were saying online about Burger King and its toys; and (2) what concerns, opinions, or complaints were being expressed by customers and other interested parties (which included parents, children, toy collectors and visitors to general interest sites). Appropriate online brand and CRM tactics were developed. A dynamic and innovative online campaign was launched to spread the word about the toys to at-risk groups and to correct erroneous consumer information about the recall. As part of the campaign, Burger King's customers were engaged in real-time dialogue about the potential dangers of the toys.

Results: Phase 1 reached an estimated 30 million targeted consumers online and helped contribute to the largest toy promotion in US history. Phase 2 provided Burger King with intelligence that enabled the brand to understand customer attitudes in real-time, and to adjust its online brand and CRM services in response. The recall efforts were so successful – and so positively received online – that they earned praise for Burger King in the *Washington Post* and from Associated Press and Reuters.

As the case studies illustrate, online monitoring and business intelligence is all about asking and responding to diverse yet always strategic questions. The answers provide businesses with actionable data that can be either short- or long-term in focus. For example, early in a new marketing campaign, online intelligence can deliver information regarding promotional elements that are alienating consumers, which could prompt the revision of creative material or other aspects of the campaign. Conversely, research can provide information on secondary marketing elements that resonate with customers to an unanticipated degree, which could be built up for the duration of the online campaign. Longer-term strategies include providing advice regarding product development (in light of campaign feedback or research on competitor brands), suggesting future campaign ideas, and offering more general trend analysis and business predictions.

Conclusion: the bottom line

It's a truism that competition among consumer brands has hit an all-time high, and that demand for innovative marketing and CRM strategies is more crucial than ever. But it's a truism worth repeating because corporations are increasingly hard-pressed to find ways to maintain their market share, never mind grow it. Getting ahead of the curve in a climate of fierce competition means knowing what consumers, competitors and other stakeholders are saying about clients, their products and brands. It also means knowing how to respond quickly and effectively to what's being said and understanding the implications of that buzz in the short run and over the long haul.

There's been a huge increase in the use of online monitoring and business intelligence services over the past six years, across a widening breadth of industries. More and more companies understand that if they represent a major brand, or if their products and services touch consumers and other stakeholders across communities, they need to pay attention to what's being said – or what's not being said – on the Internet. That need is only going to grow as high-speed access increases, as online ad spending continues to rise and as companies strive to cultivate best practices.

Having and holding onto a competitive edge has never been more important than it is today or than it will be in the foreseeable future. The good news is that CEOs love to know what stakeholders are saying about them and their corporations. The even better news is that for brands to maintain and grow market share, and for product development to thrive, companies absolutely need to know what people are saying about them.

Enter online monitoring and business intelligence - it will help you listen, help you communicate and most of all, make sure you keep your competitive edge.

Takeaway points

- The Internet provides an unrivalled opportunity to listen to stakeholders, 24/7.
- Business intelligence analysis is less about gathering data and more about drilling down into and interpreting data.
- If business intelligence doesn't benefit a client's bottom line, there's not much intelligence to it.
- Internet market research is unfiltered and honest, but this can cut both ways – the Web is a forum for positive feedback and consumer-to-consumer buzz, while also a vehicle for unfounded rumour.
- Businesses need to find out where their customers live online and listen and talk to them in these places, in real-time whenever possible.
- Online monitoring and business intelligence enables organizations to listen to and communicate with consumers, which helps them to build or maintain a competitive edge.

Reference

1 Burzinski, T. (2004) *Business Intelligence: Today's Decisions, Yesterday's Data*, 2 March, devx.com.

9

Changing the game

Steve Curran

President, Pod Digital

Television ratings continue to decline, personal video recorders enable viewers to zip past TV ads, and commercial-free satellite radio is shaking up the radio industry. Consumers are demonstrating with their remote controls and their wallets that, given a choice, they would much prefer not to have their entertainment experience interrupted, thank you very much.

For brand marketers relying on capturing the attention of an increasingly impatient and distracted consumer, there are major shifts taking place in technology and audience behaviour that are turning the business of advertising into a whole new game - literally.

Games are a wildly popular medium and represent a fertile new territory in which to engage customers in ways that are impossible with traditional advertising. Coca-Cola, Nabisco, Nokia, Procter & Gamble and many other international brands have been using games for years as an important component of their integrated marketing mix. More and more smart marketers are waking up to the fact that games and game-like interactions are the ultimate integration of brand and entertainment, and present an unbridled opportunity to maximize the benefits of connected marketing strategies.

In the past few years, a growing number of companies have started to rely on the power of 'advergaming', a genre that sets out to exploit commercially the unique characteristics inherent in computer gaming.

There are two main types of advergames: those that make use of in-game advertising via devices such as product placement (Tony Hawk's skater game, played on Xbox, PS2, or the PC is one of the best cases in point); and online branded games, specifically created to support a brand or product. In some cases, the gameplay will revolve around the product itself, as in the case of a US Army recruitment game discussed later in this chapter.

The former are capable of creating word of mouth and buzz, especially if they're topical, such as *FIFA 2006*; but it's the latter that have the most potential when it comes to connected marketing. Beyond the simple appeal of the stickiness of these promotional tools to inspire lengthy and repeat interaction, they can also generate valuable customer data, and enormous traffic through online forward-to-friend activity.

Whatever the style of advergame though, they provide a way of integrating product awareness into the entertainment experience itself without detracting from or interrupting enjoyment of the experience. Not that this is anything new; products have been strategically placed in television shows and films for years. The difference now is that, with games, marketers and advertisers can get away with completely designing entertainment around a brand - and at a minimal cost compared with devices such as product placement. There are also opportunities for definitive and precise measurement of who is interacting with the entertainment and for how long.

It's these inherent qualities of advergaming that make it such an exciting medium for marketers and advertisers. This chapter illustrates this in more detail and shows how you can maximize the potential of advergaming.

Computer gaming: even bigger than the movie business?

The advergaming business alone is expected to grow to a billion-dollar industry by 2008. In the same period, the market for cell-phone games is expected to grow to US$1.7 billion. The video game market generates billions of dollars more revenue than Hollywood every year. ACNeilsen predicts that within just four years the film industry will be only one third the size of the computer games market.[1]

If you want to see an early indication of this, in its first weekend on sale, the computer game *Halo 2* generated sales of over US$125 million, while *The Incredibles*, the number one film in the country opening the same weekend, grossed US$70 million.

But to exploit the computer games sector fully, it needs more than an awareness of how big the potential market is. Marketers must appreciate what makes gaming so different from other media.

Interactivity beats passivity every time

It's not just the popularity of games that makes them so appealing as an advertising medium; it's the way consumers are involved in a game that offers unique advantages. Marketers must leverage this engagement to get their messages across, or even to go as far as changing brand perception.

With games, consumers are engaged in an interactive versus a passive experience, a lean-forward versus a lean-back experience. With a television ad, the best-case scenario is to get 30–60 seconds of a consumer's attention and interest. With a branded game, an advertiser has the very real potential of getting hours of undivided, proactive attention. Instead of trying to compete for a few seconds of an unengaged customer's attention, games can immerse the consumer in a world designed by an advertiser. These worlds can combine brand associations with fantasy worlds of wish fulfilment, challenges or quests, athletic feats, humour, irreverence, or other attributes relevant to a particular brand message or identity.

The ultimate in social networking

Gaming is a social medium. It offers opportunities for users to challenge each other and to recommend games to their friends. Some branded games are multiplayer, a platform that enables peers and/or strangers to interact in a social setting. The potential for embedding marketing messages and using connected marketing techniques to spread those messages is huge.

Inform, educate and entertain

Games have always provided an opportunity to involve, educate and inform people. They are entertaining, thought-provoking, require focused attention and, in many cases, are social exercises. From our earliest education, we learn best in the form of games and social play. Teachers make games out of seeing who can answer a question first, who can find what's wrong with a picture, or getting us to form teams to solve puzzles. Serious games are used as vital tools with an educational purpose to teach everything from how to make war, create health and wellness, or vote in an election, to how to build an economy, operate on a heart, or fly an airplane.

Crossing demographic borders and gender gaps

Gaming now crosses all demographic boundaries. While the perception exists that gaming is the domain of testosterone- and pizza-fuelled teenage boys, more than 50% of game players are 35 or older, and mostly female, particularly online. In fact, according to a recent study,[2] women comprise the largest segment of the online game-playing audience. Clearly, these women are the ideal target audience for brand marketers online. They are upscale, educated, professional women who also use the Internet to interact socially, research product information and shop.

Every new platform is a boon for marketers

Today, gaming opportunities for brand marketers exist on every major digital platform, including televisions, digital phones, PDAs, PCs, the Internet, console and hand-held games. With every new rollout of a digital platform, from interactive TV to PDAs, games rank among the most popular applications, and are pushed forward by the manufacturer or service provider as a showcase feature that sells hardware.

The growing benefits of advergames

Advergames offer the opportunity of mining consumer data that is unheard of in traditional media. Advertisers have found that, in return for the value of the game-playing experience, the audience is far more likely to opt-in to sign up for future promotions and products, or share personal information with an advertiser seeking a richer database. Campaigns that integrate advergames will routinely get 50–75% of the participants choosing to opt-in to future messages or promotions. For example, Honda's contest to win a CR-V (see p. 136) enabled the car manufacturer to gather valuable data about potential customers.

Other advertisers have found that invaluable data can be gathered by offering players in-game choices related to a product. The resulting data can be used to create sophisticated psychographic profiles of customers in order to build better products and promotions in the future.

Branded online games can be produced for a lot less than the cost of games designed for the console, PC, or handheld markets. While PC and console games could take years and millions of dollars to create, a branded online game can be created in months, even weeks, with the cost of entry as low as US$5000–15 000.

Finally, the distribution costs of branded online games make advertisers drool. If a game is well targeted, has real play value, is innovative and doesn't beat players over the head with the branding or hard sell, the players themselves will become the distribution channel and buzz generators via forward-to-friend emails, blogs, chatrooms and word of mouth. Without the limitations of an expensive media spend, games can live on the Internet for a long time, virtually free. So build it well, and they will play.

It looks certain that the advergaming phenomenon will continue to develop, so investing now in how to best use the medium is a budget well spent. New platforms and gaming techniques are emerging all the time. The 'ilovebees' promotional game for *Halo 2* and the MiniCooper hoax site, among others, can be credited with helping to pioneer the trend towards alternate reality gaming (ARG) or immersive gaming – a potentially huge vehicle for connected marketing strategies. ARG draws the gamer into a fantasy world that becomes entwined with their own reality via devices such as hoax websites, treasure hunts and pre-recorded phone calls.

In the UK in the 1980s, confectionery manufacturer Cadbury staged a pioneering branded treasure hunt via a printed book that sent people off to dig for golden eggs (in connection with its famous Cadbury's Creme Egg brand). A few years earlier author Kit Williams sent people in search of a golden hare if they could crack a set of clues buried in illustration and rhyme in his book *Masquerade*. Both books[3] generated massive word of mouth and buzz. ARG takes the same idea, but has the advantage of the Internet with its ability to generate even more word of mouth and connect communities of people who have a common interest.

Who's playing the game and how?

As already stated, there are broadly speaking two subsets of advergames: games that make use of some kind of in-game advertising such as product placement; and branded promotional games, which are built specifically for the brand. In the past few years there have been some notable examples of advertisers tapping into the power of both types of advergame. Let's look at some of these examples in more detail.

In-game advertising and marketing

This is similar to buying a 30-second ad slot in a television programme, or buying hoardings at a sporting event - except that the medium is a computer game that plays on one of the popular platforms such as Xbox, PS2, or PC. By riding on the back of the game, advertisers can get their message out to a target audience.

Buzz is often created, especially if there is a topical link in the real world. For example, FIFA's branding can be found on the promotional hoardings featured in the computer game *FIFA 2006*, which in the run-up to the football World Cup 2006, will be the focus of a great deal of gameplay and buzz. Other less topical examples include the McDonald's brand being featured in *The Sims*. This technique is still capable of generating some buzz, but on a limited scale (there will be some kudos attached to eating at McDonald's in *The Sims*, so it is likely to be talked about).

Branded promotional games

This is where a game is actually created by the brand itself to support and promote the brand - it's sometimes referred to as advertainment. Often the brand or product will form an integral part of the gameplay. This doesn't have to be the case, but the brand will always be a central element of the game rather than riding on its back. The aim is for the game to be so compelling that it gets passed around the Internet or on mobile phones as the buzz increases and its popularity is communicated via word of mouth.

The best examples of this kind of game are found among the avalanche of games found on the Internet. Often a sweepstake-style mechanic provides a huge incentive to play, while in other cases it's simply because the game is excellent. Driving games for automotive companies are good examples, because the player is totally immersed in a brand experience - even better if players can also challenge and invite their friends to play.

Branded promotional games are a far more powerful connected marketing tool than games that simply feature in-game advertising. As well as building brand, creating awareness and generating response, branded promotional games also enable marketers to collect valuable data on customers. The following examples show how some branded games are designed to achieve one or more of these objectives, while others excel on a number of levels and can also have a direct effect on sales.

Case studies

M&M's Flip the Mix at http://mms.kewlbox.com/: This game formed part of a publicity campaign to announce M&M's new colour, in a battle between red and green. It also helped to build brand and the game scored high as a piece of entertainment in its own right: it was played eight million times for an average of 45 minutes to one hour per player.[4]

Kraft Food's NabiscoWorld at www.nabiscoworld.com: This games compendium continually reinforces brand messages by drawing more than three million visitors a month to its website. It features over 80 different games, each incorporating a different Kraft brand.[5]

Nike at www.nikesoccer.com: The sports clothing and equipment manufacturer is one of the masters of the branded game, using games throughout its website and as standalone microsites to promote basketball, soccer, track and field, and many other sports. Nike's games focus on both real world skill development and playing the game itself. They often feature cameos from top athletes in order to increase interest in the sports that will pay off in equipment sales over time. The games reinforce brand image in an entertaining and engaging way, and create buzz that is now self-perpetuating because they are well executed and have become known as the online games to play within the sport genre.

US Army recruitment at www.americasarmy.com: One of the most ambitious and controversial advergames was created by the US Army as a recruiting tool. Called *America's Army,* it proved to be a highly successful response generator. The game enables players to join the army, go through basic training and fight terrorism. The game cost about US$6.3 million to develop, and was designed with input from top games designers and military strategists. It factors in such detail as players' breathing, movement, stance and mastery of weapons. Within weeks of the release, more than one million people downloaded the game.[6] For better or worse, with more than three million registered players the game is one of the most popular games on the Internet, causing some to argue that the game is too effective in de-sensitizing civilians to war and violence. Still, the US Army was so inspired by the results of the game, it followed up with an upgraded CD release and has announced plans for the creation of a games and simulations studio.

BBC's Test the Nation at www.bbc.co.uk/testthenation: To encourage viewers to tune into its *Test the Nation* series of TV programmes the BBC runs quizzes on its website. This activity also builds brand when the programme is not on the air. For its National IQ Test programme, it also made the test available via a whole range of platforms including the Web, WAP, mobile phone, interactive TV, and even pen and paper.

Chrysler: The car manufacturer has used a variety of custom-branded games involving its products in order to increase brand recognition and purchasing intent among consumers. These games are also among the best examples of branded games generating sales enquiries and actual sales. Chrysler *Golf* resulted in a 33% increase in purchase intent, while 1000 of the 383 403 players of the Jeep 4x4 *Trail of Life* game have bought Jeeps in the last 18 months.[7] Awareness of Dodge brands was up 27.6% among Internet users who had played one of the games, and other results included a purchase intent of 19.6% for Dodge brands, and a boost in awareness of 24.7% of all Daimler–Chrysler brands.

Jeff Bell, Chrysler's VP of marketing, spoke at the 2004 Jupiter Media Advertising Forum about the importance of getting your brand into the consumer's 'consideration set' – the four or five brands the consumer has narrowed their purchase down to – as early as possible. He stressed that gaming was among the marketing channels that help to get brands into popular culture and give the consumer experience of the brand, as well as 'top of the head' awareness.

Ford's Race On the Moon: Ford Motors used an advergame to promote the Ford Escape SUV. The game involved a race that featured the Escape and encouraged players to forward an email to friends asking them to beat their best score. There was a 40% response rate to the initial email, and an 80% response rate to the tell-a-friend emails. Ford also collected data such as email addresses and vehicle colour preferences.[8]

Honda's CR-V game: The Japanese automotive company created a slick 3D racing online advergame, designed around Honda car models and engineered to play like an arcade game, even on a slow-speed connection. The game functioned as a market-research tool, asking players to submit their age, address, occupation and hobbies before playing. Honda rewarded registrants by entering them in a contest to win a Honda CR-V.

The game yielded a 30% registration rate, and over 90% of registered users played the game for an average of four and a half minutes. Based on registration data, Honda was able to gather information about players' hobbies, the types of cars they own, the age of their vehicles and their car preferences.[9]

Miller Lite's Virtual Racing League: By collecting codes from packs of beer, players could tweak the cars in their online motor racing team and race stock cars against their friends. This advergame showed how an interface can be formed between the real and virtual worlds, and is significant because it demonstrated how brands can link online activities directly to actual purchases.[10]

Want to play? Four good reasons why

The previous examples demonstrate various advergames in action. Now let's look at the basics of why advergaming is so good at facilitating successful connected marketing initiatives.

1. One of the most powerful reasons to use games for advertising is that a game is far more likely to be sent from peer to peer than is a traditional advertisement or commercial message. Whereas many video clips that get passed on via the Internet lean towards an edgy sensibility, or even a shock value, games can be designed to be appropriate for a much wider range of advertisers and audiences. (That said, if you want to take an edgy and more 'out there' approach, there's probably no better medium.)
2. A proven tenet of email marketing - and fast becoming one of mobile phone marketing - is the power of forward-to-a-friend. People are much more likely to open a message from somebody they know and trust, such as a friend, associate, or family member. Digital promotions of games via MMS, SMS, microsite email functions, or plain email marketing, can facilitate this type of peer-to-peer interaction and generate a level of third-party endorsement that is unheard of in traditional media.
3. Forwarded-from-a-friend messages generally have high open rates. And a challenge from a friend is an even more irresistible method of getting recipients to interact with the content. A message demands to be opened, whereas a challenge demanding to be taken up is even more compelling. You have to interact with it. You might ignore, or read but not respond to, a joke, image, or clip sent to you - via email, SMS, MMS, etc. - but to ignore a challenge from a friend is the equivalent of admitting defeat. Most likely, the recipient will play the game until they have had an opportunity to secure bragging rights.
4. A friend can say 'Have you seen that TV ad?'. But unless this conversation is taking place online, you might never see the advertisement at all. When a friend challenges another to play a branded online game that has a 'challenge a friend' feature, the means and opportunity are provided for the recipient to participate in the promotion right away. This kind of immediacy is hard to achieve with any medium outside the Internet.

Getting in the game: the options

For a marketer looking to tap into advergaming opportunities, a variety of options exist that depend on the usual suspects of time, budget and threshold for risk. Games are not right for every marketing problem or solution, and it's helpful to look at the options available in the marketplace when deciding where the opportunities cross-sect with the promotional goals. Here are a few of the options available:

- **Have a custom game created for you:** These are online games that have been designed around the unique features or benefits of a product, service, or brand and can be located on the brand's website, or on a standalone microsite. The advantage of having it on the main brand website is that it draws traffic to your entire Web presence. The advantage of using a campaign microsite is that you can focus the attention on specific calls-to-action, such as forward-to-a-friend, registration, or discrete product promotion.
- **Brand an existing game, or build a promotion around it:** Many advergame companies have libraries of games that can be re-skinned and rebranded, for a price. The advantage of branding an existing game is the lower cost of entry as you are renting the game on (generally) a non-exclusive basis. The game has a proven appeal and its rebranding can be turned around quickly with minimal production needs. The downside is that these offerings tend to be generic and not very innovative, product-relevant, or buzzworthy.
- **Advertise in game portals such as Pogo, Gamesville or Yahoo!'s game section:** These sites attract a large loyal audience that spends a lot of time online playing games. Simple interactive games in ad banners have proven effective at dramatically increasing click-through rates, and can be a good way to test the waters to see if games work with your product.
- **Seek out branding opportunities with game developers or publishers:** Opportunities to push advertising into game environments help to defray the increasingly expensive production costs for games publishers and, when the advertising is included organically in the content of the game, it can enhance the game experience. In one study,[11] 70% of gamers said that realistic brands make the game more authentic. Tony Hawk's skate games look more genuinely representative of real-world urban environments with realistic-looking billboards than they would if the billboards had simple generic messages such as 'drink soda'.

Ironically, for a game that featured a context that needed advertisements to seem real, it would be a fake ad that would draw more attention than a real one, simply by virtue of the fact that it seemed out of place. Advertisers should be aware though that there is a fine line between in-game advertising having the desired affect and it merely adding to the clutter and being little more than wallpaper. A company called Massive has been formed to focus on ad-serving in video games. One of Massive's approaches is to make the passive ads dynamic within PC games if the player is connected to the Web when they play. This approach ensures that billboard-style ads are relevant and contextual, based on the player's criteria, such as location. While effective, this is an example of better targeting rather than true connected marketing.

In short, a logo added to a generic game that has little relevance to a brand creates little value. A custom game, on the other hand, with an original concept that organically weaves the promotional value into the entertainment experience, can provide great value - but it's far more expensive to create. Like anything else, you get what you pay for.

Creating a branded game

While games offer an ideal opportunity to involve your audience with a product, service or message, as with any medium, you'll want to make sure you've done your homework in order to ensure the best results.

It's not enough for a branded game to be popular if it's not also reaching the desired target audience, if it fails to gather information about that audience, or if it fails to meet the basic objectives of the promotion. Here are some helpful steps for creating and promoting a branded game.

Define your objectives

Is the objective to create awareness, generate buzz, educate, generate site traffic, look cool, or build loyalty? The objectives should be feasible goals for the outcome of the campaign. After defining the objectives, you can sometimes know whether or not an advergame approach is even the correct way to go. Many times, it isn't.

Define a call to action

Based on your objectives, what is the most realistic direct response you can expect out of the promotion? The most entertaining and informational driving game about a car is not going to convert to immediate car sales, but it could get somebody to sign up for a test drive, download a PDF, or ask for a brochure to be sent. As defined by the objectives, not every advergame needs to have a call-to-action outside of getting people to send it to their friends. But many of the more successful advergames have provided valuable net results for marketers looking for deeper profiling information on customers, product feedback, leads for sales follow-up, or yes, even conversion to sales.

Define the audience

This can take two steps: defining the audience of your brand, and then defining that audience in the context of games and entertainment. For example, if you find out that the audience for an online card store you are trying to promote is primarily women between the ages of 35 and 50, and the games they prefer are puzzles, card games and word games, that gives you a logical starting point for your creative process. Men on the other hand prefer 'twitch' or action games, simulations, aggressive games and dark humour. Is your audience sentimental? Jaded? Are they game-literate? Or do you need to focus on only basic types of interaction. Are they broadband users who can tolerate big downloads? And, most importantly, what would engage and inspire them to forward it to a peer, and where does that answer cross-sect with what helps to meet your objective?

Find a hot button

One of the critical components to the creation of a buzzworthy online advergaming campaign is finding a cultural hotspot. A topic that is hot in the public's eye presents an ideal opportunity to create an interaction that takes the game beyond being simple entertainment for entertainment's sake. Look for an idea that a newspaper or magazine writer will drool over because it lends itself to inclusion in coverage of a topic that has been getting a lot of attention lately. A good example of a hot button and the column inches that can be generated is the Condomi 'Size Him Up' campaign discussed in Chapter 5.

One campaign, 'Whack-a-flack' (see www.whackaflack.com) by e-tractions, had the objective of getting the attention of journalists who get bombarded by messages from PR flacks on a daily basis. The game made fun of the growing rift between journalists and PR professionals – it enabled journalists to take out their frustrations on PR people by bombarding them right back with their own press releases. It was a good-natured but on-the-nose spoof of the industry and was topical because at the end of the dot.com bubble, journalists were tired of being hounded with aggressive hype. The results were outstanding: the promotion was written up in the *Wall Street Journal*, *Adweek*,

BrandWeek, PR Week and many other publications, and it's hard to find a journalist or PR person working at that time who did not see it.

Another online campaign, created for a new line of skin care called Dr Comenge (see www.drcomenge.com/apothia), picked up on the hot button of botox and face-lifts gone awry. It enabled users to give the classic beauty of Mona Lisa a humorous and horrific makeover. Once the player was finished giving her collagen breast implants, acid peels and Michael Jackson's nose, the message about the benefits of the brand's non-abrasive, natural beauty-enhancing skin care was an easy segue.

Keep it simple

All good creative people want to create an entertainment concept that nobody has seen before. But with games, particularly promotional games where you have to grab a player's interest and draw them in quickly, you have to balance the innovation of your concept with gameplay that is simple to figure out. There have been some great original games at the centrepiece of a promotion that have required too much investment on the part of the audience when it came to how to play. In general, if people aren't familiar with what they are supposed to do, they aren't going to invest the time to learn. The balancing act is to keep the interaction simple, and the ideas big. A well-executed trivia game is likely to get a better audience response than a great original game with complex play.

Take a risk

This is the slipperiest slope of all when it comes to connected marketing campaigns in general. If the creative material is too safe, it probably isn't that interesting. Most successful connected marketing campaigns were either launched under the radar, were used by companies whose brands were built on taking creative leaps with marketing (e.g. Virgin, Nike), or at the very least had their fair share of detractors wondering if the campaign was right for the brand. If you want your advergame to reap the benefits of connected marketing and 'go viral', you have to create something that's going to be sent from peer to peer. Bear in mind that these are usually things you would only find online (or possibly on cell phones as that medium advances), and they therefore have a novelty that works in that context. A bawdy joke that would flop at a party suddenly seems hysterical when it lands in your inbox during a boring conference call, or on your mobile phone on the train home. The worst mistake to make is to make your advergaming promotion seem like, well, advertising.

Skip the 'Buy Now' button

Again, unless it was outlined as a realistic objective, it's a mistake to assume the only measurement of value for a branded game is how many sales it resulted in from players clicking through to buy. If direct sales are the only measurement of the success and value of your advergaming campaign, it's best to look for another marketing opportunity. The online game-playing audience is tolerant of sponsored games or promotional games to a point, providing the interaction has value. But avoid stepping over the line between integrating a brand and making a shameless pitch. The former feels smart. The latter feels desperate.

Find the right developer

Look for a developer with a portfolio of work that best matches the criteria of what you've outlined for the objectives of your game. Not all game developers create their games the same way. Some branded game developers only work with proprietary technology that requires plugs-ins to be downloaded to your browser, and this might be a problem for some audiences. Some developers are great at creating games based on games they have already built (in the game business, these are referred to as 'engines') but might not be best to hire for a custom concept. Conversely, if your marketing objectives allow for a game that can be customized around a pre-existing game engine, you can save a lot of time, money and worries about testing with a game that has already been battle-tested and is simply being re-skinned with new graphics.

Plan for success

Never underestimate the power of connected marketing when it comes to anticipating your bandwidth needs for an online branded game. Make sure that your website hosting plan can accommodate the potential for big spikes in traffic, otherwise you can get hit with a whopping additional charge for unplanned bandwidth usage. Find out how much your hosting plan can accommodate and work through the bandwidth requirements with your developer and server host. It's much better to overestimate than underestimate, as nothing will kill the success of a game quicker than slow downloads or poor performance.

Make it measurable

Make sure your advergame has built in the necessary tracking abilities to measure not only the game's popularity, but also the actions taken within the game towards a desired outcome. How long did the visitors play? How many were repeat visitors? Where did the game players go after they hit the link from the game to the brand's website? Did they purchase? The more information you can track and report, the more you can learn from the campaign to build on in future promotions. At a snapshot level, the same tools that measure website traffic (e.g. Webtrends, Urchin) can measure traffic to the game. However, most games are developed in Flash, and these tools can't report on user activity inside of Flash. To get more detailed reporting, many game developers have their own tools for tracking and reporting that they can build into the game. Make sure you ask for a demonstration, or to see an example of reporting.

If one of your advergaming campaign objectives is to build a database of customer information, make sure that the information is being recorded in a format that will be easy for you to use, or to reconcile with a larger database. If you've asked for permission to contact in the future, make sure that you can, and do!

The future of advergaming: multiplayer interactivity makes it massive

Looking forward, the opportunities for branding and exploiting connected marketing strategies inside of games will multiply as games become even more interconnected via online communities and social experiences.

Massive multiplayer games are reshaping the computer games industry, connecting players around the globe in virtual worlds such as Sony's *Everquest* or *Final Fantasy*. In 2004, the virtual character of a player who had completed 74 levels of *Final Fantasy* was sold to another player for US$1200. Spider Venom, a valuable item to have in the game *Everquest,* has sold for US$700 on online gaming accessories websites. These are indicators of the fanatic loyalty and value players place on their identities in these worlds. In *The Sims Online* from EA, the characters now live in a branded world, where they can purchase McDonald's franchises and sell the products to other players earning virtual money. Eating the food increases the players standing in the game.[12]

The realm of immersive or alternate reality gaming is also radically altering the gaming landscape. *Chasing the Wish* and *Urban Hunt,* created by one of the world's leading ARG creators Dave Szulborski, have been critically acclaimed. The former combines an online game with a real-life treasure hunt, while the latter draws people into what at first seems like a reality TV show. The viral and word of mouth potential of these type of products is huge as communities unite via weblogs, forums, or even face-to-face. The potential to deliver marketing messages is also significant, as 'ilovebees' and the high profile MiniCooper *Robot* hoax site have started to demonstrate.[13]

Now, with the low cost of game console units combined with high-speed broadband Internet connectivity, all major video game consoles such as Nintendo's GameCube, Sony's Playstation and Microsoft's Xbox are rolling out multiplayer modems and games with multiplayer features.

Similarly, all major phone manufacturers and service providers are increasingly offering community-oriented functionality and content, such as tournaments for top-scores, head-to-head combat, and chatting with other players. JAMDAT Mobile hosts about 25 500 multiplayer match-ups everyday, mostly from just three hit games titles, and many more are on the way. Sprint developed *Game Lobby,* a central forum on the phone through which mobile game players can meet, with an online leaderboard so that players of hit cell phone games such as *Bejeweled* can compete with users of ATT Wireless's game room. Nokia's N-Gage includes access to N-Gage Arena, an online community where people can play against each other online, chat and post on message boards, and receive tips and community news. Los Angeles-based Tomo Software Inc. is expected to launch a mobile game that is part reality TV, part *Sims,* and employs the social networking approach of websites such as Friendster.com. Players compete in games, with other players deciding the winner. Players communicate through online journals or by sending each other messages.

The sheer number of mobile digital devices out there and the increasing leaps in technology - from 3G to WiMax - make the mobile market a potentially fertile area for advergames. It's already being exploited by the likes of games developer Capcom, who in 2004 came up with an elaborate cellphone hoax, the *T-Virus,* to drum up interest in the computer game *Resident Evil.* While this was a hoax to promote a game rather than a game itself being served via mobile, it starts to demonstrate the potential benefits of this channel for marketers.

Conclusion

Games as a connected social experience will continue to reshape how we interact with friends, family, entertainment and marketers. Interconnected entertainment communities will offer ideal audiences and channels for branded games, sponsorship and promotion. They combine the fastest growing form of entertainment within the ultimate venues for connected marketing. Amazingly, advertisers are only really starting to take notice of the power and influence that advergaming is having on our culture.

For marketers there is much to be excited about: the way advergames can create buzz, how they bring together communities of people, encourage multiplay and stimulate massive word of mouth, and all for a fraction of the cost of some traditional marketing channels. But is this enough?

Like any marketing medium, advergaming will ultimately be judged on its ability to benefit the brand. And while we're right to be excited by its potential, we should also be mindful of some inherent dangers. It's great that games can be designed for multiplay, but beware that these communities and games can take on a life of their own that marginalizes the brand's involvement. The Budweiser catchphrase 'Whassup?', for instance, became completely unconnected with the brand. Numbers can also be deceptive. Sure, you can demonstrate that lots of people around the world are playing a specific game – and it's therefore a way of generating low-cost awareness – but what if a high percentage of those playing live in a country where the product or brand isn't available?

MarketingSherpa.com, an online publisher of marketing news and case studies, did its own viral test that illustrates this point perfectly. It launched a game called *TortureASpammer.com* in autumn 2001, which has had a million plays. However, 95% of the players were outside the desired demographic of professional email marketers (in fact, the game proved very popular with Eastern European teens).[14] Wastage is a huge potential problem for connected marketers; they need to come up with ways of better targeting, as well as mechanisms for linking the gaming experience directly to purchase. A good example of this is Miller Lite's *Virtual Racing League* mentioned earlier.

As well as providing some return on investment in a relatively short space of time, such devices also exploit connectivity – encouraging competition and irresistible challenges between players – which is ultimately what connected marketing strategies are all about.

There's nothing wrong with using advergaming to create low-cost awareness, particularly if you can show that you are entertaining and engaging potential customers. But if creating awareness is the only rationale then there's a danger of advergaming becoming just another random alternative marketing technique. So showing tangible business benefits such as improved intent to purchase or recommendation rates, or ideally sales, helps give advergaming more credibility. Chrysler (see p. 136) has done the advergaming industry a huge service by amply demonstrating that its branded games have contributed to the bottom line. Expect more statistics like this to emerge as the advergaming market matures.

Branded games and entertainment are still relatively untamed territory on the media landscape, and the opportunities exist now for smart marketers to take advantage of an uncluttered medium that delivers great results for their brands. As more and more marketers learn the power of advergames and crowd inboxes, phones and websites with gaming efforts, the novelty factor is sure to diminish. But like any creative medium, the best examples will always break through and will travel the world with the click of a button labelled 'forward'.

Takeaway points

- Games present marketers with huge potential for advertising opportunities, which is already being exploited via in-game advertising and branded promotional games.
- Games have an inherent connectivity that makes them powerful and beneficial as a marketing medium.
- Games can create buzz and help generate low-cost awareness, but more importantly can provide tangible business benefits, such as sales leads, an increased intent to purchase and actual sales.
- Gaming gives brands hours of proactive, undivided attention from consumers, but research is needed to demonstrate how this can be leveraged so that marketing investment in games becomes related to ongoing brand benefits, not just a game's popularity.
- It isn't just teenage boys who play games. More than 50% of game players are 35 years +, and women comprise the largest segment of the online game-playing audience.
- The entry-level cost of a branded game could be as little as US$15 000 but it may demand more serious investment to achieve ROI and ongoing brand benefits.
- Advergaming isn't just a technique for business-to-consumer brands. 'Whack-a-flack' shows that if you get the context right and target your audience with the right hot button, then it can work well in a business-to-business context, too.
- The development of Alternate Reality Gaming, mobile technologies and interactive TV widens the scope for advergaming and connected marketing evolution and opportunities.

Notes and references

1 Lindstrom, Martin (2004) 'Playing the Brand Game', *ClickZ Experts*, 23 November, archived at http://www.clickz.com/experts/brand/brand/article.php/3438511.

2 Digital Marketing Services (2004) *Casual Gaming Report* for AOL, 11 February, archived at http://www.dmsdallas.com.

3 Shaw, Don (1983) *Cadbury's Creme Egg Mystery*. Bourneville: Cadbury; Williams, Kit (1979) *Masquerade*. London: Jonathan Cape.
4 'Advertisers use online games to entice customers', *Washington Post*, 29 November 2004.
5 'The latest marketing trend makes the consumer a player inside the commercial', *Newsweek*, 29 November 2004.
6 'U.S.Army spending top dollar; America's Army reaches $6.3 million . . . so far', *IGN.com*, 3 June 2002, archived at http://pc.ign.com/articles/361/361291p1.html.
7 *Electronic Gaming Business*, 2 (16) No. 16 (11 August 2004).
8 'Major brands play for attention', *CNET News.com*, 3 May 2001.
9 'More than a game', *Fast Company*, May 2002.
10 Saunders, Christopher (2001) 'Offline brands turning to advergaming', *ClickZ News*, 27 March, archived at http://www.clickz.com/news/article.php/725181.
11 Activision and Nielsen Entertainment (2004) *Video Game Habits: a Comprehensive Study of Gamer Demographics and Habits in US Households*.
12 'Buying into the unreal', *Mail and Guardian Online*, 6 August 2004, archived at http://www.mg.co.za/articlePage.aspx?articleid=134364&area=/insight/insight_online/.
13 Szulborski, D. (2005) *This Is Not a Game: A Guide to Alternate Reality Gaming*, available as an e-book or in print on-demand format from www.lulu.com. Two sample chapters can be downloaded at www.immersivegaming.com.
14 'Viral advertising in 2005: Top 7 tactics, how-tos, and measurement data', *MarketingSherpa.com Special*, 6 April 2005.

10

Blog marketing

Andrew Corcoran,[*] Paul Marsden,[†]
Thomas Zorbach[‡] and Bernd Röthlingshöfer

[*]Lincoln Business School; [†]London School of Economics/Associate Director, Spheeris; [‡]CEO, vm-people

What is blog marketing?

Let's start with a simple definition: blog marketing is the use of weblogs to promote a brand, company, product or service, event or some other initiative. A weblog, or 'blog' for short, is a frequently updated personal or collaborative website in the form of a diarized journal containing opinions, information and weblinks that reflect the interests and personality of the author(s) (see Blog buzzword briefing box). Because blogs are often themed, covering news and views related to a particular industry or product category, they attract readers interested in that area - and therefore can represent a useful targeting opportunity for marketers.

Although blogs have been around since the mid-1990s - even if the actual term 'blog' was only coined in 1999[1] - marketing interest in blogs is a far more recent phenomenon. The commercial interest in blogs as a marketing tool has only grown with the recent rise in the popularity of blogs, which in turn has been fuelled by new easy-to-use, low-cost blogging software. From geeky beginnings, by mid-2005 blogging had become a mainstream pastime - with the number of published blogs exceeding 10 million, and growing by one million every month.[2]

Blog buzzword briefing

Blog (n): Short for 'weblog', a frequently updated personal or collaborative website in the form of a diarized journal containing opinions, information and weblinks that reflect the interests and personality of the author(s)

Blog (v): to blog – the act of posting an update to a blog

Blogger: Someone who has a blog and blogs – also the name of Google's free blog hosting service

Blog marketing: The use of weblogs to promote a brand, company, product or service, event or some other initiative

Audioblog: A blog in audio format – also known as a 'podcast' because the audio file can be downloaded to an iPod

Blogosphere: Term used by bloggers to describe the sum total of blogs on the Web

Faux blog: A fake or false blog authored by someone different from whom they say they are (also known derogatively as 'flogs')

Moblogging: Posting blog content on-the-go from a mobile device

RSS: A Web technology (short for Real Simple Syndication) that enables people to receive automatic electronic news alerts from their favourite blogs

Photoblog: Online diarized journal in photo format

TypePad/Movable Type/LiveJournal/WordPress: Popular blog software

Vlog: An online diarized video journal

Three emerging approaches to blog marketing

So how can marketers profit from blogs as a marketing medium to promote a brand or company? Whilst blog marketing is still in its early experimental phase, three distinct approaches appear to be emerging. First, marketers can seek endorsements (aka blogvertorials) on popular, opinion leading third-party blogs. Second, you can set up your own 'business blogs' (sometimes called brand blogs or corporate blogs) to directly or indirectly promote your product or brand. And thirdly, marketers can engage in the controversial practice of creating 'faux blogs': fake or false blogs of happy but imaginary clients, customers or consumers.

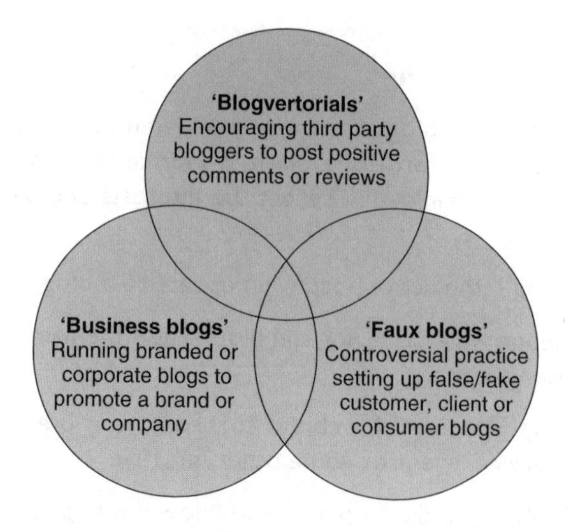

Figure 10.1 Blog marketing: three solutions

Blogvertorials

At its most elementary, blog marketing is simply an extension of classic PR – maintaining good relations with high profile 'bloggers' (people who blog) about subjects relevant to what you sell. Just as you would invite journalists and editors to openings and previews, offering them inside scoops and latest news and sending them samples, demos and other free goodies, it can pay to extend this corporate hospitality to key bloggers. One key to success with these 'blogger outreach' initiatives is to conduct them in the spirit of transparency, partnership and openness; a number of brands have had their fingers burned by negative PR when it has come to light that they were paying bloggers to shill their products or services (see Raging Cow example on p. 151).

Case Study 1: Nokia

Nokia recently seeded its new camera phone with a number of 'leading edge' bloggers in Finland, asking them to experiment with the camera function. Although Nokia did not explicitly ask the bloggers to write about the telephones in their blogs, the majority chose to do so. Nokia found that the bloggers' coverage of the new phone generated significant Web traffic for them on the Nokia site, with four of the blogs featuring in the top 15 traffic generators.[3]

Case Study 2: Dr Pepper – Raging Cow

In 2003 the soft drink manufacturer Dr Pepper offered a group of young bloggers incentives to discuss the new milk-flavoured product called Raging Cow. Dr Pepper asked the bloggers not to mention that they had been briefed – but the word got out. The lack of transparency behind the Dr Pepper blog campaign caused furore and outrage among the blogging community, who called for a boycott of the product.[4]

Business blogs (brand blogs or corporate blogs)

In addition to seeking blogvertorials on third-party blogs, companies can set up their own business blogs to promote a company or brand. Of course, to compete for readers with the other 10 million plus blogs (not to mention other websites), a business blog has to offer compelling, constantly updated, content. This is often achieved by inviting key employees, typically high-profile, opinion-leading executives or board members, to regularly blog their thoughts, musings and opinions. The challenge is to keep readers coming back for more with fresh content that offers breaking news, insider insights, and interesting opinions.

Case Study 3: Stonyfield Farm

Stonyfield Farm, a US organic dairy company, has created a blogging 'Cow'munity' to better communicate with their customers and their employees.[5] Made up of a series of themed blogs, the 'cow'munity' promotes Stonyfield Farm through a fun and interactive dialogue with readers that offers behind-the-scenes industry news, campaigns for healthier school meals, information on women's health issues and baby nutrition. Rick Bruner, president of Executive Summary Consulting and the operator of BusinessBlogConsulting.com,[6] summarizes the blogging initiative: 'Stonyfield Farm's weblogs are a great example of how a large, well-known consumer brand can use the simplicity and personal voice of weblogs to keep customers informed, build human relationships and reach new customer prospects who may not be familiar with the brand'.[7]

Case Study 4: INSCENE Embassy

INSCENE, a German fashion label aimed at 18- to 25-year-olds (a competitor to The Gap, H&M and Esprit) uses blogs as a brand building tool. As part of an integrated campaign to give the brand a cooler image in 2003, the company's marketing agency, vm-people, recruited four hip German trendsetters living in Tokyo, New York, London and Berlin to work as trend-spotting bloggers. Each with their own independent blog site linked to the main INSCENE website, the bloggers were tasked with regularly posting the latest fashion news and trends they spotted in their respective cities. Collectively, the blogs were known as the INSCENE Embassy, and quickly became popular online points of reference for fashion-forward Germans with the initiative generating a significant number of column inches in the fashion press.[8]

Faux blogs

Faux blogs or 'flogs' as they are derogatively known, are promotional blogs written to appear as if they are authored by real clients, customers or consumers. Often used to stimulate interest for new products, faux blogs fall into the controversial area of stealth marketing - marketing to people without them realizing they are being marketed to. As with underhand payments to bloggers in return for endorsements, faux blogs are not without risk - getting found out can result in negative PR.

Case Study 5: Sega – Beta 7

To stimulate interest in a soon-to-be-released Sega software videogame title, the company's advertising agency Wieden & Kennedy created a faux blog that was authored by an imaginary videogame tester called Beta 7. After having being sent a preview copy of the new *ESPN NFL Football 2K4* game, Beta 7 posted his review on his blog, telling readers that it was so extreme it triggered blackouts and fits of violence. Anxious to prevent the software from reaching the stores, Beta 7 used his blog to launch an online campaign to get the game banned – a sure-fire way to stimulate sales![9]

Case Study 6: McDonald's – Lincoln Fry

In 2005 the fast food brand McDonald's ran a faux blog campaign featuring an imaginary individual who had found a French fry shaped like the head of Abraham Lincoln. The blog received over 2 million hits and was then linked in to a McDonald's Superbowl commercial. When it later became known that the Lincoln Fry blog was indeed a faux blog, McDonald's reported that this added a second lease of life to the ad campaign.[10]

What can blog marketing achieve?

Although blog marketing is a relatively new addition to the marketing armoury, it has already proved itself to be a versatile tool. Susie Gardner, author of *Buzz Marketing with Blogs*,[11] summarizes six marketing objectives that can be fulfilled with blogs:

- **Generate interest:** Nike Art of Speed blog. As a brand-building exercise to engage people with the Nike brand with content rather than advertising, the sportswear brand ran a multimedia blog called 'the Art of Speed' showcasing the work of 15 filmmakers for 20 days. Posts included everything from background information to the filmmakers to general discussions about the concept of speed.
- **Drive action and sales:** ActiveWords. Getting favourable reviews on high-profile blogs can accelerate product adoption. Buzz Bruggeman, one of the founders of software company ActiveWords, says more than 50% of the company's trial software downloads are the result of comments and links in blogs.
- **Create goodwill:** Microsoft. Blog marketing can generate brand affinity among readers by giving a brand a human face. For example, Robert Scoble, technical guru with Microsoft, runs a very popular blog called Scobleizer written in an honest, open and personable style that helps humanize the software behemoth.
- **Establish expertise:** Stark & Stark. Blog marketing can be used to position a company as the leading authority or expert in a particular area. For example, the New Jersey injury lawyers Stark & Stark run a specialized blog that covers news in traumatic brain injury law.
- **Dialogue with customers:** FastCompany. Blog marketing can be used to supplement marketing monologue with genuine dialogue between a company and its target market. For example, FastCompany, a business

Objective	Blogvertorials	Business blogs	Faux blogs
Generate interest	✔	✔	✔
Drive action/sales	✔	✔	✔
Create goodwill	✘	✔	✘
Establish expertise	✘	✔	✘
Customer dialogue	✘	✔	✘

Figure 10.2 Blog marketing approaches and objectives

publication covering new business ideas runs a blog called FC Now in which it invites readers to contribute to the blog and discuss ideas for future coverage with FastCompany staff.

- **Dialogue with employees:** *Variety* magazine. Blogs can be used for internal marketing purposes to promote initiatives, change or simply to share information. For example, entertainment magazine *Variety* runs an internal 'Have You Heard' collaborative blog, to which any staff member can post, designed to keep employees across the business up to date on breaking industry gossip, news and internal affairs.

Blog marketing approaches and objectives are summarized in Figure 10.2.

Advantages of blog marketing

Done right, blog marketing can offer a number of intrinsic benefits over other marketing communications. In a nutshell, *blog marketing can be seen as a faster, cheaper and better way of promoting a brand or company*. *Faster,* because blog publishing and updating is instantaneous, simple to do and involves no programming skills: if you can type you can blog, which saves time with respect to the typical tedious messing around with IT departments or web agencies. *Cheaper*, because blogs use media and publishing services that are virtually free and bloggers can be significantly less expensive to woo than spoiled journalists used to lavish gifts and hospitality. And *better*, well, blogs are better for a number of reasons:

- **Blog marketing has viral potential:** Blog content, including third product endorsements, can spread rapidly and infectiously over the Web because bloggers pick up on each other's opinions, copying and pasting each other's posts into their own blogs, whilst bloggers and readers alike can use Web technology called RSS to receive automatic electronic news alerts from their favourite blogs (see previous Blog buzzword briefing).

- **Blog marketing is measurable:** Blogs are repositories of online word of mouth, but unlike traditional word of mouth that disappears without trace once the words have been uttered, blogs leave a digital trace. In other words, blogs make word of mouth measurable. For example, IBM uses an online word of mouth measurement technology called Web Fountain to not only capture brand buzz on blogs, but also to match it to sales data: In the case of books, blog buzz was found to have predictive utility in forecasting sales.[12] As well as empowering marketers to dynamically measure word of mouth for a brand or company, blogs enable marketers to trace where and how word of mouth spreads. A technology called Trackback[13] developed by blogging software company Six Apart enables the owners of blogs to find out the extent to which their postings are being cited in other blogs, thereby measuring how 'infectious' the blog is in spreading word of mouth.
- **Blog marketing gives marketing a human face:** Blogs are personal journals reflecting the personality of the author, and therefore adding a human dimension to a business – allowing readers to see the people behind company and those advocating it. For example, the popular blogs of Robert Scoble, 'technical evangelist' for Microsoft and Bob Lutz, vice chairman of General Motors, give their respective corporate giants a personal face.
- **Blog marketing puts credibility back into the marketing mix:** In an era where people are increasingly sceptical of traditional interruptive advertising, dismissing overt commercial messages as propAdganda and corporate spin-wash, blogs represent a refreshing and credible source of information. Readers are more likely to believe information in an opinion-leading third-party blog than in an ad, whilst the informal style of avoiding sales-speak and overt promotion in business blogs enhances the credibility of the medium.
- **Blog marketing replaces marketing monologue with marketing dialogue:** Because blogs offer an integrated facility for readers to leave comments and post feedback, they open a genuine channel of dialogue, rather than traditional monologue, between a business and its target market. For example, through General Motors' Fastlane blog, readers can converse directly with senior executives and company vice presidents, sharing opinions and engaging in conversations.[14]

Blog marketing: not just for big business

This chapter has provided many examples of how big businesses with big brands are making blog marketing work for them. But blog marketing is not just for big businesses. Bernd Röthlingshöfer, author of top 250 Amazon bestseller *Advertising with a Small Budget*, describes in his own words how he used blog marketing to promote his latest book:

Between 21st and 29th September 2004 I conducted the first ever 'virtual book tour' in Germany to promote my new book *Advertising with a Small Budget*. A virtual book tour can be compared to a regular book tour – with the exception that instead of touring book stores, you visit blogs and websites. The idea is simple: you contact bloggers and offer them an exclusive interview with you as author of a new book which they can post as a transcript, audio-cast, or podcast to their blog, inviting readers to ask questions, which you as author agree to answer. It's a win–win situation for both parties – bloggers get fresh exclusive content that can attract new readers, the author gets free publicity.

So, what did my 2004 virtual book tour achieve? Across the different blogs I toured, 11 856 readers joined in, triggering a jump in sales that put my book at 222 in Amazon's bestseller list! And the tour cost me virtually nothing. Which other channel enables you to reach such a number of interested and involved readers, and generate sales at no cost? The virtual book tour also benefited the bloggers who hosted the tour by generating increased visitor numbers: (Gastgewerbe Gedankensplitter: 35.1% increase = 2331 additional readers, Ideenbloggerin: 48.6% increase = 125 additional readers, Werbeblogger: 78.1% increase = 268 additional readers, Notizblog: 176.0% increase = 220 additional readers, MEX-Blog: 14.1% increase = 96 additional readers).

So, in sum, what are the advantages of a virtual book tour?

- Higher reach than a regular book tour
- Lower cost
- No travel logistics involved
- Harnesses the viral potential of blogs
- Direct links to online shops
- Improved search engine presence
- Sustainable web presence

Conclusion: the future of blog marketing

Is blog marketing just a passing fad in the faddish world of marketing, or is it more significant, representing a structural change in how businesses connect with their markets? It's too early to tell - but we think blog marketing will have an increasingly important role to play in marketing over the next few years. Why? Because blog marketing helps marketers replace monologue with dialogue, interruption with engagement, and control with collaboration.

Blogs represent a watershed in the democratization of media production - empowering ordinary people to become media producers, not just media users. With network communication technology evolving so fast, it will not be long before blogs can rival traditional media in terms of multimedia experience. Already we are seeing the rise of 'audioblogs' (aka 'podcasts', so called because they can be downloaded onto Apple's ubiquitous iPod) and 'vlogging' (video blogging) that involve, respectively, publishing audio and video diaries online. And then there is 'moblogging', the new trend of on-the-go blogging from mobile phones. Perhaps we can envisage a time in the not-too-distant future when blogs will become fully fledged 'personal communication channels' that compete with traditional media channels by offering constantly updated streaming content. If communication theorists are right in insisting that power lies not, as Marx suggested, in the ownership of the means of production, but in the ownership of the *means of communication*, then blogs are the writing on the wall for the old command and control dogma of traditional interruptive marketing.

Takeaway points

- Blog marketing uses weblogs to promote a brand, company, product, service, event, or some other initiative.
- A weblog, or 'blog' for short, is a frequently updated personal or collaborative website in the form of a diarized journal containing opinions, information and weblinks that reflect the interests and personality of the author(s).
- There are three main blog marketing solutions:
  - Seeking blogvertorials on third-party blogs
  - Setting up proprietary 'brand blogs' (also known as corporate blogs)
  - Commissioning 'faux blogs', fake blogs written by imaginary happy clients, customers, or consumers.

- Faux blogs, like other forms of stealth marketing, run the risk of creating negative word of mouth if they are found out to be fake.
- Blog marketing can be effective in achieving a number of marketing objectives: generating interest, driving action and sales, creating goodwill, establishing expertise, and stimulating dialogue with customers or employees.
- Blog marketing has a number of advantages over other promotional marketing tools: blogs give a business a human face, blogs replace marketing monologue with interactivity and dialogue, the information in blogs has viral potential to spread over the Internet, blog readership is measurable, and blogs are considered to be independent and therefore a trustworthy source of information.
- By replacing interruptive marketing messages with engagement, and control with collaboration, blog marketing is likely to have an increasingly important role to play in the future of marketing.

Notes and references

1 http://en.wikipedia.org/wiki/Blog, accessed 15 June 2005.
2 Intelliseek Blogpulse http://www.blogpulse.com, total identified blogs @ 17 June 2005 12 457 790, growth figure from Baker, S. and Green, H. (2005) 'Blogs will change your business', *Business Week* 2 May.
3 Corcoran, A. (2005) Blogging for PR, CIPR Conference Working Paper.
4 http://www.businessblogconsulting.com/2004/06/raging_cow_the_.html, accessed 12 June 2005.
5 http://www.businessblogconsulting.com/adverblogs/, accessed 15 June 2005.
6 http://www.businessblogconsulting.com/, accessed 14 June 2005.
7 http://www.stonyfield.com/AboutUs/MoosReleases_Display.cfm?pr_id=47, accessed 14 June 2005.
8 Campaign run by vm-people www.vm-people.de.
9 http://www.beta-7.com/blog/archives/2003_09.html, accessed 12 June 2005.
10 Ibid., p. 15.
11 Gardner, S. (2005) *Buzz Marketing with Blogs for Dummies*. New York: Wiley Publishing Inc.
12 'Blogs will change your business', *Business Week* May 2005.
13 www.sixapart.com, accessed 16 June 2005.
14 http://fastlane.gmblogs.com.

Part Two

Connected Marketing Principles

11

Word of mouth: what we really know – and what we don't

Greg Nyilasy

Henry W. Grady College of Journalism and Mass Communication, University of Georgia

If you're a marketing professional in the 21st century, word of mouth seems to be one of those buzzwords you can't live without. Trade books and articles often tout word of mouth as one of the 'next big ideas' for the troubled marketing communication industries. The main reason is that traditional advertising just doesn't seem to work that well anymore. Advertising experts and their clients are becoming more and more concerned with audience fragmentation, ad clutter, consumer annoyance and avoidance of advertising. Industry prophets talk about the need for a paradigm shift in marketing communications: permission marketing instead of intrusive advertising, pull instead of push, one-on-one instead of mass messages, relationship marketing instead of isolated communication events, integration of promotional and entertainment content instead of a clear identification of the advertiser. In the midst of these cataclysmic changes, word of mouth seems to represent hope and one of the guiding lights.

But what does word of mouth really mean? Do we really understand the term and the phenomenon behind it? As happens with marketing practice, word of mouth is yet another technical phrase that is defined differently depending on whom you ask. Is it the reputation of a product, or companies trying to secretly spread rumours about themselves? Is it PR, or is it consumers talking about products? And what about other related terms such as viral marketing, buzz campaigns, opinion leaders, consumer evangelists, alphas, Influentials(SM), etc.? Is there a consensus in the meaning of such terms, or are we hopelessly lost? Further, what do we really know about the word of mouth phenomenon? How does word of mouth work? How does it influence consumers? What are the factors that would make it more effective? What causes it and what effects does it have?

To be able to answer these questions, one possible tactic is to survey the available academic literature on word of mouth. Academic research serves as the key source of legitimization for any occupational practice, so this is where questions about knowledge should ultimately be asked. It's academic research that provides the theoretical base for any occupation trying to achieve the much-respected position of being a 'profession'.[1] This idea is generally applicable to marketing even though many marketing practitioners would disagree with it, and even though historically there has been quite a bit of antagonism between practitioners and academic researchers in this field. Practitioners think that academicians are irrelevant, old-fashioned and too theoretical; academicians blame practitioners for sloppy thinking, logical inconsistencies, frequent use of unreliable information and an overall anti-intellectual stance.[2]

Discussions about word of mouth could be a fine example to illustrate the truth to both sides of this argument; instead this chapter takes a more constructive path. It summarizes what is known about word of mouth at the level of academic rigour, and holds up this knowledge to assess what is directly useful for marketing communication practice. First, it reviews how academic researchers define word of mouth, since without a clear understanding of the concept itself it is hard to make any claims about it. Second, it summarizes what we really know about how word of mouth works, enumerating the known causes of word of mouth behaviour as well as its effects. Finally, it assesses the implications of academic knowledge about word of mouth for marketing practice.

The hope is that, after all, advertising and marketing are not that different from other occupations which, by using solid knowledge bases, have successfully become professions. It is not only possible but also necessary to conduct meaningful conversations between theory and practice. As is the case with other information-based occupations, our survival depends on this dialogue. And if word of mouth is part of a new paradigm for marketing communications, why not start with better relations between academia and practice?

A 50-year review

For those of us who think that word of mouth is something brand new, it's perhaps surprising that there has been a lot of scholarly research published on word of mouth, and a lot of it goes back in time quite a distance. Fifty years of research has produced ample evidence that word of mouth is an important factor in consumer information search, consumer decision-making, the diffusion and adoption of innovations, the flow of mass media messages to audiences, as well as consumer dissatisfaction and complaining behaviour.

In fact, the literature is so extensive that a number of articles and monographs had already set out with the objective of reviewing and structuring it.[3] While very useful, these summaries do not capture everything that is known about the subject. Some of them were simply published too long ago, others summarized only a single facet of all there is to know about word of mouth, yet others were limited to a single theoretical frame. An up-to-date, comprehensive review is in order.

The articles reviewed for this chapter were identified by using the EBSCOhost and ABI/Inform research databases. Only peer-reviewed academic publications were included, the likes of the *Journal of Advertising*, *Journal of Advertising Research*, *International Journal of Advertising*, *Journal of Marketing*, *Journal of Marketing Research*, *Journal of Consumer Research*, *Advances in Consumer Research*, and *Journal of the Academy of Marketing Science.* These journals represent the storage houses of academic marketing knowledge, whose reliability and validity are safeguarded by the rigorous peer-review process. The references in the initial pool of research articles were followed up. This literature search scenario resulted in over 150 items. The articles were read, definitions of word of mouth and related concepts were collected and compared, and finally, findings about the causes and consequences of word of mouth were tabulated.[4]

Let us turn to the definition of word of mouth used in this chapter first, since without a clear understanding of the concept itself it is impossible to study it, measure it, or make any claims about it. What does this term really mean?

In search of a definition

The phrase 'word of mouth' has been used in everyday English for a long time. According to the most reliable source of English language etymology, the *Oxford English Dictionary*, the first written occurrence of the term dates back to 1533. The dictionary defines it as 'oral communication', 'oral publicity', or simply 'speaking', in contrast with 'written and other method[s] of expression'.[5] As the etymological notes testify, this meaning has been stable over the centuries.

More recently, in marketing and communication literature word of mouth has taken on a more restricted meaning. It refers only to interpersonal communications about *commercial entities*. While the phenomenon of people talking about products had clearly been known throughout the first half of the 20th century, the concept of word of mouth only became a scientific term after the rise of positivist communication research in the US after World War II.[6] A useful summary of this early research on word of mouth offered the following definition:

> Oral, person-to-person communication between a receiver and a communicator whom the receiver perceives as non-commercial, concerning a brand, a product or a service.[7]

The definition consists of three essential parts. First, and in agreement with the common use of the term, word of mouth is *interpersonal communication*. This element sets word of mouth apart from mass communication (such as advertising) and other impersonal channels available for consumers (e.g. third-party sources of consumer information such as *Consumer Reports*). Further, the code of this type of communication is language. So, other, less tangible forms (non-verbal communication, or information exchange arising from the imitation of others) do not qualify for the label 'word of mouth' in themselves. They might accompany word of mouth, but they are not essential to it.

Second, the content of word of mouth communication from a marketing perspective is *commercial*. The message is about commercial entities, products, product categories, brands and marketers - or even their advertising.[8] This restriction in the meaning of the term underscores that word of mouth is a technical term appropriated for marketing, consumer behaviour and mass media. While in everyday language we might use the phrase 'word of mouth' for any kind of interpersonal communication, or in the meanings of 'hearsay' or 'rumour', word of mouth in marketing refers to talk about brands, marketers and advertising.

Third, even though the content of word of mouth communication is commercial, the communicators are *not motivated commercially*, or at least they are *perceived not to be*. They don't talk about brands because they are employees of the company, or receive any incentives from it. They talk at their own will. It's important to emphasize that this part of the definition is phrased in *perceptual* terms. It's enough that the communicator is *perceived* to be unbiased - (s)he does not necessarily have to be so. In this case, perception is reality.

It's not only theoretically possible but well-known practice for certain marketers or their agents to mask themselves as non-commercial sources of information while being financially motivated. Arndt cites some anecdotal evidence for such marketing tactics dating back to the 1930s ('whispering campaigns' conducted by 'professional rumor mongers').[9] Similarly, current trade sources often describe conscious efforts to manage word of mouth.[10] While these messages do not come from independent, unbiased sources, they are still labelled as word of mouth, if their communicators are *perceived* as non-commercial.

The early marketing and communication literatures on word of mouth often use the phrase 'word of mouth *advertising*', referencing this Janus-face nature of word of mouth. Word of mouth is commercial in content but non-commercial in perception. 'Word of mouth advertising', however, is quite a misnomer, since it seems to imply that word of mouth is always intentional, it fits into someone's marketing plan and it is inherently deceitful. This is simply not the case, nor does the above definition require such an interpretation. Word of mouth is a naturally occurring phenomenon of consumer behaviour, and it *may or may not* be induced by the conscious efforts of marketers. This is a crucial distinction, as we will see in the implications section of this chapter. For now it's enough to say that word of mouth in academic literature is not equated with efforts to manage it. Buzz is not equal to buzz marketing. Word of mouth is simply commercial talk among consumers, none of whom is perceived to be associated with marketers.

The definition analysed above has proved an enduring one. Most of the articles reviewed for this chapter used it almost verbatim. In other words, there is a strong consensus in academic marketing research about what word of mouth means. The phrasing might differ a little, but the definitions are essentially the same. Word of mouth has been defined as 'interpersonal communication between a perceived non-commercial communicator and a receiver concerning a product or service',[11] 'opinions sought from personal sources',[12] 'hearing about a product or service from friends',[13] 'product related conversation',[14] 'interpersonal interaction that does not involve personal selling',[15] 'interpersonal information exchange about a product, service or retailer',[16] 'informal communication directed at other consumers about the ownership, usage or characteristics of particular goods and services and/or sellers',[17] 'interpersonal communications in which none of the participants are marketing sources',[18] and 'the act of telling at least one friend, acquaintance or family member about a satisfactory or unsatisfactory product experience'.[19]

While there is some apparent variation between those definitions, it is mostly due to omitting one or two parts, and not substantial disagreements. The authors either leave out the requirement that the interaction should be non-commercially motivated or that the content of the messages should be about products, services, or marketers. However, there are no irresolvable logical contradictions between these definitions and no author challenges previous conceptualizations of word of mouth. Even in the cases where some prongs are left out, the context of the definition clarifies that all of the above three parts are required. Further, in a good number of studies (especially among those published more recently) there is no conceptual definition offered whatsoever. One possible reason for this might be that authors have felt that the foundations of the concept are firm. Other studies simply cite earlier definitions verbatim without discussion.

Some articles emphasize the fact that word of mouth is not necessarily positive, it does not necessarily praise the product, service, or marketer. Accordingly, 'negative word of mouth' (NWOM) has been defined as 'interpersonal communication among consumers concerning a marketing organization or product which denigrates the object of communication',[20] 'complaining to friends and relatives',[21] or 'telling friends and relatives about [a] dissatisfying experience [with a product or service]'.[22] In contrast, 'positive word of mouth' (PWOM) is conceptualized as 'product-related information transmitted by satisfied customers'.[23] These definitions are again easy to integrate into the general one discussed above, provided one keeps in mind the limited sphere of applicability. No study explores the possibility that NWOM and PWOM are fundamentally different. Instead, they are thought of as identical, merely varying between negative and positive endpoints. There is some evidence, however, that NWOM is more potent than PWOM.[24] (And see Brad Ferguson's case study in Chapter 12)

In short, we can conclude that the definition of word of mouth from a marketing perspective is solid in academic research. There is a widespread consensus that word of mouth is a naturally occurring fact of consumers' lives, who talk about products, brands, marketers, or their advertising among themselves, and by default they do not consciously serve a marketing strategy while doing that. Word of mouth is 'Oral, person-to-person communication between a receiver and a communicator whom the receiver perceives as non-commercial, concerning a brand, a product or a service.'[25]

Having clarified the conceptual foundations of word of mouth, it's time to investigate how word of mouth communication works, what causes it and what effects it has.

How word of mouth works

As defined already, word of mouth from a marketing perspective is essentially interpersonal communication, the exchange of information between communicators and receivers about a commercial topic. Most studies dealing with word of mouth focus on *either* the communicator *or* the receiver side of the interaction.

The difference between these two groups of studies goes beyond that of perspective (i.e. whether they look at and measure word of mouth from the standpoint of the communicator or receiver, respectively). They look at two different *facets* of word of mouth. One group investigates the acquisition and processing of product-related information (receiver-oriented studies), while the other examines information provision (communicator-oriented studies). The distinction between these two different facets of word of mouth is known as 'input' (receiver) versus 'output' (communicator) word of mouth.[26]

The distinction between input and output word of mouth lies in the psyche of the communicator and receiver. While the interaction between the communicator and receiver is one event, the psychology of the people taking part in it differs, based on whether they are talkers or listeners. As the literature testifies, different psychological processes are present in the mind of the communicator and the receiver, and these mental processes in turn are influenced by different factors (variables). Although the analysis of communicator and receiver behaviours and cognitions need to be separated for their analysis, naturally, input and output word of mouth are not independent of each other. Not only do they constitute the same verbal exchange (being 'one' from a linguistic standpoint), but also they are impossible to conceptualize without each other. An input by definition pre-supposes an output source. Nevertheless, this distinction is a crucial one in trying to analyse and understand word of mouth interaction, and it provides a useful tool to disentangle the complexities therein.

It's also important to note that the difference between input and output word of mouth is not identical with the question of who *initiated* the communication exchange. The focus on the psychology of the receiver does not necessarily mean that it is the receiver who started the exchange, by asking for product information. Conversely, explaining what processes occur in the communicator's head does not carry the assumption that (s)he wanted to talk. Both can be either active or passive in word of mouth interaction events. The only assumption is that 'giving' and 'receiving' advice, opinions, experiences, etc. are different mental processes.

Unit of analysis	Main focus of study	
	Antecedents to word of mouth (causes)	*Consequences of word of mouth (effects)*
Receiver of communication (input word of mouth)	**QI:** 'Why do people listen?' *Related variables*: external information search, product category (perceived risk), type of relationship with source (tie strength)	**QII:** 'The power of word of mouth'. *Related variables*: key communication effectiveness variables (awareness, attitude change, behavioural intention, purchase behaviour)
Communicator (output word of mouth)	**QIII:** 'What makes people talk?' *Related variables*: opinion leadership, satisfaction/dissatisfaction, promotional activities/direct influence of advertiser	**QIV:** 'What happens to the communicator after the word of mouth event?' *Related variables*: cognitive dissonance, ego-enhancement

Figure 11.1 Four areas of word of mouth literature

A second approach commonly used for structuring academic literature is whether the investigated concept is a cause or a consequence in the analysis. Are we looking at what causes the occurrence of a phenomenon (and what causes *more* of it)? Or conversely, are we interested in the effects it has once it has occurred (and if it has stronger consequences than other things)? To put it in academic language, we are either interested in the *antecedents* or the *consequences* of the concept we are looking at.

The above two distinctions (input versus output word of mouth, and whether we look at the antecedents or consequences of word of mouth) will be used to structure the otherwise quite complex literature on word of mouth. Figure 11.1 illustrates the matrix formed from the two dichotomies. Cells in this matrix represent the typical areas of the academic word of mouth literature.

Quadrant I: Antecedents of input word of mouth

Studies that deal with the antecedents of word of mouth and focus on the receiver try to identify factors that influence the likelihood that consumers will use (seek out or be exposed to) word of mouth. In this research area, word of mouth is contrasted with other possible sources of consumer information (advertising and other marketer-dominated sources). Factors are sought to predict the choice among these information sources. Word of mouth is therefore an important part of what is commonly called 'external information search' in marketing research (i.e. something that goes beyond marketer-dominated sources, such as advertising).[27] As noted earlier, input word of mouth is not necessarily consciously sought out. However, studies focusing on the antecedents of input word of mouth do not make this distinction, and deal only with word of mouth initiated by consumers taking part in conscious information search.

Early research proposed that one of the crucial predictors of the occurrence of input word of mouth was *perceived risk*, the subjective assessment of potentially negative outcomes of using the product.[28] These early studies and a more recent replication[29] found empirical support for the hypothesis that the more risky the consumer perceives the purchase decision to be, the more likely he/she would be exposed to word of mouth.

Higher levels of perceived risk is one of the important characteristics of services as opposed to products, according to 'services marketing theory.'[30] Risk is one of the reasons why consumers of services use word of mouth sources more often than purchasers of products, a proposition that has further empirical support.[31]

The characteristics of the commercial entity and its subjective perception are not the only factors that influence the likelihood of a word of mouth event. A different line of research, using 'network analysis' methods, suggests that the kind of social relationships the receiver has with potential communicators (preceding the communication event) is another important predictor of whether word of mouth will occur. Relying on the 'strength of ties' theory,[32] sociometric research in marketing found that the stronger the relationship between the members of a particular social network, the more likely that these strong-tie relationships will be used for word of mouth communication.[33] In other words, there is strong empirical evidence (also replicated recently[34]) for the common sense notion that if word of mouth occurs, it tends to be between close friends and relatives rather than superficial acquaintances (although 'weak ties' are also important, especially for the spread of word of mouth between overlapping social networks).[35]

Quadrant II: Consequences of word of mouth for the receiver

This quadrant has to do with the power of word of mouth. Studies in this area prove that word of mouth is a more potent communication form than paid-for messages. Articles examine the effects of word of mouth once consumers have been exposed to it. Early communication research focusing on the effects of mass media and advertising found that word of mouth was a much more important factor influencing brand awareness and favourable attitude change than paid-for messages. The 'two-step flow' theory of media and advertising effects (still dominant in US mass communication research) stated that instead of direct effects, advertising messages were channelled through 'opinion leaders' - individuals influential in social networks - and it is only through their output word of mouth that communication effects can occur.[36] In Figure 11.1, the two-step flow theory belongs to both Quadrants II and III, since it looks at both what causes the communicator to talk and the effects of word of mouth on the receiver. As of today, this relatively older theory remains the most comprehensive model explaining word of mouth.

Although not directly referencing the two-step flow theory, marketing researchers have also found considerable support for the hypothesis that word of mouth is stronger than advertising or other marketing communication forms. A host of advertising effectiveness variables has been shown to be positively related to word of mouth. Word of mouth has a positive influence on brand awareness,[37] positive attitude-change toward the brand,[38] initial and long-term 'product judgments',[39] brand evaluations,[40] service quality expectations[41] and purchase intentions.[42] This list of communication effectiveness variables basically covers the whole spectrum of potential advertising objectives. The power of word of mouth (when compared to paid-for sources) is undisputed in academic literature.

The strength of word of mouth, however, can be moderated by a number of factors, most importantly ones related to prior information about the brand. There is evidence, for instance, that favourable predisposition toward the brand moderated the effects of word of mouth on product evaluation.[43] Similarly, highly diagnostic information such as prior impressions about the brand and the presence of extremely negative information also influence the extent to which word of mouth is related to communication effectiveness variables.[44] Another study reports that a favourable brand name reduces the persuasiveness of NWOM.[45] On the other hand, the perception that the communicator is an expert source enhances word of mouth effectiveness.[46]

While the persuasive power of word of mouth versus advertising has been firmly established in academic literature, much less is known about the reason *why* this is the case. Although not explicitly stated by the two-step flow theory, the characteristics of the 'primary group' ('strong ties' in network analysis terminology) are implied to be part of the explanation. Communicators that come from a primary group (family and friends) are more trustworthy and credible than impersonal or weaker personal ones. Another reason suggested by the two-step flow theorists is that *normative influence* (conformity to influencers and group norms) is also at play when word of mouth (*informational influence*) is passed along. The two are not separable - in fact it might be that conformity is behind the power of word of mouth.

'Attribution theory' serves as another possible explanation for the strength of word of mouth. According to this theory, people tend to try to figure out why others do what they do. The tentative answers (the 'causal attributions') they come up with, in turn, determine how they relate to the observed person. In the case of word of mouth the implication is that if the receiver thinks that the communicator is unbiased, the causal attribution is that the communicator really believes in the message; so then the message is accepted by the receiver. Conversely, the attribution for paid-for messages is that the motivation of the source is economic, not genuine evaluation. Consequently, the message is received sceptically. This difference between causal attributions in the case of word of mouth and paid-for messages may explain the power of word of mouth - there is partial evidence to prove this case. The findings of one study[47] suggest that the receiver's causal attribution about the motives behind the communicator's behaviour is a key mediator in the NWOM–brand evaluation relationship. While word of mouth is not contrasted with advertising in the study, attribution processes are shown to be present, mediating the relationship between different NWOM scenarios and brand evaluations.

'Accessibility-diagnosticity theory' provides a different explanation for the power of word of mouth. According to a study testing this theory[48] the 'vividness' of information (unique to word of mouth due to its oral and interactive nature) is what generates favourable responses from consumers receiving it. Since vivid information is more 'accessible' than impersonal messages, receivers are more likely to use it when formulating product judgements (it's more 'diagnostic'). The study reports evidence that this is the case.

In sum, there is no question in academic literature about the effectiveness of word of mouth. Less empirical proof is offered to explain why this power exists. The trustworthiness of primary groups, the causal attributions receivers make about the communicators, and the perceived vividness of word of mouth are all explanations that, while compelling, need more empirical support.

Quadrant III: Antecedents of output word of mouth

The literature reviewed in this area investigates the factors influencing the likelihood and extent to which communicators engage in positive or negative word of mouth. The two-step flow theory proposes that mass media messages are not directly influential, rather they are filtered through opinion leaders.[49] Opinion leaders process the information first, since they tend to be more frequently exposed to mass media. So the opinion leadership trait can be understood as a knowledge-based antecedent to output word of mouth.[50]

This implicit proposition has generated a large body of research trying to identify opinion leaders, a line of work relatively independent from the core word of mouth literature. The rationale for this type of investigation has been that the identification of opinion leaders is the key for marketers to be able to manage word of mouth. In the framework of the two-step flow theory this has meant that opinion leaders should be targeted by advertising campaigns. However, there is clearly a potential to influence them directly, a tactic that has become much more widely used since the 1990s.

Opinion leadership studies have mostly dealt with the psychological and demographic profile of opinion leaders.[51] No general demographic predictors have emerged for opinion leaders, mainly because they can be very different people depending on the product category in question. Psychographic or sociological variables are again very poorly correlated with opinion leaders, although social leadership shows a positive relationship.[52] The only characteristic that seems to consistently predict opinion leadership is enduring involvement with the product category.[53] Opinion leaders are interested in the product category, it is significant for their lives for various reasons, and they spend a lot of time with these products. It is because they are involved with the category that they talk about it so much.

The academic literature has also identified other types of influencers who spread the word. There are 'market mavens' who have general marketplace expertise that is broader than the product category-specific knowledge of opinion leaders. There are 'innovators', consumers who are first to try new products (ahead in the 'diffusion of innovations' curve) and sometimes (but not necessarily) talk about these products. There are 'purchase pals' who even accompany you to the store, giving useful information while shopping. And finally there are 'surrogate consumers' who charge you for their information provision services about products. Attempts have been made to profile these influencers as well, but the results are mixed.[54]

Both involvement and opinion leadership are relatively stable characteristics of consumers spreading the word about a product. There are, however, less enduring, more incidental factors influencing one's propensity to engage in output word of mouth. This explains why people who cannot be categorized as opinion leaders also spread word of mouth quite frequently.

One of these incidental factors is product usage. A single incident of using the product might be enough to trigger word of mouth behaviour. In fact, some studies showed that even purchase *intention* was enough to create output word of mouth.[55] The intervening variable in the brand usage–output word of mouth relationship (the reason why it happens) is *satisfaction* or *dissatisfaction*. Numerous studies have shown that satisfaction causes an increased likelihood of engaging in PWOM[56] and that dissatisfaction causes NWOM behaviour.[57]

An important finding in the dissatisfaction literature is that the easier consumers can complain to the company, the lower are the chances that they will spread NWOM.[58] The efforts the marketer makes to set things right can also help. A number of studies show that 'perceived justice' can deter consumers from NWOM,[59] while satisfaction with the service recovery process and outcomes can generate PWOM.[60]

A number of other antecedents have been identified, including micro-level variables such as: surprise as an emotional state,[61] consumer participation in the service,[62] promotional efforts,[63] incentives and deal proneness,[64] as well as macro-level variables: market type[65] and culture.[66] These studies, however, have largely been one-shot attempts to introduce a new variable, and have lacked firm theoretical bases.

One of the most promising areas of word of mouth research lies in the 'Antecedents to output word of mouth' quadrant, because of the topic's practical significance and the relatively underdeveloped state of this line of enquiry. It's especially important to discover antecedent variables that are *controllable* by marketing professionals - factors they can do something about. Academic study striving for relevance among practitioners ('problem-oriented marketing research') should identify more manageable antecedents to output word of mouth.

Quadrant IV: Consequences of word of mouth for the communicator

While it is conceivable that word of mouth has an effect on the communicator - not only on the receiver of the communication - not much research has investigated this possibility. To the author's knowledge the only studies dealing with word of mouth's effects on communicators are motivational analyses[67] from the years when motivation research was in vogue in the United States. Although the forms of 'self-motivation' identified in Ernest Dichter's famous article on word of mouth are not dependent variables in a hypothesized relationship; they deal with the consequences of word of mouth on the communicator *after* the event has taken place.

The most important effects are ego-enhancement and the reduction of cognitive dissonance. A word of mouth episode on the one hand might reassure the communicator that (s)he has made the right purchase decision. Just by talking about it, the communicator can get rid of negative feelings associated with cognitive dissonance. On the other hand, the communicator might also feel good about him/herself as (s)he helps out a fellow human being. (S)he might also entertain the thought that (s)he was knowledgeable and competent in something. These forms of ego-enhancement can be powerful emotional effects on the communicator.

These qualitative studies, however, are not comparable methodologically to statistical hypothesis testing. Future research could clarify if these mental processes in fact take place in the communicator after word of mouth episodes.

Executive summary and implications for marketers

The purpose of this chapter was to provide an overview of what is known about word of mouth in an academic manner. As is always the case with human knowledge, there is more that we don't know about word of mouth, than we do know. In certain areas there is, however, a solid consensus among marketing researchers. Let's summarize the highlights of this knowledge base and its implications for marketing practice:

1. Word of mouth is a naturally occurring behaviour of consumers – there is nothing mystical about it. It is simply communication among consumers about products, brands and advertising, who do not normally assume that their conversation partner is motivated by anything else than wanting to offer help. So, before running to the first buzz practitioner on the corner, brand managers should assess the amount and quality of word of mouth they *already have* (as most likely, they already have a lot of it). Consumer talk can be induced by buzz marketing methods, but traditional marketplace activities (such as advertising) result in a lot of word of mouth automatically. Monitoring it is the first priority.
2. Word of mouth can be negative, and unfortunately negative word of mouth is far more influential than positive word of mouth. This is why it's even more important to monitor word of mouth. Undiagnosed negative word of mouth that spreads widely among consumers can have irreversible and lethal effects for any brand.
3. People listen to word of mouth because it's part of their normal information search about brands when scanning the marketplace. For this reason, there's no way you can fully control word of mouth; there is too great of a need for it – a need for unbiased, unsponsored information. The promises of certain buzz, word of mouth, or viral

marketing practitioners who claim an unlimited capability to orchestrate, influence, or manufacture word of mouth are to be handled cautiously.

4. Consumers tend to seek out and listen to word of mouth more when the purchase decision is risky. Managers of 'high involvement' (higher priced, more complex, more personally relevant) brands should pay extra attention to word of mouth as it will be a significant factor in the purchase decision.
5. Marketers of services should be especially wary of the influence of word of mouth. The intangibility, unpredictability and often rather consequential nature of services motivate most consumers to seek out advice before choosing a provider. Some form of word of mouth management tactics should be part of the marketing communication mix of all service providers.
6. People listen to their close relatives and friends, or others they perceive as respectable experts. While weaker ties (acquaintances, strangers) might help the overall spread of the message through society, consumers *rely on* information coming from close ties. If the objective of the campaign is attitude change or maintenance, buzz marketing solutions using weak-tie consumer evangelists should have a limited role. If the objective is to increase brand awareness, such schemes can have a role in the marketing communication mix. This is especially true for new products.
7. Word of mouth is a potent communication form, and it is more potent than any marketing communication techniques that identify a sponsor. On all known communication effectiveness variables, word of mouth scores better than advertising. Does this mean that brand managers should dismantle their traditional marketing communication programs? No. Here are the reasons:
   (a) Advertising induces word of mouth. As one of the most influential communication theories suggests, advertising is a natural step in communications flowing from marketers to consumers.
   (b) Word of mouth is inherently uncontrollable by its nature, so marketers will never have the same access to it as they have to sponsored marketing communication channels.
   (c) Brand image has a strong moderator effect on the influence of word of mouth. People who have a prior impression about a brand are more difficult to influence by word of mouth. Tactics attempting to manage word of mouth directly therefore should be *a part* and not *a replacement of* traditional marketing communication mixes. This is true even if advertising and other forms of sponsored messages are known to be losing some of their effectiveness today.

8. There are three different explanations in academic literature for the power of word of mouth:
   (a) Word of mouth communicators are better trusted as being parts of 'primary groups.'
   (b) They are not attributed motives to sell.
   (c) The way in which they communicate is vivid.
9. A potential implication of these word of mouth characteristics is that marketers might want to *emulate them in their paid-for messages*. If the traits behind word of mouth are trustworthiness, the non-identification of the sponsor and vividness, why not create paid-for marketing communications that simulate these? While such tactics have been known for quite a while (as the existence of testimonials, advertorials and creativity in advertising shows), what word of mouth proves is that they are very effective, and they continue to be so. Marketers should also go beyond the above-mentioned tools (e.g. the *genre* of the testimonial, or the *medium* of product placement), and try making these three word of mouth characteristics broader principles of their overall communication with their consumers.
10. The implications of attribution theory for word of mouth can represent a serious ethical dilemma for marketers. If attribution theory is right, the perception of whether or not the source of the message is associated with the marketer will dictate the extent to which the message is believed. If it comes from the company (or somebody associated with it), it's handled with scepticism (and is less effective); if it comes from a friend, it's more readily believed (and is more effective). If this is true, in certain buzz marketing schemes (or perhaps in all of them) where word of mouth is directly influenced by paid-for agents acting as if they were unbiased 'friends', there is a conflict between economic interest and ethical considerations. In such scenarios, it doesn't pay for the agents to be transparent, since the attribution that transparency causes – ('this agent acts on behalf of the marketer and therefore is not reliable') – ruins the very advantage word of mouth has over advertising, or other paid-for messages. On the other hand, the ethical notion of the 'informed consumer' would require full transparency. Marketers should be aware of the ethical problems attribution theory represents, and should assess its potential consequences. As of today, there is only partial evidence that attributions about communicator motives are activated with word of mouth, but there is very strong support for attribution theory outside marketing, in general psychology literature.
11. Opinion leaders (in trade lingo: alphas, influencers, consumer evangelists and enthusiasts, etc.) remain one of the key drivers of the word of mouth phenomenon. Opinion leaders have an enduring involvement in the product category, and that's what makes them talk. They are

like good, reliable engines: they hum and buzz constantly. You can count on them - and hope they are not in your competitor's car. Marketers attempting to influence word of mouth about their brands should start with opinion leaders. Unfortunately, there are no general demographic or psychographic predictors of opinion leadership. Profiling should be category- and market-situation-specific. Both mass media and direct or below-the-line communication forms should be aimed at them. If opinion leaders in a category are well defined, and especially if they are a small and homogeneous group, buzz marketing tactics trying to influence them directly will have a good chance of being successful. Brand managers shouldn't forget about other types of influencers such as market mavens, innovators, purchase pals and surrogate consumers either.

12. Attention should also be paid to *incidental* word of mouth, as average consumers (not just the alphas!) might also open their mouths. Not surprisingly, satisfied consumers will say good things, and dissatisfied consumers bad things. Although product/service quality is mostly a given, there is something brand managers and their agents can do to prevent negative word of mouth even if dissatisfaction is constant. Let them complain. Make complaining to the company as easy as possible, and consumers will let off steam this way, instead of complaining to their friends. Second, make the impression that the company is willing and able to set things right, as 'perceived justice' again reduces negative word of mouth.
13. The temporary emotional state of being surprised tends to motivate people to talk about the product this surprise is associated with. The implication of this finding is that it pays doubly for marketers to be creative in their marketing communications (whatever channels they use). Not only does fresh, creative, surprising communication cut through the clutter to generate sufficient attention, but it also makes the recipients talk about the message (or even pass along the *creative material itself* in the case of 'viral marketing').
14. Incentives offered for spreading the word do work. Brand managers should find ways to build promotional programmes that reward active engagement in word of mouth. These 'tell-a-friend' programmes are effective, despite the fact that the motivation of the communicator might become transparent. Sampling of new products has a similar word of mouth-generating indirect effect.

Despite the many things we know about word of mouth, and the implication we can draw from this knowledge, there are significant white spots on our knowledge map. There is no academic knowledge available, for instance, on the effectiveness of the many forms of buzz and viral marketing campaigns, the conscious and organized efforts to influence the occurrence or content of word of mouth directly. Research should continue in these areas to produce relevant, consistent and reliable knowledge for marketers seeking to understand and manage word of mouth. This closer partnership between academic knowledge and marketing practice will also help increase the credibility of marketing as a profession.

Takeaway points

- Word of mouth is oral, person-to-person communication between a receiver and a communicator whom the receiver perceives as non-commercial, concerning a brand, a product or a service.
- Word of mouth is important because it is much more effective than sponsored messages.
- The main reason for seeking input word of mouth is the perceived risk of the purchase decision.
- The main causes of output word of mouth are opinion leadership and satisfaction/dissatisfaction.
- The main effects of input word of mouth are higher levels of awareness, attitude change and purchase behaviour.
- The main effect of output word of mouth is that communicators feel better about themselves.
- Marketers should be aware that word of mouth occurs naturally and should monitor it closely.
- Marketers can influence positive word of mouth through targeting messages at opinion leaders, incentives and product sampling; and prevent negative word of mouth by creating platforms for consumer complaints with the company.
- Marketers may also stimulate word of mouth by the services of buzz and viral marketing specialists. Marketers should be aware of the ethical questions some of these practices may pose.

Notes and references

1 Abbott, Andrew D. (1988) *The System of Professions: An Essay on the Division of Expert Labor*. Chicago: University of Chicago Press; MacDonald, Keith M. (1995) *The Sociology of the Professions*. London: Sage.

2 Kover, Arthur J. (1976) 'Careers and Noncommunication: the Case of Academic and Applied Marketing Research', *Journal of Marketing Research*, 13 (Nov.): 339–344; McQuarrie, Edward (1998) 'Have laboratory experiments become detached from advertiser goals? A

meta-analysis', *Journal of Advertising Research*, 38 (Nov.-Dec.), 15-25; Preston, Ivan L. (1985) 'The developing detachment of advertising research from the study of advertiser's goals', *Current Issues and Research in Advertising*, 8 (2): 1-16.

3 Arndt, Johan (1967) *Word of Mouth Advertising: A Review of the Literature.* New York: Advertising Research Foundation; Buttle, Francis A. (1998) 'Word of mouth: understanding and managing referral marketing', *Journal of Strategic Marketing*, 6 (Sep.): 241-254; Gelb, Betsy and Johnson, Madeline (1995) 'Word of mouth communication causes and consequences', *Journal of Health Care Marketing*, 15 (Fall): 54-58; Price, Linda L., Feick, Lawrence F. and Higie, Robin A. (1987) 'Information sensitive consumers and market information', *Journal of Consumer Affairs*, 21 (Winter): 328-341; Richins, Marsha L. (1984) 'Word of mouth communication as negative information', *Advances in Consumer Research*, 11 (1): 697-702; Rogers, Everett M. (1962) *Diffusion of Innovations.* New York: Free Press.

4 Articles dealing only with offline word of mouth were considered for this review. Academic literature on online word of mouth is virtually nonexistent. There is some evidence that the power of word of mouth applies to online settings as well, but overall, very little is known about online word of mouth. Interested readers, nevertheless, should consult: Alon, Anat, Brunel, Frédéric and Schneier Siegal, Wendy (2002) 'Word of mouth and community development stages: towards an understanding of the characteristics and dynamics of interpersonal influences in Internet communities', *Advances in Consumer Research*, 29 (1): 429-430; Bickart, Barbara and Schindler, Robert M. (2001) 'Internet forums as influential sources of consumer information', *Journal of Interactive Marketing*, 15 (3): 31-40; Henning-Thurau, Thorsten and Walsh, Gianfranco (2003) 'Electronic word of mouth: motives for and consequences of reading customer articulations on the Internet', *International Journal of Electronic Commerce*, 8 (Winter): 51-74; Kiecker, Pamela and Cowles, Deborah (2001) 'Interpersonal communication and personal influence on the Internet: a framework for examining online word of mouth', *Journal of Euromarketing*, 11 (2): 71-88; Godes, David and Mayzlin, Dina (2004) 'Using online conversations to study word of mouth communication', *Marketing Science*, 23 (Fall): 545-560; Schindler, Robert M. and Bickart, Barbara (2002) 'Characteristics of online consumer comments valued for hedonic and utilitarian shopping tasks', *Advances in Consumer Research*, 29 (1): 428-429; Ward, James and Ostrom, Amy (2002) 'Motives for posting negative word of mouth communications on the Internet', *Advances in Consumer Research*, 29 (1): 429.

5 *Oxford English Dictionary* (1989) 2nd edn. Oxford: Oxford University Press.

6 Brooks, Robert C. Jr (1957) '"Word of mouth" advertising in selling new products', *Journal of Marketing*, 22 (Oct.): 154–161; Katz, Elihu and Lazarsfeld, Paul Felix (1955) *Personal Influence: The Part Played by People in the Flow of Mass Communications.* Glencoe, IL: Free Press; Lazarsfeld, Paul Felix, Berelson, Bernard and Gaudet, Hazel (1948) *The People's Choice: How the Voter Makes up His Mind in a Presidential Campaign.* New York: Columbia University Press.

7 Arndt, *Word of Mouth Advertising*, p. 3.

8 Dichter, Ernest (1966) 'How word of mouth advertising works', *Harvard Business Review*, 44 (Nov.-Dec.): 147–166.

9 Arndt, *Word of Mouth Advertising*, p. 3.

10 Bond, Jonathan and Kirshenbaum, Richard (1998) *Under the Radar: Talking to Today's Cynical Consumer.* New York: Wiley; Rosen, Emanuel (2000) *The Anatomy of Buzz: How to Create Word of Mouth Marketing.* New York: Doubleday; Salzman, Marian L., Matathia, Ira and O'Reilly, Ann (2003) *Buzz: Harness the Power of Influence and Create Demand.* Hoboken, NJ: Wiley.

11 Webster, Frederick E. Jr (1970) 'Informal communication in industrial markets', *Journal of Marketing Research*, 7 (May): 186–189.

12 Martilla, John A. (1971) 'Word of mouth communication in the industrial adoption process', *Journal of Marketing Research*, 8 (May): 173–178.

13 Traylor, Mark and Mathias, Alicia (1983) 'The impact of TV advertising versus word of mouth on the image of lawyers: a projective experiment', *Journal of Advertising*, 12 (4): 42–45, 49.

14 Still, Richard R., Barnes, James H. Jr and Kooyman, Mark E. (1984) 'Word of mouth communication in low-risk product decisions', *International Journal of Advertising*, 3 (4): 335–345.

15 Reingen, Peter H. and Kernan, Jerome B. (1986) 'Analysis of referral networks in marketing: methods and illustration', *Journal of Marketing Research*, 23 (Nov.): 370–378.

16 Higie, Robin A., Feick, Lawrence F and Price, Linda L. (1987) 'Types and amount of word of mouth communications about retailers', *Journal of Retailing*, 63 (Fall): 260–278.

17 Westbrook, Robert A. (1987) 'Product/consumption-based affective responses and postpurchase processes', *Journal of Marketing Research*, 24 (Aug.): 258–270.

18 Bone, Paula Fitzgerald (1995) 'Word of mouth effects on short-term and long-term product judgments', *Journal of Business Research*, 32 (March): 213–223.

19 Halstead, Diane (2002) 'Negative word of mouth: substitute for or supplement to consumer complaints?', *Journal of Consumer Satisfaction, Dissatisfaction and Complaining Behavior*, 15: 1–12.

20 Richins, 'Word of mouth communication as negative information'.

21 Singh, Jagdip (1990) 'Voice, exit, and negative word of mouth behaviors: an investigation across three service categories', *Journal of the Academy of Marketing Science*, 18 (Winter): 1–15.

22 Blodgett, Jeffrey G., Granbois, Donald H. and Walters, Rockney G. (1993) 'The effects of perceived justice on complainants' negative word of mouth behavior and repatronage intentions', *Journal of Retailing*, 69 (Winter): 399–428.

23 Holmes, John H. and Lett, John D. Jr (1977) 'Product sampling and word of mouth', *Journal of Advertising Research*, 17 (Oct.): 35–40.

24 Burzynski, Michael H. and Bayer, Dewey J. (1977) 'The effect of positive and negative prior information on motion picture appreciation', *Journal of Social Psychology*, 101 (April): 215–218.

25 Arndt, *Word of Mouth Advertising*, p. 3.

26 File, Karen Maru, Cermak, Dianne S.P. and Prince, Russ Alan (1994) 'Word of mouth effects in professional services buyer behaviour', *Service Industries Journal*, 14 (July): 301–314.

27 Beatty, Sharon E. and Smith, Scott M. (1987) 'External search effort: an investigation across several product categories', *Journal of Consumer Research*, 14 (June): 83–95; Bloch, Peter H., Sherrell, Daniel L. and Ridgway, Nancy (1986) 'Consumer search: an extended framework', *Journal of Consumer Research*, 13 (June): 119–26; Kiel, Geoffrey C. and Layton, Roger A. (1981) 'Dimensions of consumer information seeking behavior', *Journal of Marketing Research*, 18 (May): 233–239.

28 Arndt, Johan (1967) 'Perceived risk, sociometric integration, and word of mouth in the adoption of a new food product', in Donald F. Cox (ed.), *Risk Taking and Information Handling in Consumer Behavior.* Boston: Division of Research Graduate School of Business Administration Harvard University, pp. 289–316; Arndt, Johan (1967) 'Role of product-related conversations in the diffusion of a new product', *Journal of Marketing Research*, 4 (August): 291–295; Cunningham, Scott M. (1967) 'Perceived risk as a factor in informal consumer communications', in Cox, *Risk Taking and Information Handling*, pp. 265–316.

29 Hugstad, Paul, Taylor, James W. and Bruce, Grady D. (1987) 'The effects of social class and perceived risk on consumer information search', *Journal of Services Marketing*, 1 (Summer): 47–52.

30 The 'services marketing theory' is the idea that a whole separate set of marketing rules are to be applied for services, see Murray, Keith B. (1991) 'A test of services marketing theory: consumer information acquisition activities', *Journal of Marketing*, 55 (January): 10–25.

31 Beatty and Smith, 'External search effort'.

32 Granovetter, Mark S. (1973) 'The strength of weak ties', *American Journal of Sociology*, 78 (May): 1360–1380.

33 Brown, Jacqueline Johnson and Reingen, Peter H. (1987) 'Social ties and word of mouth referral behavior', *Journal of Consumer Research*, 14 (Dec.): 350-362; Reingen and Kernan, 'Analysis of referral networks'.

34 Wirtz, Jochen and Chew, Patricia (2002) 'The effects of incentives, deal proneness, satisfaction and tie strength on word of mouth behaviour', *International Journal of Service Industry Management*, 13 (2): 141-162.

35 Granovetter, M. (1973) 'The strength of weak ties', *American Journal of Sociology*, 78 (6): 1360-1380.

36 Katz and Lazarsfeld, *Personal Influence*; Lazarsfeld, Berelson and Gaudet, *The People's Choice.*

37 Udell, Jon G. (1966) 'Prepurchase behavior of buyers of small electrical appliances', *Journal of Marketing*, 30 (Oct.): 50-52.

38 Bone, 'Word of mouth effects on product judgements'; Day, George S. (1971) 'Attitude change, media and word of mouth', *Journal of Advertising Research*, 11 (Dec.): 31-40; Reynolds, Fred D. and Darden, William R. (1971) 'Mutually adaptive effects of interpersonal communication', *Journal of Marketing Research*, 8 (Nov.): 449-454.

39 Bone, 'Word of mouth effects'; Herr, Paul M., Kardes, Frank R. and Kim, John (1991) 'Effects of word of mouth and product-attribute information on persuasion: an accessibility-diagnosticity perspective', *Journal of Consumer Research*, 17 (March): 454-462.

40 Laczniak, Russell N., DeCarlo, Thomas E. and Ramaswami, Sridhar N. (2001) 'Consumers' responses to negative word of mouth communication: an attribution theory perspective', *Journal of Consumer Psychology*, 11 (1): 57-73; Wilson, William R. and Peterson, Robert E. (1989) 'Some limits on the potency of word of mouth information', *Advances in Consumer Research*, 16 (1): 23-29.

41 Webster, 'Informal communication'.

42 Charlett, Don and Garland, Ron (1995) 'How damaging is negative word of mouth?', *Marketing Bulletin*, 6: 42-50; Still, Barnes and Kooyman, 'Word of mouth communication in low-risk product decisions'.

43 Wilson and Peterson, 'Limits on the potency of word of mouth'.

44 Bone, 'Word of mouth effects'; Herr, Kardes and Kim, 'Effects of word of mouth on persuasion'.

45 Laczniak, DeCarlo and Ramaswami, 'Consumers' responses to negative word of mouth'.

46 Bone, 'Word of mouth effects'.

47 Laczniak, DeCarlo and Ramaswami, 'Consumers' responses to negative word of mouth'.

48 Herr, Kardes and Kim, 'Effects of word of mouth on persuasion'.

49 Katz and Lazarsfeld, *Personal Influence.* Lazarsfeld, Berelson and Gaudet, *The People's Choice.*
50 Richins, Marsha L. and Root-Shaffer, Teri (1988) 'The role of involvement and opinion leadership in consumer word of mouth: an implicit model made explicit', *Advances in Consumer Research*, 15 (1): 32–36.
51 Feick, Lawrence F. and Price, Linda L. (1987) 'The market maven: a diffuser of marketplace information', *Journal of Marketing*, 51 (Jan.): 83–97.
52 Myers, James H. and Robertson, Thomas S. (1972) 'Dimensions of opinion leadership', *Journal of Marketing Research*, 9 (Feb.): 41–46; Robertson, Thomas S. and Myers, James H. (1969) 'Personality correlates of opinion leadership', *Journal of Marketing Research*, 6 (May): 164–168.
53 Richins and Root-Shaffer, 'Role of involvement and opinion leadership'.
54 The five types of influentials are compared and contrasted by Kiecker and Cowles, 'Interpersonal communication and personal influence'. For *market mavens*, see Feick and Price, 'The market maven'; *innovators*, Midgley, David F. and Dowling, Grahame R. (1978) 'Innovativeness: the concept and its measurement', *Journal of Consumer Research*, 4 (March): 229–242; *purchase pals*, Hartman, Cathy L. and Kiecker, Pamela L. (1991) 'Marketplace influencers at the point of purchase: the role of purchase pals in consumer decision making', in Mary Gilly and Robert Dwyer (eds), *AMA Educators' Proceedings*, vol. 2, American Marketing Association, pp. 461–469; *surrogate consumers*, Solomon, Michael R. (1986) 'The missing link: surrogate consumers in the marketing chain', *Journal of Marketing*, 50 (Oct.): 208–218.
55 Arndt, Johan (1968) 'Selective processes in word of mouth', *Journal of Advertising Research*, 8 (Sept.): 19–22; Holmes and Lett, 'Product sampling and word of mouth'.
56 Gremler, Dwayne D., Gwinner, Kevin P. and Brown, Stephen W. (2001) 'Generating positive word of mouth communication through customer–employee relationships', *International Journal of Service Industry Management*, 12 (1): 44–59; File, Cermak and Prince, 'Word of mouth effects'; Mangold, W. Glynn, Miller, Fred and Brockway, Gary R. (1999) 'Word of mouth communication in the service marketplace', *Journal of Services Marketing*, 13 (1): 73–89; Swan, John E. and Oliver, Richard L. (1989) 'Postpurchase communications by consumers', *Journal of Retailing*, 65 (Winter): 516–533.
57 Blodgett, Granbois and Walters, 'Effects of perceived justice'; Halstead, 'Negative word of mouth'; Richins, 'Word of mouth communication as negative information'; Richins, Marsha L. (1987) 'A multivariate analysis of responses to dissatisfaction', *Journal of the Academy of Marketing*

Science, 15 (Fall): 24–31; Mangold, Miller and Brockway, 'Word of mouth in the service marketplace'; Maxham, James G., III (2001) 'Service recovery's influence on consumer satisfaction, positive word of mouth, and purchase intentions', *Journal of Business Research*, 54 (Oct.): 11–24; Singh, Voice, exit, and negative word of mouth behaviors'; Susskind, Alex M. (2002) 'I told you so!: Restaurant customers' word of mouth communication patterns', *Cornell Hotel and Restaurant Administration Quarterly*, 43 (April): 75–85; Swanson, Scott R. and Kelley, Scott W. (2001) 'Attribution and outcomes of the service recovery process', *Journal of Marketing Theory and Practice*, 9 (Fall): 50–65.

58 Blodgett, Granbois and Walters, 'Effects of perceived justice'; Blodgett, Jeffrey G., Wakefield, Kirk L. and Barnes, James H. (1995) 'The effects of customer service on consumer complaining behavior', *Journal of Services Marketing*, 9 (4): 31–42; Richins, Marsha L. (1983) 'Negative word of mouth by dissatisfied consumers: a pilot study', *Journal of Marketing*, 47 (Winter): 68–78; Singh, 'Voice, exit, and negative word of mouth behaviors'.

59 Blodgett, Granbois and Walters, 'Effects of perceived justice'; Blodgett, Wakefield and Barnes, 'Effects of customer service'.

60 Maxham, 'Service recovery's influence'; Maxham, James G., III and Netemeyer, Richard G. (2002) 'Modeling customer perceptions of complaint handling over time: the effects of perceived justice on satisfaction and intent', *Journal of Retailing*, 78 (Winter): 217–301; Susskind, 'I told you so!'.

61 Derbaix, Christian and Vanhamme, Joelle (2003) 'Inducing word of mouth by eliciting surprise: a pilot investigation', *Journal of Economic Psychology*, 24 (February) 99–116.

62 File, Karen Maru, Judd, Ben B. and Prince, Russ Alan (1992) 'Interactive marketing: the influence of participation on positive word of mouth and referrals', *Journal of Services Marketing*, 6 (Fall): 5–14.

63 Day, 'Attitude change'.

64 Wirtz and Chew, 'Effects of incentives'.

65 Webster, 'Informal communication'.

66 Money, R. Bruce (2000) 'Word of mouth referral sources for buyers of international corporate financial services', *Journal of World Business*, 35 (Fall): 314–329; Tiong Tan, Chin and Dolich, Ira J. (1983) 'A comparative study of consumer information seeking: Singapore vs U.S.', *Journal of the Academy of Marketing Science*, 11 (Summer): 313–322.

67 Dichter, 'How word of mouth works'; Stuteville, John R. (1968) 'The buyer as a salesman', *Journal of Marketing*, 32 (July): 14–18.

12

Black buzz and red ink: the financial impact of negative consumer comments on US airlines

Bradley Ferguson

Founder, Intrinzyk/Founder emeriti, Informative, Inc.

How much is a recommendation worth?

There is growing acknowledgement of the financial value of consumer-to-consumer recommendations for a brand. Reichheld[1] and others have tracked and measured the relationship between market success and product advocacy (or recommendation activity) for various industries. Running in parallel to the belief that positive word of mouth is good for business is another belief that negative word of mouth can have an equally powerful but harmful impact on business success. The effects of negative word of mouth have not received nearly as much attention as the effects of positive word of mouth. And more importantly little work has been done to compare the financial value of positive recommendations to the financial value of negative recommendations. These are important questions to consider because the marketplace is a mixed bag with every brand having strong advocates and many times equally strong detractors. I recently looked closely at these issues within the context of the US domestic airline industry and posed the following questions:

- What is the financial value to an airline when one of its customers says something **positive** about it to an acquaintance?
- What is the financial value to an airline when one of its customers says something **negative** about it to an acquaintance?
- Does a positive or a negative statement have equal financial impact or is one more beneficial and the other more harmful?
- Can the economic value of customer recommendations be quantified?

Looking to the skies for an answer

In January 2005, Informative conducted a study designed to answer these questions. One thousand people who had flown on a named group of US domestic airlines - American Airlines, Continental Airlines, Delta Airlines, Jet Blue Airlines, Southwest Airlines, United Airlines and US Airways - during the previous year were surveyed on their flying habits and their recommendation activities. Data was collected on their elite flying status, standard demographics, online check-in and boarding pass experience, and recommendation behaviour.

Respondents were first asked whether negative or positive comments have a bigger impact on their purchase decisions.

> Which of these would have the greater impact on your decision to fly or not to fly with an airline?
>
> □ A positive comment from a friend or acquaintance about their flying experience with that airline
> □ A negative comment for a friend or acquaintance about their flying experience with that airline

People were roughly split on their perception of the impact on their flying choices of positive and negative comments. 54% reported that positive comments by an acquaintance had a greater impact on their decision to fly with a particular airline, and 46% believed that negative comments had greater impact.

In general, the marketing community tends to focus on the value of positive buzz as a potential driver of customer acquisition and brand promotion. This focus on the positive benefits of consumer word of mouth behaviour is a natural tendency. I certainly like to think that positive comments have a greater impact on my decisions than negative comments. In fact, the respondents to the survey reported that the two factors that had the biggest impact on their airline choices were comparison-shopping on the Internet (38%) and personal recommendations from an acquaintance (42%). But, although personal recommendations have a clear impact on choices, the question is whether positive comments have a greater impact than negative comments. This distinction has not yet been made with regard to word of mouth marketing effects.

The US national election in 2004 was viewed by many as the most negative in recent memory. Why would hard-nosed politicians with limited budgets invest so much in negative advertising if positive comments had greater impact on behaviour? Although we like to believe positive comments have a greater impact on our decisions than negative comments, there are reasons to believe that within certain industries or contexts negative comments have a more pronounced effect on our behaviour.

Survey respondents were asked about their recommendation activities for the seven airlines, in order to measure the number of positive and negative comments for each airline during the previous year.

> Have you made a very **negative** comment to a friend or acquaintance about your experience flying with an airline in the previous year? (Yes/No)
>
> □ (If Yes) Tell us which airlines you made a **negative** comment about to a friend or acquaintance and to how many people you spoke
>
> Have you made a very **positive** comment to a friend or acquaintance about your experience flying with an airline in the previous year? (Yes/No)
>
> □ (If Yes) Tell us which airlines you made a **positive** comment about to a friend or acquaintance and to how many people you spoke

Results

Figure 12.1 and Table 12.1 illustrate the key finding of the airline study. The chart plots the relationship of (1) the percentage of negative consumer comments made about an airline by consumers who consider the airline their primary airline, and (2) the operational profits of that airline as of Q3/2004. The bubble size of the chart indicates the revenue for each airline in Q3 of 2004.

Let's make a few observations. The first thing to note is that Jet Blue and Southwest were the only airlines in the study that were profitable in Q3/2004 and they were the airlines with the lowest negative comments among respondents who considered them their primary carrier. JetBlue

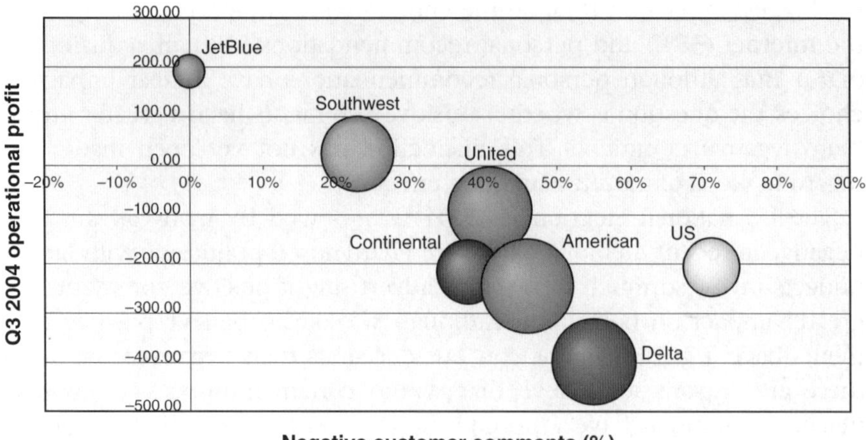

Figure 12.1 Negative word of mouth and operational profit for US domestic airline industry, 2004

and Southwest were also the airlines with the highest percentage of positive comments. The single exception is Continental that squeaked by Southwest with 93% positive comments. The second observation is that the airline industry is in dramatic upheaval and undergoing structural transformation from low-cost carriers such as JetBlue and Southwest. The result is that all the traditional network airlines are losing money at historic levels.

Table 12.1 Analysis of negative word of mouth and operational profit for US domestic airline industry, 2004

	A	B	C	D	E
Airline	*Q3/04 Operating profit (US$ m)*[a]	*% Negative comments*[b]	*Q3/04 Passenger revenue (US$)*[a]	*% Positive comments*[b]	*Ave. % profit Q3/03 – Q3/04*[a]
JetBlue	191.30	0%	311 600 000	1.00	13.12
Southwest	23.00	23%	1 570 000 000	0.92	9.04
United	−91.00	41%	2 028 900 000	0.75	−6.68
US Air	−206.10	71%	969 400 000	0.65	−6.74
Continental	−216.40	38%	1 144 600 000	0.93	−6.80
American	−244.80	46%	2 453 500 000	0.88	−8.22
Delta	−399.00	55%	2 041 900 000	0.81	−9.02

Source: Columns A, C and E, data from the Bureau of Transportation Statistics (BTS) of the US Department of Transportation, 16 December 2004; columns B and D, data from the Informative study conducted January 2005

Negative word of mouth predicts poor profit

The really interesting finding here is that the percentage of negative comments is highly predictive of operational profit[2]($R^2 = -0.84$), while positive comments are not very predictive at all. While many people might have anticipated that negative comments are as predictive as positive comments, few people would expect negative comments alone to be predictive of operating profit. The correlation of negative comments is even higher ($R^2 = -0.86$) against the five-quarter average of operational profits (column 'E') than it is with operational profits in Q3/2004 (column 'A'). This makes sense because the five-quarter average contains data over a longer period of time and smoothes out any volatility that might impact a single quarter.[3]

The strong negative correlation between operating profit and negative comments suggests that as the percentage of an airline's customers who are making negative comments about them increases the airline's operating profits decrease. Conversely, as the percentage of an airline's customers who are making negative comments about them decreases the airline's profits increase. This correlation illustrates a very interesting dynamic that appears to be at work in this particular industry.

If we view the word of mouth communication among airline customers as a media channel then we are saying the greater the frequency and reach of the negative comments about an airline on that channel the greater the negative impact on an airline's profits. This is literally what is happening in this industry. With traditional media channels, advertisers specify in advance the frequency and reach of their advertisements and then pay accordingly. With the word of mouth channel there is good news and bad news. The good news is that you do not have to pay to use the channel for advertising. The first piece of bad news is that you cannot specify the content, frequency, or reach of the advertising. The second piece of bad news (at least for the airlines studied in the current market) is that negative word of mouth advertising is much more predictive of profits than positive word of mouth advertising.

Show me the money

At this point we can answer two of our original questions:

- Does a positive or a negative statement have equal financial impact or is one more beneficial and the other more harmful?
- Can the economic value of customer recommendations be quantified?

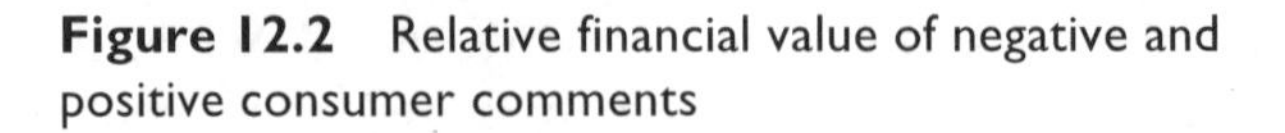

Figure 12.2 Relative financial value of negative and positive consumer comments

Not only are negative comments more predictive of operating profit than positive comments but their impact can be quantified. Negative comments have about 2.4 times more financial impact than positive comments within the airline industry. Every 1% reduction in negative comments co-varies with US$4 050 464 of operating profit; every 1% increase in positive comments co-varies with US$1 709 300 of operating profit.

The term 'co-varies' has been used because there is not yet a proven causal relationship between negative comments and operational profits. However, the percentage of negative comments definitely mirrors operational profits for the airlines within this study. To summarize, it appears *for this industry at this time* more financially beneficial to reduce negative consumer comments about a brand than it is to increase positive consumer comments about a brand.

Why negative word of mouth is linked to poor profitability

It's time to poke holes in this analysis in order to test these conclusions. We have not yet presented a theory about the causal processes that could explain the correlation of negative comments and operating profit. The competitive story illustrated in Figure 12.1 was what Clayton Christensen terms 'the disruptive transformation of an industry by low-cost entrants'.[4] In Christensen's model of disruptive innovation:

- Incumbent players compete with one another by creating high-margin services for demanding customers.
- Low-cost entrants enter the competitive picture, meeting the needs of the under-served, less demanding segment of the market.
- Low-cost entrants develop product and business models that meet the requirements of less demanding customers at a price that the incumbents cannot match, and still operate profitably.

- Low-cost entrants gradually improve their service and meet the needs of a larger share of the market while maintaining low prices.
- Incumbents are faced with a competitive dilemma:
  - Lower prices below the level that their business model can operate profitably
  - Retreat up-market to battle for high-margin demanding customers

This is precisely the financial dynamics of the airline industry depicted in Figure 12.1. In the current market the incumbent (traditional network) airlines are being pressed to maintain services by their demanding business customers and are being pressed to lower prices by their less demanding leisure customers. The incumbent airlines are struggling to operate profitably at the low end of the market while maintaining high levels of service for their profitable business travellers. Unfortunately, given the dilemma, airlines are having a difficult time meeting the demands of either of these groups and as a result are generating negative comments from both.

In this competitive environment:

1. It makes sense that the incumbent airlines would generate more negative comments than their low-cost competitors who have less demanding customers.
2. It also makes sense that the low-cost entrants' ability to operate profitably at a much lower price point would generate higher positive comments from their less demanding customers.

This model of disruptive innovation gives a plausible account for why there might be a negative correlation between negative consumer comments and operating profits among the airlines we studied. But how does this information help marketers in the airline industry do their job? How does understanding the dynamics of word of mouth marketing help marketers build a successful airline, or a successful marketing campaign for that matter? If we take seriously the notion that word of mouth is a media channel with frequency, reach and content that has a direct relationship to profitability then airline marketers have several approaches open to them.

Marketing implications

You can't manage what you don't measure. The first thing an airline marketer should consider is to track and measure the word of mouth content about their airline: (1) track the number of negative comments its customers are making to their acquaintances about the brand; (2) identify the concerns being expressed by its consumers; and (3) wherever possible, address these concerns to decrease the negative word of mouth. Decreasing negative comments is literally like money in the bank for an airline. Addressing the causes of negative comments might mean changing in some way their product or service offering. But it may be that their customers are misinformed about or unaware of some aspect of the airline's business. In this case some kind of informational marketing campaign utilizing traditional media could help decrease the negative word of mouth.

Another approach is to consider countering negative word of mouth by allocating marketing budget to target the major sources of word of mouth marketing effects. Not all consumers are equally effective in the communication of their attitudes throughout the network. Marketing segmentation approaches have tended to segment consumers by different demographic or product use attributes, but have generally ignored the impact people have on others' purchase decisions. But if you look at recommendation behaviour and other psychological and social attributes, some consumers have much greater impact on word of mouth activity than others. Borrowing terms from the traditional media channels you could say some people have greater frequency and reach than others. Further, some consumers have greater impact on the purchase decisions of those around them than the general population.

Marketers are developing segmentation profiles to identify people with various roles in the transmission of messages through the social network. For example, I have developed a segmentation profile for influencers[5] that has proven successful in identifying people who have greater frequency, reach and impact on purchase decisions than the general population. The profile measures people along two variables that are highly predictive of recommendation behaviour and impact. It utilizes the Affective Communication Test[6] (ACT) developed by Howard Friedman at the University of California and a variable I created to measure the relative size and geographical distribution of a person's social network.

A great number of people have been profiled over the past two years, and we can conclude that influencers have several attributes of interest to marketers. First, they are very active recommenders. In many industries and concerning many different product categories, they have been shown to exhibit recommendation frequencies from two to six times greater than the general population. They also have been shown

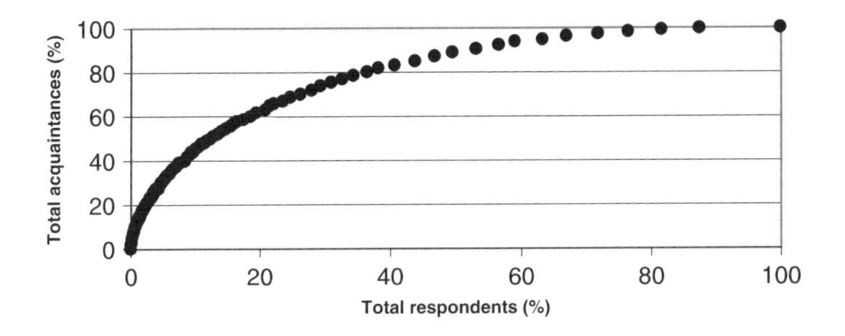

Figure 12.3 Pareto analysis of acquaintance distribution in population

to know on average four to five times the number of people as the general population.

A Pareto analysis on influencers shows that 20% of the population account for about 68% of all the social acquaintances of the population studied. It turns out that the distribution of social acquaintances does not follow a normal distribution but a power law distribution. This same kind of phenomenon has been identified in links to websites on the Internet and other places susceptible to network effects.[7] In power law distributions (unlike normal distributions) there can be extreme deviations from the norm. Since physical height among humans is normally distributed you might very well meet someone who is 7 feet tall but you will never meet anyone who is 70 feet tall. On the other hand, wealth is distributed according to a power law in many cultures. If the mean yearly income is approximately US$40 000 in the US, you can run into people who have yearly incomes of US$400 000, US$4 000 000, or even higher.

Besides having greater frequency and reach, influencers have been shown to have a greater impact on the purchase decisions of those in their social network. The impact on purchase decisions is difficult to measure in most social contexts, but we approached the problem from a different angle to approximate the effect. Insurance agents for a large US insurance agency were profiled then grouped by influencer segment, using a quadrant analysis in which people are grouped by the following four categories.

- Expressive/Social (high ACT/high social connection)
- Expressive/Private (high ACT/low social connection)
- Reserved/Social (low ACT/high social connection)
- Reserved/Private (low ACT/low social connection)

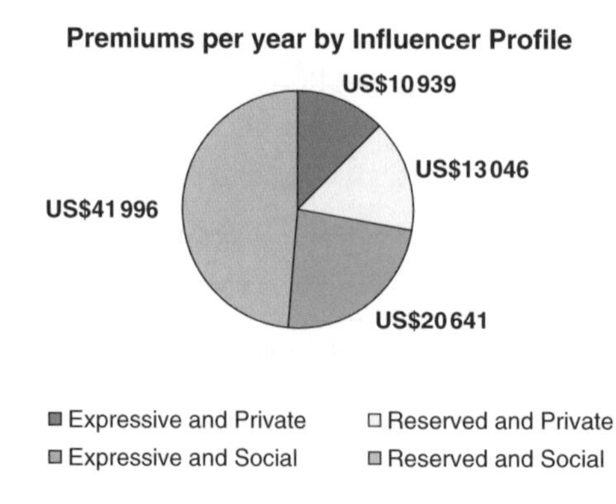

Figure 12.4 Insurance premiums per year by influencer profile

The results (Figure 12.4) showed that the influencers with a B+ in emotional expressiveness and an A in social connection were two to four times more productive in selling premiums on the insurance products under examination.[8] The influencer segmentation is very predictive of sales success and, by analogy, of the impact that influencers have on the purchase decisions of those in their social networks. An influencer is defined as a person in the top 20% of the population in both affective communication skills and social connection. This typically yields about a 10% incidence rate in the general population. However, as in the case above, we optimize the configuration for the desired behaviour and the greatest incidence rate. We have found in certain industries and product categories, such as insurance sales and recommendation of consumer electronics, you need to dial down the emotional expressiveness to yield the best results.

Influencers create an inordinate amount of the word of mouth effects that drive the financial impact in the US airline industry. They actively recommend either positively or negatively the brands they love and hate to an extended social network. In addition, they are natural born salespeople who are very capable of influencing the rest of the population's purchase decisions. Marketers can leverage this valuable segment by using an inordinate portion of their marketing budget to target influencers with special promotions that generate trial, adoption and advocacy and result in positive word of mouth recommendations.

Conclusion

As we have seen, negative consumer comments can be reflective of financial success in certain industries. The way we talk to one another about a brand is very reflective of the attitudes that drive purchase decisions. It is important for airline marketers to consider that the airlines that are having the greatest financial success in the current market are also receiving the fewest negative comments and the most positive comments from their customers. The way customers talk about brands can be viewed as an emotional bank account. Every day customers are making emotional deposits and withdrawals through the things they say about brands to their friends. These emotional deposits and withdrawals are just as reflective of financial success as the deposits and withdrawals in your actual bank account.

Takeaway points

- Word of mouth can be viewed as a media channel that has a measurable frequency, reach and impact on purchase intent similar to traditional media channels.
- In order to manage word of mouth, marketers first need to track and measure it.
- It is important to understand the dynamics of negative as well as positive word of mouth for a given industry and market.
- For the airlines considered in this study, negative word of mouth is inversely proportional to profitability – as negative word of mouth increases, operating profits fall.
- The financial impact of negative and positive comments can be quantified.
- Word of mouth marketing should be done systematically, and programmes should focus on influencers and opinion leaders who have the greatest reach (large social networks), the highest frequency of recommendations, and the greatest impact on purchase intent.

Notes and references

1 Reichheld, F.F. (2003) 'The one number you need to grow', *Harvard Business Review*, 81 (12): 46–54.
2 Operational profit in the airline industry is defined as RASM (revenue per available seat mile) minus CASM (cost per available seat mile).
3 The author presented some initial findings of this study to the 12th Ohio State University Airline Symposium in Honolulu in January 2004.
4 Christensen, Clayton M. and Raynor, Michael E. (2003) *The Innovator's Solution.* Boston, MA: Harvard Business School Press, pp. 32–35.

5 Ferguson, Brad (2003) Understanding Influence Communities: Analyzing, measuring, and accelerating word of mouth communication. Advertising Research Foundation National Convention.
6 Friedman, Howard S., Prince, Louise M., Riggio, Ronald E. and DiMatteo, M. Robin (1980) 'Understanding and assessing nonverbal expressiveness: the Affective Communication Test', *Journal of Personality & Social Psychology*, 39 (2): 333–351.
7 For an important discussion of the implications of power law distributions see Barabasi, Albert-Laszlo (2002) *Linked: The New Science of Networks*. Cambridge: Perseus Publishing.
8 The premiums listed are not the total premiums for these agents per year but the total for a line of insurance products that concerned our clients.

13

Myths and promises of buzz marketing

Stéphane Allard

Associate Director, Spheeris

Myths and misinformation abound in the world of buzz marketing. People say it's cheap. It's not; it's cheap*er*. Others say word of mouth, viral and buzz marketing are all the same. They're not; they're different. Some say you have to have an exceptional product to harness the word of mouth effect. You don't. Critics say you can't measure buzz. You can. Others say that word of mouth marketing is risky. In fact, it's no more risky than any other type of marketing.

This chapter dispels some of the common myths associated with viral, buzz and word of mouth marketing, and guides you through the jargon jungle to find the connected marketing strategy that's right for you.

Myth 1

The myth: Buzz, viral, street and stealth marketing are all the same.

User comment: 'My viral email campaign is coupled with an undercover marketing campaign that will create buzz in Internet forums.'

Reality: Word of mouth marketing techniques differ and shouldn't be confused with other alternative marketing techniques. They don't serve the same objectives.

Not all connected marketing techniques are the same

	What's different about it?
Word of mouth marketing	Umbrella term for marketing practices which aim to make consumers talk about a brand
Buzz marketing	Using a special 'hook', event, or promotion to get consumers and the media talking about a campaign
Viral marketing	Creating branded Internet materials or websites that consumers enjoy sharing with their friends, usually by email
Influencer marketing	Identifying and involving the most influential consumers in a target market to turn them into brand advocates
Evangelist marketing	Involving your most loyal customers to turn them into brand advocates
Street marketing	Reaching and interacting with consumers directly in the places they frequent offline
Stealth/Undercover marketing	Marketing below the threshold of awareness (for example: hiring actors to spread positive brand messages in public spaces, passing them off as real consumers)

As for every fashionable new marketing technique, it's highly tempting to add a 'viral' here and a 'buzz' there. A brand new field, some weak definitions and you end up with general confusion as to what the activities really are and what to expect from them. Terms become completely interchangeable and everything, including old practices, become viral or buzz campaigns.

But buzz is not viral is not street is not stealth marketing. Table 13.1 shows one interpretation of some of the marketing techniques you may come across in our brave new world.

Each technique uses different levers to get different results. So you need to choose the technique that's appropriate for the objectives you want to achieve. The first step is to ask yourself some questions:

1. *What are your objectives?* Do you want to drive brand or product awareness? If so, your activities don't necessarily need to result in the consumer endorsing your product or brand. You want consumers to recognize and remember it. Or do you want to push sales, adding credibility to your marketing mix? If so, you'd be better focusing on activities that put a strong emphasis on getting consumer endorsement for your brand or product.

	Drive brand awareness – weak endorsement	Push sales – strong endorsement
Message spreads from consumer to consumer (word of mouth marketing related)	Viral marketing Buzz marketing	Influencer marketing Evangelist marketing
Personal brand contact with each individual consumer (non-word of mouth marketing related)	Street marketing	Undercover/stealth marketing

Figure 13.1 Objectives in connected marketing

2. *Do you want the message to be spread from consumer to consumer? Or would you prefer to create personal brand contact with each prospect individually?* Street marketing and stealth marketing get your brand noticed, but they aren't focused on stimulating consumer-to-consumer conversations. Alternatively, with viral, buzz or word of mouth marketing, the point is to relinquish a degree of control of the message so it spreads within consumer networks.
3. *Are you ready to take the risk of possible consumer backlash?* Undercover or stealth marketing can come under high scrutiny and can be perceived as unethical. If you're exposed, your brand's image may suffer a great deal.

Figure 13.1 provides an overview of the different connected marketing practices categorized by objective. It will help you to choose the right activities for your organization, according to your objectives. The remainder of this chapter focuses on the marketing activities in the first field of the diagram: activities that encourage consumers to spread messages to each other. From now on, we'll refer to these activities as word of mouth marketing.

Myth 2

The myth: Only great products benefit from word of mouth marketing.

User comment: 'Unfortunately I'm not Apple. My products aren't that sexy.'

Reality: Even the most unlikely products can benefit from word of mouth marketing.

Let's face it: if you have a stand-out product such as the iPod or the VW Beetle, you can expect greater results from word of mouth marketing campaigns. And undoubtedly, promotion of high-involvement products is also largely driven by word of mouth.

But if you're selling, say, bags of flour, it doesn't mean word of mouth marketing can't be used to your advantage. Management consultancy McKinsey estimates that roughly 70% of the whole economy, including the flour industry, is influenced by word of mouth.[1]

The first thing to consider is that your product may be more interesting than it first appears. Are Post-it Notes cool? Is Google sexy? Are Eastpak backpacks incredible? Not at first glance. Yet they all grew popular thanks to word of mouth. There are people out there who are waiting for your product (let's hope so, otherwise, don't even bother launching it); people who will greatly benefit from your offer. Find them, develop relationships with them and transform them into brand advocates.

Let's continue with the example of flour. There are people who are very interested in flour because they do a lot of cooking. They want to make the best cakes and breads and are always looking for ways to improve their recipes. They are known in their social circles as cookery experts and are highly influential on that topic. People will turn to them for advice on cooking bread or cakes. Putting your flour into the hands of these influential consumers will generate word of mouth for your brand. What's more – the icing on the cake – your product information will be well received by these individuals. This is in direct contrast to most other consumers who will try to avoid it, forcing you to adopt interruptive marketing strategies such as advertising.

Now let's suppose that your product doesn't have such experts interested in it (any peapod fans out there?). Or that it doesn't slip into everyday conversation very easily (I can't remember the last time I talked about toilet paper socially). What do you do?

The ploy here is to stimulate conversations, not about the product itself, but about the marketing campaign behind the product. Buzz or viral marketing campaigns are perfect for this: they create social currency that feeds consumers' conversations ('Have you seen . . .?' 'Have you heard about . . . ?'). Talking about the campaign is a trigger to making people talk about your product at the same time. And if your product is buzzworthy because of intrinsic qualities (blue peapods or toilet paper made of real tree leaves, for example), you have a surefire winner when it comes to word of mouth stimulation.

Myth 3

The myth: Great products don't need word of mouth marketing.

User comment: 'I have the iPod-killer in my hands. I don't need to stimulate buzz, it will grow by itself.'

Reality: Even great products are drowned in the current marketing clutter. You need to stimulate and sustain the natural flow of word of mouth.

Great products wouldn't have needed word of mouth marketing at some point last century when:

- The number of competitive products was limited
- Consumer information sources were limited
- The volume of marketing communication was limited
- Product lifecycles were long.

Now consider the following information:

1. Every year, over half a million new products are launched in Europe.[2]
2. In 1965, three advertising spots on TV in the US would reach 80% of a mainstream target audience. By 2002, the same reach required 117 spots.[3]
3. It is estimated that people are exposed to approximately 3000 messages a day.[4]
4. The product lifecycle is ever-shortening.[5]

Some key words emerge from our current marketing landscape that explain the present situation: saturation and speed. Too many choices, too many sources of information, too many messages and too little time to let the market grow the demand by itself.

Will a great product find its way through this current clutter and 3000 other messages? Possibly. But when it happens, it may be too late (most of the profits on a new product are made during the early sales phase), and it may be due more to luck than intrinsic product qualities.

'Harry Potter' books[6] are a great word of mouth legend. The proof that a great product can explode simply from natural word of mouth, right? Not really. What few people know is that the first book in the series was refused by several different publishers. How could these experts have turned down such a nugget? Great products are unfortunately not always considered as such by the public at first sight. Even more interesting, recent academic research on the 'Harry Potter' phenomenon shows that 'its success may have more to do with particular attributes of the social and media network it is spread across than with any inherent quality of the book'.[7] And indeed, the publisher of the books used a highly involved word of mouth marketing campaign, spreading the book into the relevant social networks.

Another example is 3M Post-it Notes, a great product with obvious qualities. So obvious, in fact, that the traditional marketing campaign to promote it failed miserably. The product was about to be completely withdrawn from the market when a word of mouth marketing campaign saved it. (For the full story, see Chapter 1).

In a nutshell, natural word of mouth can sometimes make your product a hit. But most of the time it won't since it will be drowned out by the thousands of other messages being pumped out. It will spread too slowly to fuel the take-off of your product. What you need to do in this day and age is to stimulate this natural word of mouth flow using word of mouth marketing campaigns, even if your product is already buzzworthy.

Myth 4

The myth: Word of mouth marketing is cheap.

User comment: 'Let's drive millions of consumers to my stores with word of mouth marketing. I have a few thousand dollars to spend.'

Reality: Word of mouth marketing is not cheap, but cheaper.

Viral and cheap are often seen as synonymous. There's no denying that a few companies manage to achieve success with almost no marketing budget. Hotmail, Google and eBay are among the best examples. But do you see a common pattern here? They all exist on the Internet, their services are easy to sample and are (or were) free. Unfortunately, this is not the case for 99.9% of the companies out there. Most products and services are not 100% Internet-based and so are more difficult to sample. Most businesses only make money by selling products or services to the end user. How many customers would Hotmail have acquired initially if the service hadn't been free?

And what about the *Blair Witch Project*? It was an offline product that people had to pay to see. True, part of the success of the movie is due to word of mouth, but this didn't happen without a heavy marketing push.

Artisan reportedly spent US$10 million on distribution and promotion, three times less than the average marketing budget in the US for a big studio film (estimated to be US$25–30 million).[8] Artisan harnessed the power of word of mouth for *less* money, but not with no money at all.

How you spend your budget is also different for word of mouth marketing than it is for traditional marketing. In traditional marketing campaigns, media buying tends to consume a large part of the total budget, while creation is allocated a much smaller portion. With word of mouth marketing, you don't need to invest as much in media, but you need a realistic budget for strategic planning and creation. You want the consumer to spread the message for you (thereby reducing your media cost), but they need a very good reason to do so (so investment in planning and creation is crucial).

Unfortunately, word of mouth marketing is not the Holy Grail some expected: you need to invest if you want to get results. And, more likely than not, the more you invest the better results you will get. In some instances, you can still achieve a measure of success for a few thousand dollars, but you need to adapt your expectations accordingly. It all depends on what you define as success.

Myth 5

The myth: Word of mouth marketing is risky.

User comment: 'I don't want to run the risk that my message is distorted or hijacked while spreading from consumer to consumer.'

Reality: The heyday of message control is over.

For the past 40 years, marketing has been all about control. Control the message. Control the brand image. Control the media channels. Control the distribution channels.

There was, and still is, value in control but marketers must accept that the landscape has changed. Over a period of just a few years, everyday consumers have gained access to a variety of communication tools: email, SMS, websites, instant messaging, blogs, forums, GSM . . .

What was the cost for a consumer to spread a message to 1000 other consumers in 1980? You can factor in a few thousand dollars, as well as the cost of the time spent doing it. What is the cost in 2005? Close to zero and the time needed to execute the task is now a matter of minutes.

Like it or not, the reality is that consumers are increasingly taking control of the marketing communication landscape. That's not an opinion; it's a fact. There are already millions of consumers talking about your products in Internet forums, or expressing and spreading their opinions far and wide to their friends via email.

The movie *Hulk* lost 69.7% of its potential audience within the first few days of its release. It was the same for *2fast2furious* which lost 63%, and *Charlie's Angels* which lost 62%.[9] The reason? Consumers spreading negative word of mouth using SMS, email and websites. Never in the history of the movie industry has a phenomenon made so much impact, so quickly.

'In the middle of difficulty lies opportunity,' said Albert Einstein.[10] So either you ignore the word of mouth phenomenon (at your own risk), or you try to embrace it and benefit from it.

Consumers are already talking about your brand and products – they may badmouth them or spread positive comments, and you can't control that. However, you can exert influence on that outcome with word of mouth marketing. Ultimately, in this media-saturated world, the winners will be the brands that stimulate the flow of conversation between consumers.

Myth 6

The myth: Word of mouth marketing can't be measured.

User comment: 'Buzz is great. Problem is, you can't measure it.'

Reality: Metrics exists to measure the impact of word of mouth marketing activities on the bottom line.

One of the criticisms levelled at word of mouth marketing is its lack of measurability. We're not even close to developing a single measurement system that encompasses every kind of connected marketing technique – if that's even possible. But metrics for connected marketing are improving every day and some useful models and tools already exist. As stated previously, viral, buzz and influencer marketing are not the same things and consequently each have their own tools and metrics.

Viral marketing, due to its online nature, is probably the most measurable of all the word of mouth marketing approaches. It's not difficult to measure traffic to a specific website. Burger King's viral website,

Subservient Chicken, for example, had received 12.7 million unique visits as at December 2004.[11]

Tools also exist to track the viewing of and interaction with elements downloaded from the Web and spread by email from user to user. For example, online viral marketing consultancy Digital Media Communications has expertise in the peer-to-peer tracking of viral video clips. In 2002, the company engineered a viral campaign for Xbox's European launch and tracked over 450 000 calculable, proactive views of copies of the *Champagne* clip within just four weeks of launch.[12]

Buzz marketing is closer to PR and events marketing by its nature, so metrics used to measure success in those two sectors can also be applied to buzz marketing. Depending on what kind of buzz activity is used, you can measure the equivalent of media column inches and GRP (Gross Rating Points); track online buzz links via a specific URL; do a pre- and post-qualitative and quantitative survey; or track a specific element of the campaign (such as coupon redemption).

Influencer marketing is probably the most difficult word of mouth activity to track but it's also one of the most impactful due to the credibility of the messages it generates. An indisputable metric that can be used to measure it is based on sales volume. Compare a sales area in which an influencer campaign has run to one in which it hasn't and that will give you a good view of the impact of the campaign. For example, Procter & Gamble saw a 14% increase in volume sales in Providence, USA due to an influencer marketing campaign.[13]

Field reports and surveys are other tools available to measure word of mouth marketing results, notably in terms of reach. Ammo Marketing's 2004 campaign for Miller Genuine Draft in the USA showed that over their six-month programme, the 500 enrolled members indirectly influenced nearly 1.5 million beer drinkers.[14]

An emerging metric in the field is the measurement of online word of mouth to gain an insight into its associated offline buzz. One study discovered that the dispersion of online conversations was a fairly accurate measurement of what was happening offline.[15]

Conclusion

The field of word of mouth marketing is developing quickly and so is the credibility of its metrics, especially with academic researchers now coming on board. It's in the interest of all marketing stakeholders to keep up this momentum. However, to sustain it, we need genuine facts and figures, not myths, misconceptions and inflated expectations.

Takeaway points

- No matter what your product, there is a high probability that it will be influenced by word of mouth.
- If your product is not buzzworthy, focus on creating a marketing campaign that is worth talking about instead.
- Natural word of mouth alone isn't enough to sell products; you need to stimulate the flow.
- To get greater and more in-depth results, you need to invest a realistic budget.
- Consumers are taking control of marketing messages, but marketers can still influence the outcome of a word of mouth marketing campaign.
- Viral marketing is the most easily measurable of all word of mouth marketing activities.

Notes and references

1 Dye, Renée (2000) The buzz on buzz. *Harvard Business Review*, (Nov.-Dec.).

2 (1997) *New Product Introduction - Successful Innovation/Failure: A Fragile Boundary.* ACNielsen BASES and Ernst & Young Global Client Consulting.

3 Bianco, Antony *et al.* (2004) 'The vanishing mass market', *Business Week*, 12 July.

4 Shenk, David (1997) *Data Smog: Surviving the Information Glut.* New York: HarperEdge.

5 Gleick, James (2000) *Faster: The Acceleration of Just about Everything*. New York: Vintage Books.

6 Green, L. and Guinery, C. (2004) 'Harry Potter and the fan fiction phenomenon', *Media and Culture*, 7 (5), archived at http://journal.media-culture.org.au/0411/14-green.php.

7 Watts, Duncan (2003) 'The science behind six degrees', *Harvard Business Review* (Jan.-Feb.).

8 (1999) 'How "The Blair Witch Project" cast online marketing magic', *Digitrends* (Fall).

9 Munoz, Lorenza (2003) 'High-tech word of mouth maims movies in a flash', *Los Angeles Times*, 17 August.

10 http://www.quotedb.com/quotes/2575.

11 Bess, Allyce (2004) 'Selection of Crispin Porter shows that Gap isn't chicken', *San Francisco Business Times*, 24 December.

12 Cases study on DMC's website: http://www.dmc.co.uk.
13 Tremor presentation 'SmartMarketing' at the International Dairy Foods Association meeting, 5-7 March 2002.
14 Miller Genuine draft case ctudy, December 2004. Ammo Marketing website at http://www.ammomarketing.com/clients/miller.htm.
15 Godes, David and Mayzlin, Dina (2002) 'Using online conversations to study word of mouth communication', *HBS Marketing Research Paper* No. 02-01.

14

Buzz marketing: the next chapter

Schuyler Brown

Co-creative director, Buzz@Euro RSCG

This chapter gives you a thorough insight into buzz – the word of mouth effect, the transfer of information through social networks; and buzz marketing – the scripted use of action to generate buzz.[1]

Back in 2003, connected marketing activities such as buzz, guerrilla and influencer marketing were still emerging practices in terms of their recognition and credibility with marketers. Since then, a lot of buzz has burned through the world's social networks. One example is the concept of 'metrosexuality' that Buzz@Euro RSCG helped spread to the world. Some of the buzz about metrosexuality can be attributed to the able minds at various marketing agencies, but most of it was spontaneously ignited by consumers themselves.

Unlike other forms of marketing, which consumers can take or leave – and, more often than not, they're leaving it these days – connected marketing such as buzz activity happens with or without marketing people. The world's buzz marketing professionals could disappear tomorrow and, in some form or fashion, buzz would still be happening.

Isn't it great to be totally . . . irrelevant?

But, wait! There is something buzz marketers can do better than others: we can enhance and sometimes even direct buzz. And we can do it in a way that benefits our clients. Buzz marketers also have the advantage of being entertainers, educators, activists and content providers, rather than simply 'interrupters'. Buzz marketing, like other connected marketing activities, works because it flows through the people who actually want to hear and spread the news. It's not interrupting the regularly scheduled programming; it *is* the programming. And it's a channel to which most people want to stay tuned.

The future of marketing?

Marketers didn't just get into all this buzz stuff because they were bored with the status quo. They realized the methods that worked in the second half of the last century weren't going to have the same effect in this century. And that's because the fundamental relationship between consumer and marketer has changed.

In the beginning, marketing was an immediate business and one that required cooperation between advertiser and consumer. The Mary Kay gal knocked on your door and you had to open it or her pitch wasn't going to be very effective. In 1965, PF Flyers, a classic sports shoe manufacturer, developed a promotion with cartoon hero Johnny Quest and his magic decoder ring. A television commercial showed Johnny (clad in his trusty PF Flyers, naturally) rescuing Race Bannon from an erupting volcano. Despite the rarity of active volcanoes in the US, kids were encouraged to get their own decoder ring (just in case!) by buying the shoes and sending in for it. If you think about it, this required a lot of work for what was, in the end, a piece of plastic that took nearly three weeks to arrive - but it seems it had kids flocking to the shoe stores.

Over the past couple of decades, efforts in advertising strove for broad reach and appeal. The ad industry got big. It got bold. And ads got noticed, but maybe not understood. They came at people with a 'You're either in or you're out, so which is it?' vibe. This was cool in many cases: Yes, I want to be a Pepper! Yes, I will have it my way, Burger King! I want to sing in harmony with the world, Coke! Count me in. And not so good in other cases: If I log on to Outpost.com, I can shoot gerbils out of a cannon? Think I'll pass on that one.

Ads, like the companies they were representing, became opaque and standoffish. In looking to appeal to everyone, they ended up appealing to no one in particular. Marketers took the life out of the messages, and audiences started looking for ways to take us out of their lives. You've got to hand it to them, they've done a pretty good job - TiVo, on-demand cable, satellite radio, pop-up blockers. Consumers are now putting the burden back on marketers to court them, woo them and give them something worth listening to, looking at and talking about. It's time to rebuild the relationship between the marketers and the market.

Buzz marketing may be the key

As marketers emerge from the hands-off marketing of the past couple of decades, they are experiencing a back-to-basics approach to getting the word out about the products they are tasked with marketing. Today, that Johnny Quest promotion sounds simultaneously (1) old-fashioned and naive and (2) like the beginning of a pretty good buzz marketing idea for PF Flyers, only today that decoder ring would be TAG Heuer with a GPS device that could track your friends on their eco-travel expeditions to places where volcanoes actually exist, and Johnny Quest would be a limited-edition, manga-inspired figure available only at select stores in Manhattan and Tokyo.

It's been a while since marketers have asked so much of consumers. No longer content to let consumers soak up spiels, marketers now want them to volunteer to be active participants in spreading the word. So here marketers are, once again, standing at the door, their pink Cadillac parked in the driveway, asking for permission to come in and show off the latest line of lipstick shades. But harder even than getting in the door, they are working to make such an impression that their target will tell her friends about the products, and they'll tell their friends, and so on, and so on.

Buzz marketing (as defined earlier in this chapter) is far from science. It is art, and those practising it know just how scary that can be. Yes, we may be poised on the edge of a great era for marketing. It feels like a golden age of understanding is right around the corner. But marketers have to face down a few challenges - sorry, opportunities - before they can get to the smooth-sailing part.

Right now, the practice of buzz marketing is being defined by its successes, because it is a business in which failures die a quick and very quiet death. There are a number of people and campaigns that have come to define buzz marketing at its best. There are innumerable lessons in their stories and potentially a few glimpses into what the future might hold. Here, we'll seek to give some structure to the amorphous business of buzz marketing as it is happening today.

First and foremost, people who hope to understand the fundamentals of the buzz discipline need to know why buzz tactics have come to the fore in recent years. The simple answer is that current conditions are ideal for buzz marketing - and, in fact, make them an imperative. These conditions are grounded in five new realities that impact businesses and the lives of consumer targets.

New reality 1: Business as usual is getting very unusual

Most people would agree that Corporate America is due for a makeover. And because we are living in a time when ordinary citizens have a vested interest in the way the businesses around them make money, treat their employees and sell products, advertisers and marketers trying to put a bright gloss on these businesses have their hands full.

When my mother, an educator in Kentucky, started talking about such things as the outsourcing of jobs overseas, corporate ethics, owning stocks ('Did you buy Google?') and sustainability, I knew 'business' was now considered part of everyday people's domain, no matter what their occupations. And it should be, given the impact of business mismanagement and criminality on people's personal finances.

The average person's loss of faith in business has consequences for marketers, of course. But we may be headed towards a time when the public's revulsion with unethical business practices reaches the point at which genuine change is effected. OK, go ahead and scoff. But if you think Nike simply shrugged off consumer criticism of its labour practices, think again.

When we look at what's buzzing in business, we don't just see the bad. We're also seeing signs that today's teens and young adults in the US are genuinely interested in restoring the sheen to the American Dream. The most influential among them are socially conscious, economically responsible and interested in protecting the environment. They have made the Toyota Prius the hot car of the moment (despite a 44% increase in its global production, there are more people on the waiting list than there are cars heading to dealerships). They eat organic, vegetarian and raw foods to preserve the environment and respect their bodies. They shun big businesses, refuse to wear fur, and flock to such brands as Carhartt and Camper because they are 'well made'. They are modern-day hippies, and they have a spokesperson in the flamboyant figure of Dov Charney, CEO of clothing company American Apparel.

Dov Charney has built his company on buzz, with a predilection for shock tactics (big and small) that get people talking. He is fond of using images of scantily clad young women in his advertising and trade-show booths, often using employees as models and taking many of the photos himself. One of the company's ads has featured his own bare backside. He has spoken of consensual sexual activity with his employees and an article in *Jane* magazine described his fondness for pornography and sexual favours.[2]

He sports a 1970s-style mustache and macho-man glasses, and can be seen about Brooklyn wearing a terrycloth tennis outfit of short shorts, polo shirt and sweatband. He is capable of using the f-word as a verb, noun and adjective – all in one sentence. And, of course, he is the darling of the US media – both the business press and the popular press.

In August 2004 Charney was voted Person of the Year by apparel-industry publication *Counselor*. The choice of Charney suggests a growing level of respect for his business practices, which have been called everything from kooky to revolutionary, and a begrudging acceptance of the man. In just seven years, he has grown his company from 50 employees to nearly 2000 and has proved that you can manufacture quality apparel without outsourcing anything. His treatment of employees is nearly enlightened, with benefits ranging from salaries that are almost double the minimum wage to classes in English as a second language, immigration support, computer literacy, on-site masseuses, even bicycles for employees without cars to use as transportation to and from work. No wonder there's a waiting list of 2000 people looking for work at American Apparel.

Charney is worthy of note not just because he's an amazing buzz marketer (he does very little advertising) and self-promoter, but also because he's emblematic of his generation's determination to make money and build businesses without crossing ethical boundaries and exploiting workers. When asked about sweatshop labour, Charney rants in *Counselor*, 'Understand: I think hot young girls are sexy. Hot young girls slaving over looms for 12 hours a day in Indonesia to make US$1, however, are not.' In his view, 'innovation and social responsibility are the new American dream'.[3]

New reality 2: Great expectations, not so great budgets

You really have to be a spin artist to take the challenge of shrinking budgets in the marketing industry and turn it into an opportunity, but for buzz marketers, there really is a silver lining. We all know that Wall Street analysts and company boards are demanding more accountability from companies, particularly with regard to the way they are spending marketing dollars. Because so much of marketing ROI is intangible, it's getting harder to defend to those who watch the bottom line.

No great help are the studies – often flawed – that like to suggest marketing dollars are being squandered on ineffective ads. According to a study released in May 2004 by Deutsche Bank using customized marketing-mix analysis from Information Resources Inc., only three of 18 major brands competing in established product categories could demonstrate financial returns greater than the company's marketing investments. In the other 15 cases, the companies were spending more money than they were making. This environment leads to turbulence (constant ad agency reviews and expensive pitches for businesses that will be lost within the year) and frustration due to unrealistic expectations.[4]

One side effect of the slashed budgets, however, is not so negative, and it's where marketers can see opportunity shining through. Advertisers are looking for creative solutions that cost less and deliver more bang for their buck. Enter buzz marketing, a practice that can be substantially less expensive and substantially more effective than traditional means.

Polaroid tried this solution in late 2003. The brand was emerging from bankruptcy with a Board determined to climb out of the hole but without incurring major costs. From a broad perspective, the goal was to re-establish the brand's reputation for innovation, something that would be necessary to keep the company afloat in the competitive technology sector, and to lay the groundwork for the launch of a line of newly designed products in spring 2004. But, their immediate goal was to stem the decline of instant film sales. Compounding their financial restraints were two perceptual problems: (1) some people assumed Polaroid was obsolete, and (2) others were not really sure how Polaroid might be relevant to them in today's age of tiny digital cameras and camera-phone hype. Polaroid was a retro icon in need of an immediate infusion of relevance and currency to buoy the brand until the new products were on the shelves.

There was not enough time or money for advertising to do the job Polaroid needed. Buzz@Euro RSCG was called in to present potential alternatives to ads. The hunch was that there might be a glimmer of hope in the iconic status of the brand, so a pop-culture audit was conducted. This was a programme designed to uncover instances of a brand in the deep recesses of pop culture.

The audit uncovered the secret life of Polaroid as the darling of the creative classes: fashion designers, art-school students and photographers. The brand was being picked up by art directors and fashionistas for its retro-chic qualities – mirroring what was happening with sneakers, sports jerseys and 1980s fashion icons.

Buzz@Euro RSCG took this spark of creative life and ran with it. The most promising lead came from the world of music, in the form of a then little-known (and now world-famous) hip-hop duo from Atlanta and a song, 'Hey Ya', which featured the lyrics 'Shake it like a Polaroid picture'. The song was discovered before its release, enabling some fortuitous networking that panned out in a small sports-jersey store in Atlanta, a favourite shopping stop of the manager of OutKast.

A Polaroid/OutKast relationship was started with the simple placement of six cameras in the hands of the band for a live performance on US TV show *Saturday Night Live*. This relationship grew exponentially over the next six months. It became obvious, as the song's popularity began to explode, what the benefits to Polaroid were: that most coveted of buzz catalysts, natural and authentic endorsement by a hot celebrity. What was less obvious to the public, but key to the deepening relationship, was Polaroid's part in taking 'Hey Ya' beyond a simple pop tune and into the realm of pop-culture phenomenon. Having the band use the actual product on stage put movement and meaning into the lyrics and started 'the Polaroid dance'. Around the US, as DJs in clubs picked up the tune, people literally shook it like a Polaroid picture on the floor.

Good faith was added to the relationship, and Polaroid's profile and connection with Andre 3000 (half of the OutKast duo) was raised, by throwing two star-studded parties for OutKast in Los Angeles. Polaroid hosted the VIP lounge at both of these parties - one after the VH1 Awards and the other after the Grammys - and brought in the giant 20x24 Polaroid camera for celebrity portraits. The spectacle of the giant camera and celebrities actually sitting for party portraits in this age of the hideous paparazzi was too much for the press. The parties were swarmed. Media coverage of these parties and the Polaroid/OutKast phenomenon spanned from *The New York Times* and CNN to *US Weekly*, *The Post*, and E! Television.

The Polaroid/OutKast relationship culminated in an enormous celebration at the NBA All-Star Game the week after the Grammys. Because the band was so dedicated to the brand, 1000 cameras were able to be placed on the seats of the audience on the floor. As the players were introduced, Andre 3000 performed his high-energy 'Hey Ya', and the audience on the floor waved and snapped thousands of Polaroids for a televised event that reached in excess of 8 million people.

Beyond the OutKast relationship, Polaroid's profile was raised simultaneously within some other hip, pop culture-focused industries. Celebrity photographer Kevin Mazur used the 20x24 camera for a week during the Sundance Film festival. And fashion designer and Polaroid fanatic Cynthia Rowley was persuaded to ambush New York Fashion Week by, among other things, designing a custom Polaroid camera for all of the VIP show attendees and taking her own red Polaroid camera on stage for her final bow. This was shocking in that camera-maker, Olympus, was the primary sponsor for all of Fashion Week, and Rowley's association with Polaroid was considered very controversial. *The New York Times* covered the drama around the competing camera sponsors during a week when every designer has their fingers crossed for a simple mention.

At this point, it was clear that Polaroid had successfully ridden the buzz wave to its peak. The brand began showing up in magazine spreads and on websites, from *The DailyCandy* to Nike.com, and 'Shake it like a Polaroid picture' became the hip 'Macarena' of the season. The brand achieved their immediate goal of stemming instant film sales (in fact, instant film is in less of a decline than 35mm). The proof was also in the numbers: among those who viewed the Grammys, Polaroid had a 58% increase in relevance; among those viewing the NBA All-Star Game, there was a 133% increase in relevance. The estimated media coverage amounted to more than US$4 million in ad-equivalent value, which translated into over 125 million consumer impressions, and traffic to Polaroid.com in January and February 2004 was greater than total traffic the previous year.

Though all of this may sound costly, it wasn't. It was, however, a logistical challenge of enormous proportions and a crash course in Hollywood-style negotiations. Constant communication and fast action between Buzz@Euro RSCG, Polaroid and Polaroid's marketing partners helped maximize exposure while minimizing costs. Every dollar needed to count. In the end, the keys to success lay in the strength of the relationship that was forged with the band, the fact that Andre 3000 is a genuine fan of Polaroid and eager to promote the product in an authentic way, and the relationships the project team tapped into to gain access to celebrities and celebrity press at every turn.

New reality 3: Conversation is currency in the information economy

To say technology has changed the way we communicate would be the understatement of the century – take 'C U L8R'. In a discussion about the future of buzz marketing, it's worth making a few points about how it has changed things. We're talking about speed, reach, the lure of anonymity, the spotlight for everyman and, of course, in the information economy, the value of being the first to know.

After all, there is nothing more helpful in getting a product message spread than cyberspace. Today, the Internet, global connectivity and our 24/7 availability has put a premium on several conversational threads: (1) being the first to know, (2) having an expertise, and (3) being highly entertaining. To be an effective buzz marketer, you must understand how each of these motivations works and respect the power of the people's buzz in driving them.

Being the first to know used to be reserved for insiders. It used to be possible to track the trickle of information from those in the know to the early adopters and then to the masses. Journalists were often, if not first, second to know, and, depending on their publication schedules, could inform everyone pretty darn quick. Today, thanks to the Internet, information sometimes hits the masses before it hits the mainstream press.

In an article in the *New Yorker* regarding the Republican National Convention's arrival in New York City, Ben McGrath mused: 'The news cycle is a tricky matter these days, what with cable TV and the Internet, but in the build-up to the Convention a new complication materialized: clairvoyance. Not content with the traditional sequence of choreographed tips and leaks, some of the participants in the political game sought to surmount the limitations of time and space. Suddenly, everybody seemed to know about things before they had happened.'[5]

People love to feel they have the scoop; it's the backbone of gossip. But today, it seems just about everything and anyone is fodder for gossip - even products, brands and services. How about RBK, the line of sneakers from Reebok, created in collaboration with music producer Pharrell Williams? The line is called Ice Cream and the shoes have been nearly impossible to get. Few things build buzz faster than coveted items that are difficult to attain.

Pharrell is no stranger to buzz, having used word of mouth to gain fame for himself and his protégés. Besides being the front man of fêted band N.E.R.D., he is the brains behind Justin Timberlake's morphing from mop-top *NSYNCer to superstud R&B phenomenon.

The limited-edition approach has been used to great effect in the world of sneaker fetishists - and there are a lot of them out there. After all, with supplies limited, being the first to know is virtually the only way to guarantee you can get your hands on the product. When news of the Ice Cream line broke, sneaker stores were inundated with requests for information. Most of the callers never got to see the product, but they did get the bottom line on when the sneakers would launch in the US, what colours would be available and who would be carrying them. And for good buzz-building measure, there was the invitation-only launch party in Manhattan featuring DJ-of-the-moment Jus-Ske.

Conversational currency based on expertise, our second motivation, often involves personal weblogs (aka blogs) and subscription email services. While there are an incalculable number of blogs online, only a few have risen to prominence. These blogs can attract as many readers in a given day as some mainstream media properties. Plus, the information they broadcast tends to carry more weight with devotees because the blogs are perceived to be independent. From Gawker.com to the Drudge Report, blogs have become a significant source of information, news and gossip for many netizens.

The opportunity to take advantage of blog audiences hasn't escaped the notice of businesses. The entertainment industry has been the most visible in its exploitation of these audiences. In August 2004, Chiore Sicha, editorial director of Gawker Media, told *Billboard*: 'Film companies and music companies are seeing that 18- to 35-year-olds who are smart and have money and buy everything online are almost entirely our audience. Blogs have this shocking demographic that most magazines would kill for.' The power of blogs is evident in the music industry's love/hate relationship with them and the perceived experts who run them.[6]

The Web has provided music enthusiasts with the type of community that used to be reserved for those with a really good indie music store in their hometown. Blogs, large and small, offer gathering places where like-minded fans can share information about up-and-coming bands, new releases and news on their favourite groups. Many of these blogs are simply personal websites built by avid fans with no commercial motives. According to an issue of *Billboard* devoted to music blogs, 'The majority of blogs have low traffic but can accurately target a specific audience. A typical blogger may only have a following of 30 or so friends, but those 30 readers are likely to have similar musical tastes.' Blogs can help the music labels find their audiences and get the word out fast by spreading the right content to the right people.

As the music industry figures out which blogs work for them, they are racking up successful launches for bands that already have some underground blog-built momentum before they hit the general public and the all-important charts. Island Records tapped urban marketing company Cornerstone Promotion, which also happens to be the publisher of respected music magazine *Fader*, to help them build buzz for an act called The Killers. When their album was released in June 2004, the first single debuted at number 27 on the US Modern Rock Tracks chart. For a debut, this is good news.

Blogs are built on buzz, and they build buzz. Marketers are increasingly aware of this and are recognizing the genuine power certain blogs have to generate positive or negative talk about brands. By targeting blog owners, marketers are simply engaging in a specialized form of influencer marketing. And it works because, as Faith Popcorn told *Advertising Age*, the information is seen as 'expertise from people who can't be paid to lie'.[7] At least for now.

Periodically, you'll hear stories about companies enlisting cheap labour to monitor and populate chatrooms and post commercially motivated messages to blogs. Ethical marketers know that this crosses the line. Of course, monitoring blogs is a smart move for any marketer in charge of a brand. Over the past year, a few free tools have popped up that are worth checking out. Beyond helping you monitor the chatter about your brand and competitive brands, they are helpful in gauging the general mood and minds online. Check out Google's Zeitgeist and BlogPulse.com.

That leaves the last motivator: entertainment – definitely one of the best ways to get people to spread information online. One of the best examples of this simple premise can be illustrated through the story of an unlikely mascot for a major fast-food brand – the Subservient Chicken.

What is it about that silly chicken in garters that makes him so buzz-y? It's nearly impossible to resist the urge to tell your friends about him, to marvel as he dances and flaps around the screen. 'Is he really there?' 'Is this live?' 'Oooh, type in something dirty and see what he does!' The Subservient Chicken, created by Crispin Porter + Bogusky for Burger King, was one of the runaway hits of 2004. It was a pop culture phenomenon and the kind of viral marketing success that makes jaws drop. Jeff Benjamin, Crispin Porter + Bogusky's interactive creative director, claims not to understand it completely himself, though he does offer this explanation: 'The chicken became symbolic within our industry. It was buzz marketing happening completely correctly. We knew it was good, and we were braced for it to become popular, but the buzz happened within a matter of hours. It all happened more quickly and much more on its own than we expected.'

On closer inspection, the chicken had a few things working in his favour buzz-wise. Most important, it was something that had never been seen before. Though the technology is relatively simple, the gag comes across as a real wonder. It seems amazing that you can type in absolutely anything and the chicken will respond. This leads to a 'You gotta see this!' moment for the user. Second, the chicken is funny. People like to share humorous jokes and gags online. And, third, the chicken seemed, at least at first, to come out of nowhere, tapping into both our love of the obscure and our desire to share new information first.

This last advantage was only sort of intentional. Benjamin explains that the chicken was actually released prematurely. It was still in development when the designers sent the link to a few friends for testing. Next thing they knew, the server was on the verge of crashing due to almost-instantaneous traffic to the site. Within two weeks Benjamin's own mother called. 'At that point we realized how big it had become. When my mother called to tell me about it – and she's not even an Internet-savvy person – I knew.'

Initially, the link to the Burger King brand was quite hidden. This was intentional. Benjamin's team did not want the stunt to look heavy-handed. They wanted the phenomenon to build organically underground before the association with a fast-food giant was revealed. There is no doubt the chicken was a success from a pop culture standpoint; it still receives thousands of visits a day from 'more countries than participated in the Olympics,' says Benjamin. And as for buzz for the brand, he says, 'Burger King and the featured chicken sandwich were mentioned in the press more over the course of this project than we could have ever achieved through traditional means.'

What all of this means is good news for buzz marketers. The Subservient Chicken was a brilliant, strategic connected marketing idea, well executed to take advantage of the Internet's power. And, as Benjamin points out, this was not work for an icon such as another Crispin Porter + Bogusky brand, the Mini Cooper; this was buzz for a chicken sandwich. He concludes, 'The success of the chicken says to me any advertiser can do great things online'.

New reality 4: The balance tips from life to pop life

Imagine a scale with two balancing trays: one containing all the stuff of life, and the other containing the candy-coated components of pop life. Here are some of the things each tray might contain:

- *Life*: those moments, people, places and things that reflect and shape the consumer's identity – my family, my friends, my beliefs, my values, my hopes and fears, my needs, my emotions, etc.
- *Pop life*: those moments, people, places, and things that reflect and shape the consumer's image – my desires, my tastes, my experiences, my hobbies and habits, my environment, etc.

In past generations, the scales tipped heavily toward the essentials of life. People were concerned with survival, with the fundamental health and well-being of their families and with the attainment of their basic physical needs.

Today, most consumers have their basic needs taken care of. Of course, we are still concerned with the health and well-being of our loved ones, but what this actually means has changed. We no longer have to grow and harvest our own foods, we don't have to ration, we are further removed from the source. We can simply hop in the car and stock up at the supermarket or the fastfood drive-thru. The boundaries between want and need have blurred. Do we actually need the big-screen TV for the family room? On a fundamental level: no – but it does make movie night with the family fun.

For marketers, this shift spells a change in the way consumers are approached, especially when it comes to buzz marketing. For years, advertisers and brand strategists tried to delve deeply into the lives of consumers, looking for emotional triggers and messages that would appeal to who they fundamentally were. Remember the AT&T and Hallmark commercials of the 1970s and 1980s? They portrayed similar scenarios of tender and private moments between family members expressing their love over the phone or through the sentiments in a greeting card. These spots proved it was actually possible to make a person cry within 30 seconds. Reach out and touch someone, for real.

As consumers have become more sceptical and, frankly, numb to the blatant appeals of marketers, the attempts of marketers to find a place in the lives of consumers seem disingenuous. Let's face it, it takes more than Happy Meals to make a happy family.

That's the bad news.

Here's the good news: even as consumers shut marketers out of their lives, they are actually giving marketers more licence to play with them in their pop lives. Consumers are open to the overtures of marketers who have something to add to their pop repertoires, their entertainment, their reserve of things that are fun to experience and talk about. This shift in where consumers want marketers, what they want to hear from them and what they don't, is evident in McDonald's recent messaging struggle. This is why the Smile campaign with its overly sentimental imagery didn't work, and why *I'm lovin' it*, as sung by Justin Timberlake, did.

Consider the difference in the way people are consuming these fundamentals today:

- **Pop food:** celebrity chefs; The Restaurant; Costco; the cult of Trader Joe's; to buy or not to buy organic; George Foreman grill and Snoop Dogg's takeoff, Snoop De Grill; farm-bred versus wild salmon.
- **Pop family:** Newlyweds with Jessica Simpson and Nick Lachey; the Hollywood baby boom; Apple Martin; the Osbornes; cloning; making babies without sperm; gay marriage; the Hilton sisters; political daughters Bush versus Kerry in the pages of *Glamour.*
- **Pop religion:** Kabbalah as practised by Britney and Madonna; WWJD (What Would Jesus Do?) bracelets and billboards; 'Jesus is my homeboy' T-shirts at Urban Outfitters; the Left Behind series of best-sellers; Bikram yoga; meditation for the masses; *The Art of Happiness* by Richard Gere's friend, the Dalai Lama; Scientology.
- **Pop politics:** P. Diddy's Citizen Change; Rock the Vote; Ben Affleck at the DNC; *Fahrenheit 9/11*; *My Life* by Bill Clinton; Jim McGreevey; 'The only Bush I trust is my own' T-shirts.
- **Pop careers:** *The Apprentice*; teen moguls (e.g. the Olsens, Hillary Duff); *American Idol*; parlaying a reality TV show stint into a career in fashion or broadcasting; America's Next Top Model.
- **Pop physical health:** Globesity; Atkins, South Beach, Anna Nicole and TRIMSPA; erectile dysfunction takes over the Super Bowl; recreational Viagra; magnets, acupuncture, and those suction-cup marks on Gwyneth Paltrow's back; pomegranate juice and the next

big antioxidant uncovered in the underbrush of the rainforests; plastic surgery; The Swan.

- **Pop sexuality and romance:** *The Bachelor*; *Joe Millionaire*; Britney's 72-hour marriage; the Playgirl Channel; online personals; online porn; yummy mummies; metrosexuals; and the 'emo boy'.

Pop life is for sharing. Pop life is about the next thing and being a part of something bigger than ourselves. Real life, as it gets more scarce and occupies less time in the day, is precious and private. It does not get talked about. As long as marketers keep their tactics aimed squarely at enhancing the pop lives of consumers, consumers are not only willing to accept marketing messages, they will also take the most relevant and valuable of them and run with them.

Metrosexualmania proves the theory that if you give people the means to talk about what's happening around them, they will run with it. This is the 'Aha!' factor. Metrosexuality is a great example of pop life. It's the commercial manifestation of a fundamental shift in gender roles that has been taking place over the past decade. When it was given a name in summer 2003, having been appropriated and redefined from a term coined by British writer Mark Simpson,[8] the phenomenon exploded in the media. The idea of the metrosexual – a heterosexual guy who is in touch with his feminine side, manifesting itself in a taste for such things as pampering, manicures, grooming products and fashion – was so compelling, it caught on around the globe. Suddenly, everyone was questioning who among them would qualify as a metrosexual. In the US, a Metrosexual Dinner Party, attended by *The New York Times* and followed by a front-page story in the 'SundayStyles' section of that paper, launched a nationwide outrage and fascination with the state of gender roles.

How established is the term? In August 2004, metrosexual made it into the *Oxford English Dictionary*: 'Metrosexual: n. informal, a heterosexual urban man who enjoys shopping, fashion, and similar interests traditionally associated with women or homosexual men. – origin 1990s: blend of metropolitan and heterosexual.'

The bottom line is that if you can get into a consumer's pop life you'll find your target is far more willing to be a carrier for your buzz.

New reality 5: The importance of trends in a neophiliac society

The iconic linguist William Safire mused in *The New York Times* on the phrase '[insert colour here] is the new black', as used in fashion and to describe something that is new or very trendy. In the piece, he highlights a word, relatively new in terms of linguistics, that perfectly describes people fascinated by the latest and greatest: neophiliac. He writes: 'I cannot let "neophiliac" go by: It is a useful word coined in 1942 for "one who believes that every change is an improvement".'[9]

In 1997, *The New Yorker* published an article by a man who has gone on to become a leading pop expert on trends and the spread of influence – Malcolm Gladwell (author of *The Tipping Point*). The article[10] was a profile of a woman called DeeDee Gordon, poster girl for the new profession of trendspotting which was at the time becoming more important to image-based retailers and manufacturers in categories where competition to offer the next in cool was stiff: athletic shoes and gear, fashion, accessories, soft drinks, etc. DeeDee was working with Reebok at the time. This article described the perfect outlet for my passion for pop culture, and it changed the course of my life and no doubt the lives of thousands of other young people.

Today, my inbox is inundated with so many newsletters and newsflashes from various trendspotting outfits I can barely get through them all before the news is obsolete. As mentioned earlier, people today are obsessed with being the first to know. This can be overwhelming and, at times, a bit stressful for individuals. 'Is this jacket so last season, or has enough time passed that it is now retro chic?' For a multimillion-dollar corporation, it can be a nightmare – or, again, an opportunity.

One of the most popular trendspotting sites is a subscription-based push email called *The DailyCandy*, which was the brainchild of former *New York* magazine writer Dany Levy. She started the service in 2000 at the age of 28 and, at time of writing, is now covering New York, Chicago, Los Angeles and San Francisco. The email helps neophiliacs (mainly Gen X women) stay on top of the latest in everything from restaurant openings to local brow-shaping experts. The service has never advertised and grew almost exclusively through word of mouth. Women began forwarding the daily update to friends and soon Levy had a booming business on her hands. While *The DailyCandy* is hardly underground anymore (it now serves the mainstream rather than the trendsetters), it is evidence of the importance people place on being among the first in the know.[11]

A sign of Corporate America's growing understanding of the importance of such services as *The DailyCandy* is the multimillion-dollar investment Levy recently received from an investor group spearheaded by Bob Pittman, former COO of AOL Time Warner. In a recent issue of the *Los Angeles Times*, Levy spoke of the power of her service: 'People we've written about get inundated with orders and calls. It's good cocktail party chatter, too. It's cultural literacy, in a way. I'm not saying you have to know what the newest jeans are to be culturally literate, but a lot of people want to know these things. "Hey, did you hear the guys from Koi just opened a new place?" Or, "Did you hear about that fat-burning water that's just come out?"'

No one can afford to ignore trends today. Pop culture whims and fads are no longer only the concern of image-related brands. Absolutely everything rides on trends in our instantaneous society. Even such packaged-goods companies as Procter & Gamble (P&G) have figured that out. P&G keeps its products fresh and buzz-y by enlisting the help of over 250 000 teenage 'talkers', who seed information among their peers electronically via message boards, short message service (SMS) and email, or face-to-face in scripted conversations in such public places as Starbucks. And P&G is far from alone. Armies of teens can disseminate information for big brands and also bring information back from the field. It's a hipster intelligence network, and it helps brands stay relevant and top-of-mind.

Buzz marketers should have a strong background in trendspotting, so their recommendations are focused on the future and up-to-the-minute present rather than grounded in the past. Recommendations for celebrity spokespeople, events, promotions, product placement - whatever the strategy - should also be under the radar. And it's important to monitor constantly the shifts in consumer behaviour around the world.

This attention to the ebbs and flows of pop culture enables buzz marketers to insert their clients' brands into a rising tide. At that point, it becomes a matter of physics. Think: Newton's first law of motion: a body at rest stays at rest, a body in motion moves - and fast, nowadays.

No brand can maintain buzz indefinitely. Remember? We live in a neophiliac culture; people are always looking to talk about the next best thing. For this reason, Buzz@Euro RSCG developed a diagram that helps track the movement of pop culture phenomena (Figure 14.1).

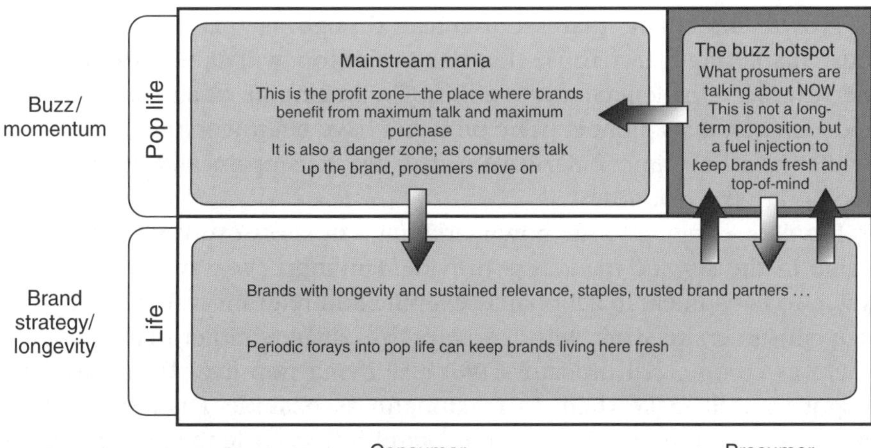

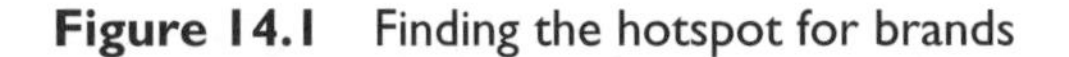
Figure 14.1 Finding the hotspot for brands

The idea is to help brands navigate the natural flow from Hot Spot to Mainstream Mania to Commodity and back to Hot Spot again. Brands that have mastered the art of riding the waves of pop culture include such companies as Apple and Nike, and such personalities as Madonna.

The more you know about pop culture and trends, the better equipped you'll be to spot rich opportunities for building buzz. Buzz is very dependent on seeding the right message with the right people at the right time. This element of timing is the trickiest part of buzz marketing and is almost completely dependent on knowing what is happening now and what will be happening next with your audience.

Conclusion: a mirror or a spark

Someone once said, 'Great advertising either reflects pop culture or ignites it'. The same is true of buzz campaigns, only more so. Buzz marketing is about inserting brands into the popular conversation, the running narrative of life in the homes and on the streets and in the clubs, schools, offices, churches – you name it.

Right now, it's still unclear what the future of connected marketing's various strands such as buzz, undercover, guerrilla and word of mouth hold. Will marketers find ways to quantify the results of their unconventional efforts? Brands certainly hope so. Will marketers be able to construct a business model for buzz that involves some efficiencies and processes? The industry would like to think so. Will marketers face a backlash as consumers tire of lazy and uninventive buzz marketing efforts that do little more than interrupt them? That one we can pretty much take for granted.

At present, the buzz discipline is defining and redefining itself as it grows. One thing that seems clear, though, is that the truly great buzz marketing teams, those that rise to the top, will exhibit three traits: perseverance, patience and flexibility. The beginning of any groundswell movement has its pioneers. The pioneers have been working at this for a while. Now it's time for the second wave to step in and apply some method to the madness.

It will be exciting to see how marketing's next era evolves. There is real value in the service marketers provide. Finding new ways to bring this value to consumers in actionable, exciting and relevant ways will encourage consumers to work with marketers as partners rather than shunning them as commercial predators. We're all living pop lives these days; we might as well make them as meaningful as possible. Finding the right mirror or spark that will help your brand reflect or ignite pop culture is the key to getting people talking – and that is the key to buzz marketing.

Forecast 1: We will see a growing reliance on proactive consumer panels and human billboards

Euro RSCG Worldwide keeps an ever-evolving list of the 50 most influential style setters in the US and abroad for inspiration in their buzz generation efforts. Think stylists rather than starlets, producers versus talking heads. They also conduct an annual study that tracks the attitudes and actions of the leading 20–30% of consumers, those people who exhibit experimental, innovative behaviours in the way they live their lives, consume and disseminate information. These are the men and women who can help predict trends poised to hit the mainstream 6 to 18 months out. Companies interested in generating buzz are beginning to keep more proactive tabs on their industries' influencers. Look at public relations agency Ketchum's recent work with FedEx as an example of how targeted influencer programmes can work. Agency and client worked together to identify 147 key industry influencers and then assigned an executive (including CEO Fred Smith) to cultivate a relationship with each. Talk about one on one.

Today, brands have mastered the art of product placement and endorsement by celebrities who are in the public eye. But these people are so removed from the day-to-day lives of most of us average Joes that their lifestyles hardly seem attainable (and often not particularly enviable). Tomorrow, we expect to see more brands recognizing the power of influence wielded by more ordinary people. How about Armani outfitting the legions of Weight Watchers meeting leaders who stand in front of millions nightly around the world? They are role models, generally with inspiring stories, and they touch a large number of people in a very immediate way. Seeding products among these influencers may, in the end, go further than a single (sometimes expensive) product placement on a television personality or a red-carpet walker with a body most of us will never see reflected in our bathroom mirrors.

Forecast 2: Brands will buy their way into the glamorous life

We are living in aspirational times. Such phenomena as teen moguls and rags-to-riches hip-hop impresarios have taught our young people that anything is possible – even at a preposterously young age. Teen girls have been known to say in all seriousness that their hobby is couture. We flip through the pages of *Us Weekly* and *In Touch* until we are so familiar with celebrity haunts, hangouts and getaways, it's almost as though we've been there ourselves. Velvet ropes, guest lists and members-only clubs abound in global cities, where the glitterati get off on being a part of the in crowd.

One of the ways brands can get into the in crowd (or at least be mentioned in the same sentence) is by orchestrating their own events and populating them with scenesters. A mention on 'Page Six' of *The New York Post*, a photo spread in *Us Weekly*, association with the It Girl or Boy of the moment – these items can bring immeasurable bounty to the brands they benefit.

Just such an association was orchestrated for Select Comfort in summer 2004 in New York state's star-studded Hamptons. Select Comfort makes beds, not a high-involvement product until you start thinking about buying one. The PlayStation Estate in Southampton was outfitted with beds for all of the guest bedrooms and two beds were used as loungers by the pool. The house was the venue for a series of high-profile Fourth of July weekend bashes, including Paris Hilton's album release party on Friday night and P. Diddy's infamous White Party on Sunday. The beds showed up in *Us Weekly* the following week, adorned by an amorous Paris Hilton and Nick Carter, and received mention in the much-read *Hamptons Diary*. Adding much to the buzz quotient was the fact that numerous celebrities, including Jay-Z and Beyoncé, spent time determining their own 'Sleep Number' compatibility as sleepover guests at the house.

Another brand that has successfully worked its way into the glamorous life is Motorola. In summer 2003, the company hawked phones through partnerships with some of the hottest DJs in the US. The handset maker's brand was in the hands of and rolling off the tongues of hipsters in New York and Los Angeles because of a calculated effort involving parties, events and product seeding among influencers. As reported by Technologymarketing.com, 'Motorola unveiled its latest multifunction wireless device, the A630, at a gathering at the trendy Hotel Gansevoort in New York. There, celebrities and style setters including Drea de Matteo, Russell Simmons and Molly Shannon were seen – and photographed – giving the new handset a trial run. Motorola will continue seeding the device with celebs and industry insiders via contacts in Los Angeles . . . [Kathleen] Finato [director of brand communications for PCS North America] said Motorola has been thinking outside the marketing box for years. "Now we're just trying to style it up more."'

Note the reference to the photographs; this takes the buzz immediately beyond the actual event and attendees, and out to the masses.

Forecast 3: Branded content makes everyone happy . . . well, almost everyone

Hallmark was a pioneer of the branded-content genre, creating sentimental television movies as extensions of their sentimental greeting cards. In just a few years, the concept of branded entertainment has matured markedly. An early example is *The Bulgari Connection*, a mystery book commissioned by the high-end jeweller. American Express invested in the creation of *The Restaurant* and *Blow Out,* two US TV shows that tout the company's open business for entrepreneurs and small-business owners. In the world of action sports, where credibility and authenticity are paramount, Vans and Quiksilver worked with surfer Stacey Peralta to create the movies *Dogtown and Z-Boys* and *Riding Giants*, respectively. The former promoted the origins of skateboarding and made Vans appear to be one of the original brands in the sport. The latter premiered at Sundance where it made a huge impact on an audience eager to learn more about the origins of big-wave surfing.

Last year, an extremely innovative and buzzworthy take on branded content was developed by Meow Mix cat food and its agency, LIME Public Relations + Promotion. In keeping with the brand's stated mission to keep cats happy, the agency and its clients came up with the concept of Meow TV. The idea was to create a fast-paced programme of cat-centric entertainment for cats 'and the people they tolerate'. The show was aired on US cable channel Oxygen and promoted with a series of events around the country, including a Hollywood-style launch party in Times Square. The concept generated more than 250 million media impressions on such high-profile outlets as CNBC, NPR, CNN, Extra!, and in *The New York Times, Newsweek, Entertainment Weekly* and *People*. Aside from being a clever idea, and on strategy, the campaign played to the age-old cat-person-versus-dog-person debate, which the media lapped up. A press release for the launch quoted Geraldine Laybourne, president and CEO of Oxygen, as saying, 'Oxygen has long been accused of being too pro-dog. Now we are giving cats equal time.'

Forecast 4: Buzz will shift its focus toward customer retention versus acquisition

While buzz marketing is an excellent tool for generating initial interest in a concept or brand at launch, or in the pursuit of new customers, expect to see more buzz marketing employed among loyal users. After all, there's a reason those dedicated users are called potential brand evangelists. Brands will begin to understand the importance of these loyalists and will launch programmes designed to keep interested parties buzzing about company news and brand activities. Apple mastered this strategy long ago. Others will take a page from their book. Look to the travel industry, especially hotel chains and possibly restaurants, for these types of buzz tactics.

Forecast 5: Limited editions will generate talk in a world where just about everything is available to absolutely everyone

There has been a growing amount of buzz about the limited-edition phenomenon. One of the best examples of this tactic used to tremendous effect is the relatively new concept of short-term retail. It's supply and demand at its most basic, with a dash of too-cool-for-school name recognition thrown in for good measure. The idea is simple: work with a brand in need of buzz and a dash of street cred; pull in a few underground, cult artists (graffiti artists, DJs and anime figurine-makers are popular); find an empty storefront in a hip, up-and-coming, but still seedy neighbourhood; open your doors for a limited time; close the doors, tear down the store and disappear forever. Years ago, this might have looked like mob activity or the work of a skilled con artist; today it makes for super buzz among the hip and trendy.

According to Vacant – a pioneer of short-term retail concepts whose clients include adidas, Nike, Puma, Ice Cream, House of Courvoisier, and Marshall Field's – the buzz for these events happens in three waves: first, among those alerted to the fact that the store is going to be happening – generally, friends of those involved, readers of certain hip magazines, and proactive consumers of hip; second, among those who either make an effort to get to the store or those who stumble in (Vacant say most of their visitors come back several times, bringing new friends each time); and third, after the fact, when those who experienced it make all those who didn't jealous.

In September 2004, downtown sneaker-and-style mecca Alife teamed up with Levi's 501 to produce a 'happening' in the form of a temporary store on Orchard Street in the Lower East Side of Manhattan. The store was open for one month only and featured approximately 15 products. Alife designed six pairs of limited edition Levi's 501s, with T-shirts and sneakers to match. Shoppers could also purchase a can of spray paint or a Zippo lighter branded with the happening's signature colours. The store was outfitted with vintage Levi's posters. It was so hip, you had to see it to believe it. Of course, that's not the point. The point is: I've seen it, and I'm telling you about it right now . . .

Forecast 6: Smaller markets will provide more fertile ground for buzz building

Bye-bye Hollywood and New York City; hello Austin, Charlotte, Woodstock and Portland. A new phenomenon is picking up steam. The idea involves the movement of urban dwellers to beautiful, rustic, rural and overall quieter small towns for quality-of-life reasons. Now that it's possible to work virtually anywhere, young urban professionals are finding the allure of wide-open spaces preferable to the rat race and gopher cubicles of the big cities.

From a buzz perspective, this makes for a lot of interesting new opportunities to get the word out about brands – in a more geographically constrained and focused way. Certain big media events are now just so overexposed from a marketing, branding and buzz perspective that they barely warrant consideration: the Academy Awards, Sundance, Fashion Week – unless you have some seriously good connections and/or mega-bucks, it's difficult to make a splash at these events. (Of course, when you do, the pay-off is enormous.) Some buzz artists are finding they have more options and can make a more immediate impact at smaller events such as the South by Southwest festival in Austin, Bonnaroo Music Festival in Manchester, Tennessee, or by targeting local hot spots in such second- and third-tier markets as Portland, Oregon, Louisville, Kentucky, or Charlotte, North Carolina.

Pabst Blue Ribbon used this strategy a few years ago to revive its working-class brew within the alternative scene in Portland. Next thing you knew, downtown hot spots were carrying the canned beer alongside their microbrews on tap. Chrysler also used this local strategy when it designed a promotion in 2004 called the Minivan Summer Games, a tribute to the Athens Olympic Games. The contest pitted regular people against one another in a series of events that tested their minivan-manipulation skills. According to *Automotive News*, more than 5000 people registered online for events that were staged in eight communities: Boston; Cheyenne, Wyoming; Detroit; Houston; Milwaukee; Raleigh, North Carolina; Gilroy, California; and Virginia Beach. Chrysler knew its customers were more likely to live in these markets than in New York or Los Angeles, so it went to where its people were, and it worked. As the buzz in the big markets gets saturated, we're certain to hear more grassroots buzz coming from the heartland.[12]

Takeaway points

- Buzz marketing, defined in this chapter as the scripted use of action to generate word of mouth, is a response to changes in attitudes to traditional marketing.
- Buzz marketing offers a creative and cost-effective solution to driving awareness, interest and demand, which can complement traditional marketing.
- The Internet enables buzz marketers to reach markets fast and directly, before information hits mainstream mass media.
- Blogging and branded entertainment are two of the most effective ways to create online buzz.
- Buzz marketing will see a growing reliance on consumer panels, human billboards, and celebrity involvement.
- Buzz marketers will increasingly use branded content, such as advertainment and sponsorship, to entertain and engage consumers.
- Buzz marketing will shift its focus toward customer retention versus acquisition.
- The perennial buzz marketing strategy, creating demand by limiting supply with limited editions, will continue to be powerful.

Notes and references

1 Brown, S., Matathia, I., O'Reilly, A. and Salzman, M. (2003) *Buzz: Harness the Power of Influence and Create Demand*. New York: John Wiley & Sons/Brandweek Books.

2 Palmeri, Christopher (2005) 'Living on the edge at American Apparel', *BusinessWeek*, 27 June, archived at http://www.businessweek.com/magazine/content/05_26/b3939108_mz017.htm; Ko, Claudine (2004) 'Meet your new boss', *Jane Magazine*, July.

3 Bell, Michele (2004) 'Dov Charney: rebel with a cause', *Counselor* (August).

4 Neff, Jack (2004) 'TV doesn't sell package goods: study shows medium fails to deliver ROI for mature brands', *Advertising Age*, 24 May.

5 McGrath, Ben (2004) 'Talk of the town: civil disobedience', *New Yorker*, 13 September.

6 Garrity, Brian (2004) 'Labels tap promo power of online commentaries, but sites linking to MP3s cause concern', *Billboard*, 21 August.

7 Chura, Hillary (2004) 'How to calculate word of mouth: Ketchum finds 200 build the buzz', *Advertising Age*, 26 July.

8 Simpson, Mark (1994) 'Here come the mirror men', *Independent*, 15 November.
9 Safire, William (2004) 'The way we live now: the new black', *New York Times*, 30 May.
10 Gladwell, Malcolm (1997) 'The coolhunt', *New Yorker*, 17 March.
11 Kinosian, Janet (2004) 'Metropolis/snapshots from the Center of the Universe; now wear this: on the Web, coolness updates are the latest thing', *Los Angeles Times*, 22 August.
12 Stein, Jason (2004) 'Chrysler brings some zaniness to launches: word of mouth marketing pays off for automaker', *Automotive News*, 2 August.

15

How to manage connected marketing

Martin Oetting

ESCP-EAP European School of Management/MemeticMinds.com

Establishing connections with clients, customers or consumers to promote positive word of mouth in the vast, invisible networks that exist between all of us can seem a complicated business. But it doesn't have to be.

Just as with any marketing campaign, it begins with product or brand differentiation. You then research the word of mouth that's already taking place. Next, decide if you need to use a tactical approach (such as a campaign that uses contagious communication), or a more strategic plan that will manage the relationship with the most influential buyers in your market. Better yet, go holistic and integrate both. Either way, remember to set objectives so your efforts are measurable, and make sure you have the right people with the right skills to implement your connected marketing strategy. And that's it.

Of course, there is a little more to it in reality, and this chapter goes through the necessary steps in more detail. With opinions and advice from many of the leading practitioners in the connected marketing field, this chapter is a blueprint for marketers who want to turn a marketing plan into a connected marketing strategy.

It begins by covering two basic issues that should be addressed before a connected marketing campaign can be developed, then it describes two basic approaches that are currently available for creating these connections. Finally, it looks at the key challenges you're likely to encounter when working on this type of marketing activity.

Forget definitions; it's all about how to manage word of mouth

There is a great deal of talk from all sorts of people in the connected marketing arena, offering all kinds of services destined to insert your brand into the word of mouth exchanges that happen a billion times a day. Does your brand need viral marketing? Should you go for a buzz marketing strategy? Or do you really need a word of mouth marketing campaign? Is stealth marketing the best solution, or rather a seed marketing approach? Do you need influencer marketing, or is it better to find out who the hubs and connectors for your brand are? What do they mean and how do you distinguish between all of them?

In the end it's all about how to manage word of mouth - offline and online - by connecting with your target market in a meaningful way. And in order to do so, we suggest not getting caught up in definitions but rather looking at underlying principles and approaches. (For some brief definitions of terms that you may find useful, see Chapter 13.)

Another factor to bear in mind is that many of the better-known connected marketing case studies deal with business-to-consumer (B2C) communications. However, connected marketing can also be applied to business-to-business (B2B) projects, as well as internal communications and business-to-employee (B2E) projects. So if you're a human resources manager wondering how to get employees enthusiastic about a new company-wide training initiative, this book is also relevant to you.

Whatever the context, before any action plan can be developed there are two issues that must be addressed in a connected marketing strategy:

1. How do I motivate word of mouth?
2. What are my customers saying today?

Decide how you want to motivate word of mouth

Differentiation

Ultimately, marketing is about differentiating your offering from that of the competition, and connected marketing is no different. If you want people to talk about your product or brand, you must give them something newsworthy to talk about - it must stand out or have unique qualities.

Even if you're operating in a highly competitive market where it becomes increasingly difficult to establish obvious differences between products, you need to find a way of distinguishing yours, such as using your brand as a symbol of value.

Differentiation provides the foundation for the word of mouth in your connected marketed strategy. You need to give your target market news that they can act upon - something they can learn about your product or brand and feel intrinsically motivated to share with other people.

Financial stimulation

Companies can also generate word of mouth that's motivated not by factors related to the product itself, but by extrinsic factors. In most cases, this is some kind of financial remuneration.

One of the most common examples of financially incentivized word of mouth techniques is an affiliate or online referral programme, such as that run by online retailer Amazon. The system is simple and straightforward: anyone who runs a website, or who is active on the Web, can recommend products from the company via the affiliate programme. The recommendations are made in the form of referrer-personalized weblinks. The referrer can publish them on his or her website, send them in an email, or post them in Internet forums. When someone clicks on such a link, the system logs the referrer, and once the referred visitor completes a purchase, the referrer will get a reward for the purchase made.

Affiliate marketing's emergence can be attributed to the trend towards cost-per-action advertising models on the Internet after the dot.com bubble burst. As companies were beginning to realize that marketing communications' primary objective should not be awareness but return on investment (ROI), these finder's fee systems seemed a great invention: people who run themed websites, blogs, or newsletters could provide a benefit, earn money and create ROI for the advertiser by connecting their visitors with products that fit the subject. Recent evidence, however, has shown that affiliate programmes can also corrupt some referrers and companies into questionable activity, especially when the products sold are bigger-ticket items.[1]

According to Aberdeen Group, an IT research and consulting group, 'Affiliates generate sales through third-party education and validation - much in the same way that word of mouth recommendations have traditionally referred business directly to movies, restaurants, merchants, doctors, lawyers and accountants.'[2]

There is one important difference though: traditional word of mouth happens between customers who inform each other of their own freewill about choices and experiences with brands or products. Introducing referral or finder fees, on the other hand, in a company and customer relationship changes the equation and creates a quasi-sales-agent character. Consequently, it can lead to a different type of word of mouth.

LinkShare, one of the largest affiliate marketing brokers, which boasts over 10 million partnerships with merchants and affiliates,[3] experienced problems with this change in company-customer relationships. Its quarterly Titanium award was given to 'the one marketing partner in The LinkShare Club loyalty program that supports the largest number of participating merchants, and drives the greatest increase in online sales for these merchants'. The result: for three consecutive quarters, every winner was later found out to have manipulated search engine results, or used questionable tactics such as cookie stuffing or spam emails, in order to get the massive click-through numbers needed to win the award. (Cookie stuffing tricks affiliate software into registering that a buyer had previously clicked on an affiliate link even if the buyer had not actually done so.[4])

Brian Clark of GMD Studios, which began the affiliate marketing analysis site ReveNews.com in 1997, comments: 'Once you empower consumers to do your selling based on a commission, you no longer control what they are doing.' In the end, he goes on to say, some affiliate programmes can lead to affiliates trying anything in order to trick search engines into displaying their referral links: 'It's a race between affiliate marketers and search engines. People will always find ways to glean money from the system.'[5]

There are other types of financially incentivized word of mouth approaches that can also damage a brand. One example is paid-for fan endorsement, a tactic frequently used by the entertainment industry. Warner Brothers Records tried to promote the band The Secret Machines on frequently read music blogs. Sending MP3 files to bloggers in the hope of publication didn't produce the desired results, so the company published its own enthusiastic 'fan posts' on the MP3 blog Music for Robots. The truth was uncovered because these postings had come from the same IP address that had previously been used to email out the promotional MPs – and the fans were not amused.[6] Examples of this type of 'stealth marketing' abound[7] – and in many cases the backlash quickly follows. (Justin Foxton goes into more detail about the risks and issues of secret marketing stunts like these in Chapter 2.)

Ed Keller, CEO at NOP World Consumer, who published a well-known book[8] on what his firm calls the Influentials(SM) consumer segment (consumers who can influence the purchase behaviour of others), comments:

> When people start asking themselves 'Is someone telling me this because they believe it, or is someone telling me this because they are a paid agent of somebody else?', there's a fine line that needs to be watched closely. Marketers who make their marketing message available to those who seek it out, who learn from it, and then start acting on it, will be successful. And those that end up trying to penetrate paid agents into the marketplace who are saying things

> they don't believe, will be looked at in the same way that people look at telemarketing today: 'You're intruding on my conversation, get out of here.' Right now, we are probably still too early in the process to know what crosses that boundary. But everybody in the industry needs to keep an eye on this.[9]

To address these issues, the Viral + Buzz Marketing Association published its first manifesto in June 2004 about rules and principles that should be respected by its practitioners when establishing consumer connections. It states, among other things: 'We strive to . . . deliver the message . . . in a way that makes it an enjoyable or valuable experience . . . We will . . . be providing a benefit to our audiences and their acquaintances and in so doing, to the brands for which we work. . . . Our goal is to foster genuine enthusiasm about brands and brand communications, which can spread through networks in a way that is enjoyed, appreciated and/or valued.'[10]

Conclusion: the importance of differentiation and trust

In order to generate word of mouth that both stems from and creates solid conversational connections between a company and its customers, a connected marketing initiative should be built on brand or product differentiation and on trust. People are interested in new and better things - connected marketing is about reaching and inspiring a given audience with a piece of information or a branding idea. But the information has to be there in the first place. When other motivating factors are introduced, there is often the risk of people exploiting the system, with a potential backlash for the brand.

Key questions: Have I identified sufficiently differentiating factors about my product or service, or the brand under which I am marketing them? Simply put: Do I have compelling news to tell?

Find out what your customers are already saying

Are online forums tearing your product apart? Are blog authors claiming your company is hurting the environment? Or maybe customers enjoyed your last TV ad so much that they're circulating it to friends as an mpeg file via email? A marketer who wants to engage and connect with customers should first of all find out what they are already saying.

Listening to the market and understanding what people are saying about your brand or product has several positive effects. First, it will help you get the feel for word of mouth – when you listen to people, you find out what is relevant to them. Second, it's a good starting point for building long-term connections with customers – talking to someone who's already speaking is better than cold calling. Third, it can help identify the movers and shakers, those opinion leaders whose recommendations play an important role in the way word of mouth spreads in a given market. And finally, it can give you a headstart in crisis management. Often, more vocal customers will voice criticism and point out problems directly to a company before these reach the media. So, if a company is attentive and responsive to the word of mouth that's already going on, problems can be spotted and dealt with at an early stage.

There are different ways of learning about the word of mouth that is already spreading. According to Jupiter Research, 'Sixty-three percent of consumers who contact customer service are tattletales (i.e. consumers who would tell others about poor customer service experiences).'[11] So a good first step is listening to consumers who contact your company. Ed Keller and Jon Berry at NOP World agree in their study on influencers.[12] According to their research, the influencers segment of consumers tend to voice their criticism more than average. So when consumers complain, treat them with respect and help them – chances are they are influencers.

Research can also be a good way to analyse and even instigate word of mouth (see also Chapter 1 and Chapter 11). Pioneering researcher George Silverman explains that for some product categories group research is the most successful word of mouth technique.[13] This is because:

1. It enables you to find out how customers are talking about your product or service peer-to-peer, in order to understand what the actual word of mouth is. Silverman specifies: 'Not only do you want to hear the contents of the word of mouth, you want to hear the sequence and the source.' By learning which sequence of arguments works best you can understand how a message must be designed to work well. Understanding which source is most credible in a given context is also important – in some cases people need an expert's advice, in others they seek testimonials from peers.
2. Research groups can be a key tool for spreading word of mouth. In focus groups, one person can often change the whole group's opinions. Standard research environments usually try to minimize this effect to get a better idea of individual opinions. Contrarily, to create word of mouth, it needs to be encouraged – and the one person influencing the others has to be the product's advocate.

In 2000, Renée Dye, associate principal at McKinsey & Company, recommended research as a means to understand how consumers influence one another. In *The Buzz on Buzz*, she says: 'Marketing researchers are thus developing new methodologies to account for customer-to-customer interactions. . . . More sophisticated techniques attempt to model how consumers interact with one another and how highly they value others as sources for information or as behavioral models.'[14]

With the proliferation of conversational or customer-generated media, it becomes increasingly possible to monitor word of mouth activity electronically on the Internet. Customers are setting up their own websites, or are publishing their opinions in forums and on blogs. Companies should try to learn as much as they can from these opinions – most importantly because negative word of mouth can spread like wildfire across the digital sphere. In a later section on blog marketing, this subject will come up again.

There are companies that can help you learn about online word of mouth. In the white paper *Rumors and Issues on the Internet*, Intelliseek's Pete Blackshaw and Karthik Iyer claim that they can help avoid massive publicity crises, quoting the example of the Ford/Firestone tyre blowouts: 'Would [they] have found themselves embroiled in lawsuits, product recalls and finger-pointing and government intervention in 2000 had they paid attention to Internet-posted consumer complaints about tire blowouts that appeared, according to Intelliseek research, as early as 1994 – long before the problems hit the nightly news, the front page and the class-action court room?'[15]

It's important to make sure, however, that you're monitoring a truly influential source. As Dr Paul Marsden at the London School of Economics points out: 'How do you truly measure word of mouth online? Unless you know whose opinion actually has any influence at all, all you are measuring is simply chatter at best and noise at worst.'[16] (New Media Strategies' CEO Pete Snyder has a long track record with online intelligence, and he presents his view on this issue in Chapter 8.)

Key questions: Am I learning about existing word of mouth yet? Am I tapping into all the channels that I have at my disposal to understand what my customers are saying today? And do I use my word of mouth intelligence to counter potential crises?

Now that we've covered the two preliminary issues – deciding what drives people's motivation for word of mouth and understanding (and possibly reacting to) existing word of mouth, we can move on to how connected marketing can be put into practice.

Spread ads, or find influencers

So what do you want to achieve? Develop an ad message so contagious that customers will want to spread it? Or build relationships with the people who have a say in the marketplace? The choice is critical because there are two fundamentally different approaches for connected marketing.

1. *Advertising in connected marketing (such as crafting messages that help connect with customers)*: This approach is tactical. It could be compared to an ad campaign or a marketing event or stunt, essentially based on some kind of advertising message, crafted and distributed to be further spread by customers. It tends to focus on creating more immediate results within a defined timespan (even though they can often carry on for much longer because by definition a word of mouth initiative of any sort involves the uncontrollable contribution of the other market participants).
2. *Relationships in connected marketing (such as strategic relationships with connected customers)*: This approach is more strategic in nature. It involves activities that seek to connect with selected influential customers and develop long-term relationships with them.

In the best cases, both approaches are combined - they can and should play off one another. And sometimes, classifying the approaches is somewhat arbitrary. (For instance, alternate reality games - covered later in this chapter - can be considered part of 'advertising'. However, they can also turn into strategic projects that establish long-term relationships with a loyal fan base. And conversely, seed marketing can be considered part of the relationships field, yet it can sometimes be executed without a very strategic side to it.) But in order to present a clearer picture of the market, some choices in terms of differentiation of approach have to be made.

> *Key questions*: Do I need a tactical campaign that communicates fairly quickly? Or do I want to establish strategic long-term relationships with ongoing benefits?

Spreading ads: advertising in connected marketing

When you do your advertising well, the customers do all your tedious and costly work – they spread the message and they might even do one-to-one targeting for you.

In *Unleashing the Ideavirus*, Seth Godin put it this way: 'The future belongs to the people who unleash ideaviruses. What's an ideavirus? It's a big idea that runs amok across a target audience. It's a fashionable idea that propagates through a section of the population, teaching and changing and influencing everyone it touches.'[17]

Marketing can make use of the connections between customers if it manages to create ideas so infectious that people simply have to pass them on to others.

In *Anatomy of Buzz*, Emanuel Rosen explains that successful word of mouth needs a product that can serve as an ideavirus by itself and that you then need to help word of mouth get started: 'Two things are needed to create buzz successfully. The first one . . . is to have a contagious product. But having such a product alone is not enough. Companies that get good buzz also accelerate natural contagion.'[18]

Connected marketing experts will beg to differ, at least to some extent. Justin Kirby, managing director of online viral and buzz marketing consultancy Digital Media Communications, says: 'If your brand, service, or product is not so one-in-a-million, you can still generate massive peer-to-peer endorsement. The challenge is that the campaign's communication agent – such as video-based advertainment content – is the element that needs a wow factor.'[19]

And that creative challenge might be a bit more demanding. Stéphane Allard of French buzz marketing agency Spheeris explains: 'In the viral and buzz field, the audience is the media. There is no way to force them to spread a message if they don't want to. So we need to be a lot more creative and at the same time a lot more aware and closer to consumers' needs than we would need to be in the advertising media field.'[20]

So the basic principle is surprisingly simple – in the words of Richard Perry and Andrew Whitaker, authors of *Viral Marketing in a Week*:[21] 'Content makes the difference between forward and delete.' The more compelling the advertising message, the more likely people will share it with their friends. That means they distribute your ads with no additional media cost. And they will spread the message particularly to those people for whom they think the message will be relevant. Unless they are a spammer or a very obnoxious person, they do not tell their friends about things they don't find interesting. In other words: they, the customers, do the targeting.

The following range of successfully employed techniques and approaches show how you can undertake advertising the connected way. This list should not be considered exhaustive; rather, it illustrates how you can present your marketing message in a way that makes it contagious.

1. Developing stories and events for brands

People love a good story. Stories are the glue that has bonded society ever since we sat around the fire together.

Some advertising agencies have looked at their existing services and developed them into new tools for connected marketing. Euro RSCG is one such example. The Buzz@Euro RSCG unit has been active for more than a year now.

Its approach to connecting brands with customers through word of mouth is built on insights from Euro RSCG's international network of trendspotters. The process involves finding the intersection between current hot topics of conversation for the target audience; the brand's perception today; and intended strategic goals. Co-creative director Schuyler Brown explains: 'We are looking for natural "insertion points" for the brand in the already-flowing current of pop culture. The fit has to be believable, authentic, but just a little left of center. We're looking for a branded storyline that not only allows consumers to continue the conversation, but enhances it.'[22]

In practice, the challenge is about developing the right story idea for the brand, creating matching events, occasions, or PR stunts – frequently connected with a suitable and credible celebrity endorsement – set in motion at the right point in time. Ron Berger, CEO of Euro RSCG Worldwide in New York, points out: 'Our entire industry these days is engaged in connecting consumers to brands by any and every means possible, and that's a reality that allows for far more flexibility, creativity and ingenuity than was possible with traditional advertising.'[23] (Schuyler's own contribution to this book (Chapter 14: Buzz marketing: the next chapter) describes, among other things, how the Polaroid brand has been successfully rejuvenated with this approach.)

In 2004, *The Oprah Winfrey Show* (seen by about 30 million weekly TV viewers) started its 19th season. For the event, the Pontiac marketing team responsible for launching its G6 model organized a brand story that made clever use of both the media and the word of mouth generated. Oprah Winfrey surprised 276 members of her audience by giving each of them a free G6. The guests were chosen from a list of people who had told the programme about their need for a new car.

A simple idea, yet the media coverage and reaction of those involved or who heard about it was considerable. Jim Bunnell, general manager of GM's Pontiac-GMC division, says: 'During the Athens Olympics, GMC ran 25 to 30 television spots for its truck products over two weeks that cost between US$7 million and US$8 million, but no one really talked about those ads. But the same budget, spent in one day, drove a significant amount of buzz for the G6 on a daytime television show.'

Also, the G6 website had 242 000 hits in the 24 hours after the show. Usually, the site gets a daily average of about 30 000 hits. (The idea did, however, produce a small aftertaste. As it turned out later, each recipient also received a tax bill of around $7000 if they wanted to keep the car. The regular price is $28 000, and the US tax authorities count the gift as income.)[24,25] On the other hand, it does seem noteworthy that despite the buzz, the car has not proven to be much of a success in terms of actual sales so far.[26]

PR agencies' primary task is to develop compelling stories around products or brands. Extending this beyond their regular target of journalists in order to get the press excited and stimulate word of mouth is something a few agencies have been doing for a long time. Graham Goodkind, founder and chairman at Frank PR, explains how to use a PR approach to creating word of mouth in Chapter 5.

2. Using live buzz marketing to enable people to encounter brands in their daily lives

Witnessing how people do things in real life can often create the most lasting impression about products or brands. By carefully planning, scripting and executing such live brand encounters, brands can make good use of these effects.

Kenneth Cole claims that he started his shoe business that way. At the beginning of the 1980s, he wanted to open a shoe store in New York and sell his own shoe designs. But he didn't have the resources that were needed to enter the market in the conventional way: either marketing at the trade show at the Hilton, or opening up a fancy showroom in midtown, not far from the show.

Instead, he went for live buzz marketing. His idea was to park a 40-foot trailer truck in Midtown and use that as his outlet – right at the show's door. But he couldn't get the Mayor's permission. Usually, the city is only willing to give out parking permits to production companies that are shooting full-length motion pictures, or to utility companies. Cole's solution: 'So that day I went to the stationery store and changed our company letterhead from Kenneth Cole, Inc. to Kenneth Cole Productions, Inc. and the next day I applied for a permit to shoot a full-length film entitled *The Birth of a Shoe Company.*'

The truck was parked at 1370 6th Avenue, right across from the New York Hilton, the same day the shoe show took place. On the side of the truck they had painted Kenneth Cole Productions. There was a director on the 'set', there were models as actresses, sometimes they were actually filming, sometimes not. And they started selling shoes directly from the truck: 'We sold 40 000 pairs of shoes in two and a half days (the entire available production),' says Cole. 'And we were off and running.'[27]

This example shows that generating live buzz can sometimes do more than just create word of mouth. If done well, it can also generate sales!

Another fairly well-known piece of live buzz marketing was the campaign Sony Ericsson organized for launching its mobile phone, the T68i, in 2002. It used a technique that has frequently been employed for alcoholic beverages and tobacco. Attractive models were sent bar-hopping in cool places as undercover agents, posing as tourists. They were briefed to address unsuspecting customers, hand them the phone and ask if they could take a photo of them. This usually sparked a conversation about the new feature of the phone - taking photos. This campaign later led to a debate about whether this kind of stealth approach is ultimately beneficial for brands. Some customers who found out that they had been marketed to in these encounters expressed that they felt uneasy about the approach.[28]

3. Harnessing viral advertising films that are like TV ads, just without the media budget

Ever wondered how to save the massive TV budgets usually needed to make your advertising idea known to the world? Viral advertising may be the solution. It not only enables you to spread the ad films without a media budget, but also enables a response mechanism, and usually comes as a welcome surprise, not as an annoying interruption.

Closer to traditional advertising than other forms of word of mouth, viral advertising is based on campaign material that the customers themselves want to spend time with online and spread to others via email.

Matthew Smith and Ed Robinson of The Viral Factory create ad films that are distributed on the Web. 'When we talk about "a viral" we mean a file that has spread on the Internet through peer-to-peer networks on an exponential curve', says Smith. These advertisements, however, do not necessarily have much in common with the classic 30-second spot because they do not 'live' in spaces that are 'exclusively reserved' for them, i.e. ad breaks. They need to be contagious enough that users will want to press the forward button.[29]

Some marketers express concern that this type of advertising carries a problem – it puts the message in the hands of the customers who can then tamper with it. According to connected marketing experts, these concerns are no longer in keeping with today's realities.

Jackie Huba, co-author of *Creating Customer Evangelists*,[30] says: 'Technology now is so simple – anyone can create their own ads. MoveOn.org asked its fan base to write and create 60-second commercials with an anti-Bush theme. They had massive submissions from people who made high-quality commercials the way you would find them on TV! These were just regular people – with a video camera, computer, and editing software. Customers can create advertising today. So if companies say "We cannot give up control and let the customers control our advertising" – it's too late for that. It's way too late.'[31]

BMW USA can certainly be considered one of the pioneers in this area. When answering a BMW briefing for a regular advertising campaign, the US agency Fallon Worldwide proposed to do things a bit differently. As opposed to creating regular advertising spots, it recommended shooting a series of short films exclusively for distribution on the Internet. Part of the idea was to hire high-profile directors such as John Frankenheimer, John Woo or Guy Ritchie to give the project more appeal and credibility.

High-quality execution and celebrity actors, such as Mickey Rourke and Madonna, coupled with exciting car scenes, helped turn bmwfilms.com into a success story – more than 13 million downloads were counted. Jim McDowell, head of marketing at BMW North America comments: 'We had no idea how successful it would be, since we were going into uncharted territory when we started the project. In the end, the project far exceeded any of our expectations.'[32]

Today, Fallon is building on this success, and has realized similar projects with other high-profile brands, the latest being Amazon.com, with its Amazon Theatre. Five short films, made by respected directors and featuring some better known actors, not only serve as entertainment on Amazon's website but also as a direct selling channel since they display direct links to areas of the site where you can buy products featured in the film.[33] (Justin Kirby's DMC has a long track record of successful campaigns with viral advertising films. His contribution to this book (Chapter 6: Viral marketing) describes in much more detail the viral marketing approach and how it works.)

4. Making online games that make people want to play with you

If you want customers to interact with your brand online for 15 minutes, and not just 10 seconds, think games. Like films, they can also spread on the Internet, being forwarded between customers who recommend them to each other.

A very popular and frequently cited example, with millions of downloads and referrals, is Burger King's Subservient Chicken by Crispin Porter + Bogusky. Highly viral, highly popular, it attracted more than one million hits on its first day online alone.[34]

One of the most successful, most expensive and most debated online games was produced for the US Army. *America's Army* is an online game that is meant to let the player experience what it means to be part of the Army. Critics say it's about making war look attractive to 13- to 21-year-olds. The game has proven to be one of the Army's most effective recruiting tools. First released in July 2002, the game had more than 16 million downloads by October 2004. Recognizing that today it takes a quality effort to be noticed among the many online games available out there, the Army is said to have spent around US$6.3 million on the first version of the game alone.[35,36]

Games can also work in a business-to-business context. The company Conference Calls Unlimited offers web and audio conferencing services. It wanted to change its marketing, in order to make it more efficient than the pay-per-click search engine advertising it was using. e-tractions, a company responsible for the development of several successful online games, designed '@work office' for the conferencing company. The game is based on a small office scene to which users can invite friends with an 'add staff' functionality. The game was sent out as a follow-up to e-traction's own Snowglobe promotion, to the e-tractions' in-house mailing list.

The emails generated a click-through rate of 24%. A third of the visitors recommended the game and invited others. Click-throughs on these referrals were significantly higher: almost 52% – proving that an acquaintance's online recommendation is usually more likely to be followed up than a commercial invitation. In terms of business objectives, a 28% click-through rate (on the number of times the game was opened) to Conference Calls Unlimited's own website was achieved, with a campaign at half the cost of its usual pay-per-click advertising.[37]

One thing should be noted about online games: there's an awful lot of online gaming out there, to the point that it has turned into a commodity. So, if this is your chosen route, some investment in terms of creative strategy and execution must be made so the game stands out from the crowd. (Steve Curran, President of Pod Digital, was also co-founder of e-tractions. He explains advergaming and its word of mouth potential in Chapter 9.)

5. Producing alternate reality games: supercharging brand communications

Earlier, we mentioned 15 minutes of brand interaction with games. What if you could get an audience to interact with your brand for weeks by connecting on the Web then travelling and seeking out locations and unravelling hints that you are giving them while trying to solve a mystery? Welcome to the world of alternate reality games - a variation on the online gaming theme, albeit quite an elaborate one.

In principle, almost any website can serve as a starting point for word of mouth activity. Strictly speaking, forward-to-a-friend functionality is already a basis for word of mouth. A well-designed teaser page with compelling content can also prompt people to forward the address to a friend.

Developing alternate reality games for brands is a way of taking the idea of a 'teaser website' way beyond its original meaning and even outside of the Internet. The concept could be said to have started online with the *Blair Witch Project*, an idea that really took momentum as a project when the people behind it realized how powerful their story idea was for generating word of mouth, particularly on the Web. By using online media - a website and a corresponding newsletter in particular - the team managed to create a personal connection with fans and build up the suspense around the film itself. The website became a resource for information about the myth of the witch and about those behind the film (filmmakers Daniel Myrick and Eduardo Sanchez) claiming to be filmmakers uncovering a mystery. By limiting access - the film was to have its premiere at the Sundance Film Festival with only one screening - they managed to generate a hype that quickly led to a bidding war for the rights to the film and, ultimately, to worldwide distribution.[38,39]

Even though the *Blair Witch Project* was about creating and maintaining the illusion of an alternate reality, the game aspect of it was not as strong as that in later projects that approached word of mouth from a similar angle.

Sean Stewart was involved in creating the online mystery world *The Beast* for the Steven Spielberg film *AI*. On his website,[40] he talks about the approach they took: 'So there was the project: create an entire self-contained world on the web, say a thousand pages deep, and then tell a story through it, advancing the plot with weekly updates, concealing each new piece of narrative in such a way that it would take clever teamwork to dig it out. Create a vast array of assets - custom photos, movies, audio recordings, scripts, corporate blurbage, logos, graphic treatments, websites, flash movies - and deploy them through a net of (untraceable) websites, phone calls, fax systems, leaks, press releases, phony newspaper ads, and so on ad infinitum.'

By spinning an elaborate mystery story based on a compelling narrative idea, and locating and promoting it on the Web and through other media, these games invite an attentive online audience to connect, work together and search for the truth, the solution, or whatever else lies at the heart of the story.

A fundamental aspect of these games is that it is usually impossible to solve the mystery alone - a collective effort is required and, in successful cases, this gets self-organized rapidly on the Internet. Fans find each other quickly through postings in newsgroups or by setting up blogs. Blogs, as a type of user-generated media, have received much attention by marketers recently and are mentioned again later in this chapter (as well as in Chapter 10).

Importantly, an alternate reality game usually doesn't inform its audience that it's artificial. The game is treated as reality, and the puppeteers pulling the strings will only give away their identity at the very end, if at all. Some critics call it a deceptive tactic because of that. Those advocating this technique see this kind of non-disclosure as a type of texture that needs to be used with caution, yet is ultimately part of the fun for players. They argue for making the distinction between fake content and 'fictional storytelling'.

Looking at the intention behind the game is one way to draw a line between the kind of deception that's likely to damage a brand and the kind that probably won't. When they're not supposed to be revealed as promotional activity, deceptive techniques can have a backlash; for example, marketers posing as independent end-users and hyping products in Internet chatrooms. When consumers find out about this, negative word of mouth is almost certain. With alternate reality games and similar approaches, the deception is either obvious, or meant to be found out in the course of the game.

More recently, there have been successful attempts to realize this kind of game with more overt branding - for instance, the MoreToSee campaign by Sharp.[41] Even though it could be considered branded entertainment, it was compelling enough to generate enthusiastic responses and involvement from consumers.

The 'ilovebees'[42] campaign for the Xbox game *Halo 2* is another recent example. It has no overt branding but is still obviously commercial - the address www.ilovebees.com was featured at the end of an Xbox advertisement. The story was introduced when the site appeared to have been hijacked with hints and information that were written in a strange language. Microsoft, Bungie Studios and agency 4orty 2wo developed and distributed an elaborate mystery plot via a wide range of media (even voicemail telephone messages) that covered the lives of six characters leading up to the events of *Halo 2*. More than 2 million people accessed ilovebees.com. Players worked together internationally, tackling each part of the story at a time, sharing information in chatrooms and splitting up the tasks until the riddle was solved.[43,44]

These projects require a lot of planning, preparation and innovative creative work (that is why it's appropriate to see a strategic component in them as well), but the reward is that they can entice large audiences to interact with a brand's communication at a highly intense level. Results: *Halo 2* had more than 1.5 million pre-orders before reaching the retail shelves, achieved sales worth US$125 million on the first day, and is considered the most successful video game launch to date.[45,46] (It should be noted, however, that industry experts thought the fact that the original *Halo* was popular also helped push pre-sales.)

Summarizing advertising in connected marketing

Online games and mysteries, films, events, celebrity endorsements, stunts and even ads - no matter which route and idea you use to make your brand message contagious, the creativity, knowledge and experience of those involved in the development determine whether or not a campaign can flop or fly.

Having a big name and an innovative creative idea, however, aren't necessarily sufficient to develop contagious advertising that helps your return on investment. Subservient Chicken (see www.subservientchicken.com) is perhaps one of the most cited online viral campaigns from the past year. But the question is, did it sell additional chicken burgers?

In the end, it's the return on investment that determines whether or not a campaign is successful. Measuring the success of a campaign (and defining beforehand what success means, in a given context) are essential elements for advertising the connected marketing way.

Key questions: In my briefing, which types of objectives should I state for my campaign? Are they based online? If so, should I aim to develop Internet-based connected advertising? Is the measurement criteria offline? If so, the campaign must generate direct results outside of the Internet.

Finding influential friends: relationships in connected marketing

Some people are more connected, knowledgeable and convincing than others when it comes to talking about what's hot and what's not. If you want your connected marketing to create long-term word of mouth effects, you'll have to find and engage with these people.

When people buy a car, they often first consult a friend or relative who knows about cars. Some people know a lot about films and can recommend a film with such an avid passion that you must see it. We often marvel at acquaintances who have contacts in just about any walk of life. These are the people that Malcolm Gladwell refers to as 'mavens', 'salesmen' and 'connectors'.[47] And they are at the centre of what connected marketing relations are all about.

In the 1940s and 1950s, studies on product diffusion were already establishing that some customers impact differently on the spread of products and information. In Everett Rogers' seminal *Diffusion of Innovations*,[48] he quotes from the first study on the subject in 1943 by Bryce Ryan and Neal Gross that focuses on hybrid corn in Iowa, US. Earlier adopters of a new product have an influence on others: 'There is no doubt that the behavior of one individual in an interacting population affects the behavior of his fellows. Thus, the demonstrated success of hybrid seed on a few farms offers new stimulus to the remaining ones.'

In other words, Ryan and Gross were witnessing and trying to understand word of mouth and imitation, triggered by those that adopted a new idea earlier than others. These early adopters also had traits that distinguished them from the masses: larger farms, higher incomes and better education. And they were more active travellers, as indicated by their more frequent travels to Des Moines, Iowa's largest city.

Ever since, diffusion research has tried to better understand what kind of triggers are needed to help an idea spread. And in many cases, the focus has been on those individuals that seem to have more influence than others. The phrases 'early adopters' and Rogers' 'diffusion curve' have long since become part of marketing terminology, and we all know about the role that trendsetters play in fashion, music, art and, increasingly, for any type of consumption.

With interest in word of mouth marketing re-emerging during the past decade, the need to better understand how these particularly influential people can be identified and reached has intensified. There are companies that offer services in this field to do just that – connect with these important people for word of mouth.

As previously illustrated, advertising in connected marketing is about designing a message in a way that makes people - any people - forward it to their peers. So the message and its viral quality are at the core of the approach. Relationships in connected marketing are, at first glance, not so much concerned with messages that spread easily. Instead, this is about looking at the people who are likely to best help a message spread and about establishing relationships with them.

Let's look at a number of current approaches and how they identify and reach these market influencers. The following examples are by no means exhaustive; they will however help to give you a general overview and understanding of this concept.

How to find your market influencers

Different approaches to finding those clients, customers or consumers who can have a *disproportionate* influence on their peers' buying behaviour can be distinguished by the number of influential types they identify and work with.

Profiling market influencers: models with one influencer type

NOP World: Influentials(SM): Ten per cent of the US population influences the rest - at least that's what research corporation NOP World (now Gfk NOP) claims. It published one of the best-known recent studies on influential customers. In *The Influentials: One American in Ten Tells the Other Nine How to Vote, Where to Eat, and What to Buy,*[49] Jon Berry and Ed Keller describe their influencers segment of consumers - the 10% of the US population who are identified via a questionnaire about people's involvement in public matters. According to the book, these people are a decisive factor influencing commercial, political and religious decisions in US society. Not necessarily at the top of government, business, finance or the celebrity elite, they are instead those with an important personal influential sphere: early adopting neighbours with active opinions and strong commitments about things that matter. They take an active interest in how things develop in their community and neighbourhood, working as, what Berry and Keller call, 'central processing units of the nation'.

NOP has been conducting research about and with influencers in the US for more than 30 years. According to Ed Keller, the company has, for the past two years, extended this research activity to other countries.[50]

Burson–Marsteller: e-fluentials®: In cooperation with NOP World (now Gfk NOP), PR agency Burson–Marsteller has developed a subset of influencers, the so-called *e*-fluentials. According to the company, they 'have exponential influence shaping and driving public opinion through the Internet and throughout the offline world. Compared with the average Internet user, *e*-fluentials are far more active users of email, newsgroups, bulletin boards, listservs and other online vehicles when conveying their messages'.[51] (Burson–Marsteller's expert on *e*-fluentials, director of knowledge development Idil Cakim, describes the approach in Chapter 7.)

Procter & Gamble's Tremor – connectors: The world's largest consumer goods manufacturer Procter & Gamble has a firm stake in the field of word of mouth marketing. Its unit, Tremor, is dedicated to managing connections with influential teen customers and using these connections to spread word of mouth. The company has recruited a group of 250 000 influential US teens, referred to as 'connectors'.

Originally, Tremor tried to identify trendsetters - those kids that determine what the next hot item is (comparable to the aforementioned early adopters). But these efforts failed since it turned out that the trendsetters were not necessarily interested in spreading the word about products, because they wanted to be - and remain - different from their peers. So the company changed its research approach and looked for trendspreaders - those that have connections that they feed news into, to help ideas or products spread.

Trendspreaders are teens who have a wide social circle. Tremor recruits them with an online questionnaire that asks, among other things, how many friends, family members and acquaintances the respondent communicates with every day. The most sociable applicants (about 10% of the respondents), are invited to join the network. The members have an average 170 names in their address books (regular teens have about 30). To them, Tremor is presented as a channel to influence companies and as an avenue to discover cool new products before their friends do.[52,53]

Informative: influencers/brand advocates: Research and marketing services company Informative run a programme called Customer Influence Marketing. It claims to combine different approaches - such as attentively monitoring customer opinions and identifying influential customers with a set of questions (including, among others, the Affective Communications Test by H.S. Friedman *et al.*,[54] which is a kind of measure of charisma). It then uses the knowledge gained to influence the brand advocates by involving them in a closer relationship with the brand.[55] (Brad Ferguson explains more about word of mouth campaigns and their return on investment in Chapter 12.)

Ketchum: influencers: Would you believe that only 200 people determine which toothbrush sells best in the whole of the US?

The PR agency Ketchum makes the assumption that, in any given product category, less than 200 people shape the buying habits of the other 300 million Americans. The agency is methodically trying to identify these influencers. Depending on the type of product, their identity can vary, but they always include financial and industry analysts, scientists, authors, academics, futurists, government officials, grass-roots organizers, advocacy groups and reporters.

Ketchum identifies them using web searches and interviews, and by tracing information back to an original source. The agency claims that in 2003, hot cereal sales were influenced by 163 Americans. According to their research, toothpaste influence is spread by dental hygienists or dating experts, while influential advice on mascara is spread by make-up artists, women's magazines, eyeglass makers and optometrists.[56]

Profiling market influencers: models with two influencer types

Emanuel Rosen: expert hubs and social hubs: Emanuel Rosen defines his set of influential customers, 'network hubs', in *The Anatomy of Buzz*,[57] he distinguishes 'regular hubs' from another category, the 'mega-hubs', the same influential voices that Ketchum is researching - press, celebrities and analysts that link to millions of other people. According to Rosen, however, these mega-hubs are the regular targets for most classic PR efforts, proven by Ketchum's focus on them. This is why Rosen does not factor them into his approach on how to generate word of mouth.

As opposed to NOP and P&G, Rosen looks at two different types: expert hubs and social hubs. The former draw their influence and authority from their own particular expertise in a given subject. They tend to specialize in a field and will predominantly want to share their opinions and views on it. Social hubs are more influential because they are more charismatic or more socially active than their peers. They are trusted because people relate to them or use them as a role model.

Rosen recommends identifying these hubs through questionnaires or by careful observation. To help, he has identified the following shared characteristics: they are ahead in adoption, connected, travellers, information-hungry, vocal and exposed more than others to the media - or, in short: ACTIVE.

Euro RSCG: alphas and bees: According to the Buzz@Euro RSCG book *Buzz: Harness the Power of Influence and Create Demand*, word of mouth spreads from trendsetters (alphas) to trendspreaders (bees) who carry it to the mainstream.[58]

Alphas, roughly 8% of the population, are powerful consumers not because of the money they spend but because of the minds they influence. They look for stimulation, disregard convention to some extent and like to take risks. They are keen on being the first to know and are the people at invitation-only events. But they are not necessarily very social - they leave that to the bees.

Bees are connected and communicative, helping to move a trend from a margin to the centre of society. They are less exclusive and more interested in others adopting something they themselves find worthwhile. Their sense of style is based on imitation - they need confirmation, knowing that what they do will be admired. Alphas and bees are highly complementary. Alphas appreciate the bees' contacts and new ideas that come with them; bees are fascinated by the alphas' sense of style and creativity.

The alphas correspond with Procter & Gamble's trendsetters; the bees with the connectors. While Procter & Gamble doesn't try to involve the trendsetters in their efforts, Euro RSCG maintain that a successful attempt at creating buzz about a brand or product should include both types in the communication.

Profiling market influencers: models with three influencer types

Malcolm Gladwell: connectors, mavens and salesmen: One of the most influential books about the new ways of spreading ideas is Malcolm Gladwell's *The Tipping Point.*[59]

As previously mentioned, Gladwell has identified three types of market influencers. Connectors are those who know everyone, people 'with a truly extraordinary knack of making friends and acquaintances'. And, as these people habitually make use of their connections to help create new constellations and acquaintances, they are, says Gladwell, needed to help spread an idea and make it a success.

Mavens are information specialists, the people we respect and trust because we know that they know more about a particular subject (what Rosen calls expert hubs). These mavens are important because 'they know things that the rest of us don't. They read more magazines than the rest of us, more newspapers, and they may be the only people who read junk mail.'

The third group are salesmen. They are people who are more capable than average of spreading enthusiasm about ideas and products. It's not necessarily the vocabulary they use that enables them to make an impact. Gladwell: 'It's energy. It's enthusiasm. It's charm. It's likeability. It's all those things and yet something more.' The most important aspect of these salesmen is that it's quite difficult to resist them, which is why they are also an important force for successfully spreading word of mouth.

Gladwell doesn't provide an approach on how to identify these consumers. But his work helps us to understand the different forces at play when ideas spread.

Communicating with market influencers

Again, there are different approaches for engaging with those who influence your market.

Matching your messages to their needs

NOP World (now Gfk NOP) doesn't identify individuals and address them directly. Instead, it develops insight from its research to help companies better design their communication efforts to match the needs of the influential audience.

Ed Keller explains:

> Any channel can become a part of what we call an Influentials Marketing campaign. You can have your advertising agency develop print ads that offer the right type of information and you place them in the right types of media . . . Or traditional public relations through media outreach could place stories in the right types of places offering the right type of content. I think it's all going to be in the execution. At the end of the day, the consumers are going to look all around them, and certain types of messages are going to become appealing to them, certain sources will seem more credible, and they will seek them out, and in trying to learn more, these sources and information will become part of their conversation.[60]

To summarize some of Keller's and Berry's key points when matching messages to needs:

- Provide reliable clear information
- When customers complain, treat them with respect and help
- A genuine community involvement helps earn influencers' respect
- Products that make life easier or less complicated are great news for influencers
- Brands that help speed up buying decisions and keep life hassle-free are appreciated by influencers who are therefore willing to pay extra for these brands' goods and services.

In *Creating Customer Evangelists*,[61] Ben McConnell and Jackie Huba outline six rules on how to adjust your communication so that market influencers feel inspired to spread the word about your brand:

- Continuously gather customer feedback
- Make it a point to share knowledge freely
- Build word of mouth networks
- Encourage communities of customers to meet and share
- Devise specialized, smaller offerings to get customers to bite
- Create a cause that customers can rally around.

According to Huba, companies often miss the chance to help customers create a community and in turn build a word of mouth network:

> I bought a Palm Treo from Verizon, but I couldn't figure out how to email with it. I called Verizon and spoke with three different guys, none of them could help me. So I went on the web to a place called TreoCentral.com – a self-organized community of Treo customers – and looked through the comments. I found recommendations for installing a third-party application, which I followed. And it worked! So effectively, the customers were smarter than the company. If Verizon had brought these customers in, affiliated them with their own website, or at least pointed their users to the forum, it would have provided them with a great opportunity to create a base of fans that would spread the word. But that didn't happen, so the only impression I get from them, as a customer, is that they're clueless about their own product.[62]

Ketchum advises clients on how to approach their influential people – how to deal with critics, how to approach ambivalent targets, and how to make them spreaders of positive word of mouth.

One of its clients, FedEx, implemented an influencer marketing programme which identified 147 influencers in the US. Within the programme, dedicated executives including CEO Fred Smith were coached in how to deal with each influencer individually. Later, the programme was extended globally, with more than 5000 influencers. Eric Jackson, VP worldwide corporate communications at FedEx, says of the programme: 'It's critical to know who these people are, where they are on issues and consistently communicate with them so we are able to get the benefit of the doubt.'[63]

However, agency and client communication with these influencers sometimes resorts to financial incentivization. In order to rally support among black families for an education reform law, the Bush administration's Education Department (a Ketchum client) paid prominent black talk show host Armstrong Williams US$240 000. In return, he was asked to promote the law on his national television show, and to influence other black journalists to follow suit. The deal eventually leaked to the press and to the blogs.[64] This shows once again that financial incentives to stimulate word of mouth can quickly become a very dangerous tactic.

Blog marketing

One way of matching messages to the needs of influential customers can be the use of weblogs or 'blogs'. 'Blog' was the Merriam–Webster Online Dictionary's Word of the Year 2004,[65] the most looked-up word on its website. It's defined as:

> Blog *noun* [short for weblog] (1999): a website that contains an online personal journal with reflections, comments, and often hyperlinks provided by the writer.

The year 2004 was particularly important for the rise of blogs, in part because bloggers (people who publish blogs) took on an increasingly visible role in the US electoral race. One example was the scandal surrounding Dan Rather and the faked documents about George W. Bush's service in the National Guard. The fact that these documents were not genuine was uncovered within minutes by the connected blogging community after the report came to light.

Blogs are not 'just' private websites. They're also part of what's referred to as user-generated media. Anyone who has something to say, about products for instance (i.e. market influencers), or about anything at all, can set up a blog. A blog can soon become part of a network of people respecting each other and each other's information sources. For example, if a company or individual wishes to cite existing information on their blog, it's not considered sufficient just to link to the interesting piece; if you have discovered it on another blog you must also credit the place where you first discovered the information.

So how does blog marketing work? Essentially, it's about providing the community with information that they find worth spreading. When a company does or says something and bloggers find out about it and consider it worth publishing, they will. The news can then spread across the blogosphere, from blog to blog. Influential consumers satisfy their hunger for information increasingly by reading blogs, as well as by publishing them.

One way to reach influential bloggers is by starting a corporate blog. If the company itself provides interesting news and information with an insider's blog, the blogging community is likely to take notice. According to Jonathan Carson of research firm BuzzMetrics: 'Blogs are a very big deal, but they aren't necessarily the appropriate medium for every brand. If the marketing calls for connecting with various key audiences in a timely, informal and community-like manner, then a blog might be appropriate.'[66]

So this approach is not necessarily applicable to candy bars, but can certainly work for items that have a higher level of customer involvement - e.g. software, or books. (Blog marketing is covered in greater detail in Chapter 10.)

Seed marketing

Tremor essentially invites its connectors to be the first to sample a product or service. The underlying principle is that these connected customers are flattered and excited to be involved in the marketing of a product at an early stage. The experience prompts them to spread the word, showing their friends that they're in the know and letting them be part of something they want to share. Procter & Gamble research with Tremor has shown that this approach can generate up to double-digit percentage increases in sales.[67]

When companies involve this kind of connected customer early on in their marketing research, they can create powerful endorsements by making the customer feel empowered and listened to. This effect is not limited to teens - Chapter 1: Seed to spread describes this approach and presents a number of examples in much more detail.

Tremor sees this kind of marketing as a two-way connection with their teen influencers: 'The way we operate is very empowering to teens, who are free to say whatever it is they want to say about the product,' says Steve Knox, Tremor vice-president of business development. 'All we do is provide the catalyst. That's the first half of our brand promise to them. The second half is we give them a voice that's heard within the companies.' Tremor's service is not exclusive to Procter & Gamble's brands - others that have used it are Coca-Cola, Sony, AOL and Toyota.[68]

Marketing to and with teens, however, is a sensitive issue. That's why Tremor claims never to ask teens directly to market the products, and it also claims to communicate clearly with members' parents about the objectives of the venture. Yet some parents have expressed dissatisfaction with the company describing itself as a channel to exchange opinions when it's actually about marketing products. Says one father: 'If they're going to try to sell things to kids, they need to make it explicit that this is a selling channel.'[69]

The next step in the Tremor evolution is already planned. A second panel is currently being assembled, focusing on mothers to help market brands that appeal to them, such as Tide, Pampers or Bounty.

Influential customers in research groups

George Silverman recommends marketing with influencers in research settings. To reach and communicate with them, he uses telephone-based research groups with respected authorities who (a) have been provided with the product in advance, (b) have experienced positive results and (c) will recommend it during the research sessions. The sessions are planned to enable the other members of the group to verify information about the product.

Silverman is convinced that direct experience is the best way to market a good product (for example, free trial and sampling), but it's often difficult to implement. Making the experience available indirectly and over the phone from a trusted source is the solution he recommends – in particular for marketing pharmaceuticals.[70] For more details about influence research see Chapter 1.

Summarizing connected relations

Some clients, customers or consumers are more influential than others. A single, accepted and successful solution for identifying who they are and how to best speak to them doesn't exist, but these trends emerge from the plethora of routes available:

1. Influencers are online. All experts agree that most influential customers are information-hungry and connected. So they spend time online and that's where you can most easily find them, through questionnaires or by listening to your most vocal customers.
2. They are not necessarily early adopters or trendsetters, but many are.
3. Influencers are often somewhat detached from the mainstream, with category-related interests and information sources.
4. They want to be heard. Giving influencers a voice within your company – such as listening to their opinions and criticism and proving to them that their opinions matter – provides them with a compelling reason to spread the word.

5. They want to be your VIPs. They have egos and like to feel special. Making sure they have information that's exclusive and providing it in a way so they can share it easily helps them increase their authority within their peer group.

Which approach is used to identify and recruit influencers largely depends on the situation a company or brand is in. Some companies might make a first step by mining the feedback they're getting from their website and looking at who it's coming from. A well-designed CRM database may contain information that's helpful for finding influential customers. A company might also be able to spot them by, as Emanuel Rosen puts it, 'careful observation'. Paul Marsden explains that sometimes it can be helpful to simply think of certain professions that fit a given brand's influencer profile. Another company may want to use the services provided by companies such as Tremor or Informative to formalize the approach and use a scaleable process.

The insight that matters most is that these customers are out there, they are active and they are vocal. And the Internet is giving them the tools to become a power that shapes the future of marketing. Or, as Jackie Huba summarizes: 'Because of the Internet, people can connect with other people and exchange marketing messages – the customers themselves are now producing as much or more content that is marketing than the company itself. So companies don't have a choice but to connect with the customers. Otherwise, the marketing simply happens without the company.'[71]

Key questions: Do I have the resources in place to spot and listen to my market influencers? Is my market research customer-centric enough so it can learn about and connect with them?

Advertising and relationships – at best, use both

At first glance, we can assume that brand messages, which are more concerned with image than with specific product features, will be best promoted via an advertising approach. On the other hand, if there are really interesting or even breakthrough features in a product, chances are that you can get your influential customers hooked.

But overall, there is no either/or. First of all, many connected marketing approaches can be part of both your advertising and relationship marketing activities. If a company develops a highly interactive and involving online advergame but doesn't track its most vocal advocates, identify their locations, nor try to engage with them in a connected dialogue, it's throwing away massive potential for successful marketing.

And if a company is brilliant at identifying its most influential customers but fails to provide them with advance, trackable online versions of its most engaging television ads (which the recipients can forward to friends), the concept of the influential customer would be lost on that company.

Companies must find smart ways to combine both advertising and relationship approaches to connected marketing and learn the needs and pitfalls involved. The relationships in connected marketing share an important element with public relations: any good PR executive will agree that maintaining relationships with journalists requires tact and an understanding of their needs and wants - you only ever give journalists what you believe will help them with their job (or life, for that matter). The same goes for market influencers - they should be seen as strategic partners who provide immense benefits to the company. So a connected marketer should provide influential customers with what will delight, surprise and enthuse them.

And a compelling connected advertising idea is one way to do just that.

Conclusion: the way forward

Brands should listen to the word of mouth that is going on among connected clients, customers or consumers. There are different reasons why people feel inclined to talk about a brand, and that product or brand differentiation - in some way, shape, or form - is usually the best motivator for genuine word of mouth.

Smart brands have already started developing connected advertising strategies and creating communication ideas for their products, which customers enjoy and want to pass on to their peers.

There are certain market influencers who play a crucial role in the way word of mouth spreads, and brands should instigate connected relationships with them so that these influencers turn into powerful advocates for the brand.

A good way to manage and benefit from connected marketing is by combining both advertising and relationship approaches.

So what lies ahead?

Measurement

Marketing has become a return on investment-driven activity. Marketing people the world over moan about financial controllers nit-picking with them, leaving no room for creative ideas. In some cases, it may be true that controllers don't understand enough about marketing to be able to make the right calls. But essentially, controllers should be advertisers' and marketers' best friends. A marketing initiative is neither useful nor successful unless it pays for itself and more.

That's why connected marketing is becoming increasingly attentive to the accountability of the business. Particularly so, because in many cases, connected marketers can now prove the effectiveness of their work - and often more thoroughly and precisely than other forms of marketing communication. Some of the case studies in other chapters of this book provide compelling proof of this fact.

Dave Evans, of US-based agency GSD&M, says: 'The next big thing in this field will be when a world-class marketer, such as Procter & Gamble, demonstrates conclusively that [connected marketing] really works. That is when the TV upfronts will fall. Attention will no longer be a commodity.'[72]

Ed Keller agrees: 'I think one of the big challenges that needs to be overcome is "What's the measure by which we know whether it's successful or whether it isn't." And I think without that measure, we will stay in the shy and narrow stage. And whenever that measure does get developed, it will help set the course.'[73]

Procter & Gamble's Tremor is already showing some exciting results. So the future is beginning.

Integrated communications

Integrated communication has been an advertising buzzword for years. But the fact is, companies can still organize advertising, PR, events, Internet campaigns, etc. as stand-alone activities. And many still do. In lots of departments and agencies, there is still too much ego involved to really get together and cooperate. According to Jon Berry: 'What events are arguing for is a highly integrated marketing approach. A lot of ad agencies that I come in contact with are still fairly fragmented. They have an Internet marketing group, or a web advertising group on one hand, and a TV advertising group on the other.'[74]

Connected marketing does not work as a stand-alone activity, as this chapter and indeed book helps prove. When creating viral advertising films, you need a viral marketing specialist, an ad agency and a production house, a web agency, a technology partner, and the PR team all around the same table. When you want to know and engage with your influencers, you have to involve your web people, CRM experts, the market research department and your PR people. And in both cases, they really have to work together - across disciplines, across departments, across companies. If they don't, the whole initiative won't work.

And things get worse when more complex projects such as alternate reality games are on the agenda. They really do require a degree of collaboration that should not be underestimated.

Marketing to minors

As mentioned already, Tremor shows that the way connected (and other) marketers interact and deal with children and teens, must be addressed. Marketing consultant Rhona Berenstein says: 'Youth marketers claim they're tapping into a pool of outspoken enthusiasts who are invited to evaluate products and services, and share their opinions (positive or negative) with their pals. But if you're an average 14-year-old who's offered a first-look at a movie trailer, or a videogame, or a free make-up sample - all of which are promoted as "the newest and coolest" - where and when, exactly, is your rational, analytic mind going to kick in?'[75]

Protecting the rights and innocence of children in these marketing efforts is an important issue that needs to be looked at and dealt with sooner rather than later. Otherwise, and rightly so, the public is likely to turn against certain connected marketing practices to the detriment of the entire field.

The brand marketers' mindset

The most demanding challenge for practitioners of connected marketing is how to change brand marketers' perceptions about what the future will look like. The connected marketing approach requires new skills and competencies - an attitudinal change about how brands approach clients, customers and consumers is required. And massive changes like these never occur easily.

The faster companies realize that they must find new ways of connecting with their target audience, in exchanges that provide meaningful conversational content, the better equipped they will be for building successful brands in the future. A rather holistic approach has recently been dubbed 'Open Source Marketing', and it's about involving marketing influencers early in the marketing plan.[76] The Blowfly case study by Liam Mulhall in Chapter 4 shows how one company has built its entire business on this principle. And Procter & Gamble, Bacardi, Coca-Cola, Pepsi, Toyota and Ford, among other major brands, have all started to take on the challenge. That proves that connected marketing is here to stay.

So we'll close this chapter with the rallying cry of the connected marketer: 'Let's connect and collaborate!'

Takeaway points

- Get to know the word of mouth that's already going on about your brand.
- Determine the differentiating factors about your brand, product, or service that you want your audience to notice and talk about.
- When developing a contagious advertising campaign, make sure you plan to measure the results as precisely as possible.
- Know your market influencers and put structures in place in order to create useful and trusting relationships with them.
- Make sure your marketing departments and agencies are experienced and integrated enough to develop connected marketing campaigns – by connecting and collaborating both internally and with your market influencers.

Notes and references

1 Edelman, B. (2004) *The Effect of 180solutions on Affiliate Commissions and Merchants*, archived at http://www.benedelman.org/spyware/180-affiliates/.

2 Aberdeen Group Inc. (September 2003) *Revisiting Affiliate Marketing: A New Sales Tier Emerges in the Digital Commerce Network*. An Executive White Paper, archived at http://www.linkshare.com/press/aberdeen.pdf.

3 LinkShare company information at: http://www.linkshare.com/about/index.shtml.

4 Edelman, B. (2004) *Cookie Stuffing Targeting Major Affiliate Merchants*, archived at http://www.benedelman.org/cookiestuffing/.

5 Interview with Brian Clark, via telephone, on 16 December 2004.

6 'Warner verärgert mit Werbung in Weblogs', 16 August 2004, archived at http://www.futurezone.orf.at.

7 For more information and a discussion on Stealth Marketing, also see: Kaikati, A.M. and Kaikati, J.G. (2004) 'Stealth marketing: how to reach consumers surreptitiously', *California Management Review*, 46 (4), and http://customerevangelists.typepad.com/blog/2005/01/exposing_stealt.html.

8 Berry, J. and Keller, E. (2003) *The Influentials: One American in Ten Tells the Other Nine How to Vote, Where to Eat, and What to Buy.* New York: Free Press.
9 Interview with Ed Keller, via telephone, on 10 December 2004.
10 See http://www.vbma.net/mission.html.
11 Daniels, D., Kanze, A., Matiesanu, C. and McGeary, Z. (2004) *Mitigating the Threat of Viral Customer Behaviour.* JupiterResearch, 13 July.
12 Berry and Keller (2003) *The Inflentials.*
13 Silverman, G. (2001) *The Secrets of Word of Mouth Marketing.* American Management Association.
14 Dye, R. (2000) 'The buzz on buzz', *Harvard Business Review*, Nov.-Dec.
15 Blackshaw, P. and Karthik, I. (2003) *Rumors and Issues on the Internet.* An Intelliseek White Paper.
16 Interview with Paul Marsden, via telephone, on 24 November 2004.
17 Godin, S. (2000) *Unleashing the Ideavirus*, Do You Zoom, Inc.
18 Rosen, E. (2000) *The Anatomy of Buzz.* New York: Doubleday.
19 Interview with Justin Kirby on 9 November 2004.
20 Interview with Stéphane Allard, via telephone, on 21 October 2004.
21 Perry, R. and Whitaker, A. (2002) *Viral Marketing in a Week.* London: Hodder & Stoughton/Chartered Management Institute.
22 Interview with Schuyler Brown on 9 November 2004.
23 *Media Alert: Buzz You - Talking About Conversational Currency.* Euro RSCG Worldwide Press Release, 10 September 2004.
24 Bodipo-Memba, A. (2004) 'Oprah's stunt is Pontiac's coup', *The Wichita Eagle*, 18 September.
25 Lepper, J. (2004) 'Oprah guests count the cost of GM Pontiac G6 give-away', *Brand Republic*, 24 September.
26 Webster, S.A. (2005) '"Oprah" buzz works no magic for Pontiac G6', *Detroit Free Press*, 22 March.
27 *The Birth of a Shoe Company as told by Kenneth Cole*, archived at http://www.kennethcole.com/scripts/aboutus/ourstory.asp.
28 Vranica, S. (2002) 'That guy showing off his hot new phone may be a shill - new campaign for Sony Ericsson puts actors in real-life settings; women play battleship at the bar', *Wall Street Journal*, 31 July.
29 Zorbach, T. (2004) 'Viral clips verdrängen den guten alten Fernsehspot', *io new management*, 7-8: 10-13.
30 Huba, J. and McConnell, B. (2002) *Creating Customer Evangelists.* Dearborn Trade, a Kaplan Professional Company.
31 Interview with Jackie Huba, via telephone, 23 November 2004.
32 Zorbach, 'Viral clips'.
33 Garrison-Sprenger, N. (2004) 'Fallon's films on Amazon.com', *Minneapolis/St Paul Business Journal*, 29 November.

34 Anderson, M. (2005) 'Dissecting "Subservient Chicken"', *AdWeek*, 7 March. Archived at http://www.adweek.com/aw/national/article_display.jsp?vnu_content_id=1000828049.
35 Ryan, J. (2004) 'Army's war game recruits kids', *San Francisco Chronicle*, 23 September.
36 Ian Hopper, D. (2002) 'Army to release computer game', Washingtonpost.com, 2 July.
37 Anderson, H. (2004) 'Fun @work: viral marketing for the office – let's have some fun, shall we?', *ClickZ*, 27 May. Archived at http://www.clickz.com/experts/em_mkt/case_studies/article.php/3359071.
38 Pierson, J. (undated) *The Blair Witch Connection*, archived at http://www.grainypictures.com/blairwitch/.
39 Kendzior, S. (1999) 'Sons of a witch', *The 11th Hour Web Magazine*, archived at http://www.the11thhour.com/archives/071999/features/blairwitch1.html.
40 Stewart, S. (undated) *The A.I. Web Game*, archived at http://www.seanstewart.org/beast/intro/.
41 MoreToSee: http://www.moretosee.com/index_flash.html.
42 See http://www.ilovebees.com/.
43 'Viral marketing reaches new heights', *GameDailyBiz*, 1 November 2004, archived at http://biz.gamedaily.com/features.asp? article_id =8247§ion=adwatch.
44 Jenkins, H. (2004) 'Chasing bees, without the hive mind', *Technology Review.com*, 3 December.
45 Oser, K. (2004) 'Microsoft's Halo2 soars on viral push', *Advertising Age*, 25 October, p. 46.
46 *Rekordumsatz für neues Videospiel von Microsoft.* Handelsblatt, 11 November 2004.
47 Gladwell, M. (2002) *The Tipping Point.* Back Bay Books (*The Tipping Point: How Little Things Can Make a Big Difference.* Boston: Little, Brown and Company, 2000).
48 Rogers, E.M. (2003) *The Diffusion of Innovations*, 5th edn. New York: Free Press.
49 Berry and Keller (2003) *The Influentials.*
50 Interview with Ed Keller.
51 See http://www.bm.com/pages/insights/efluentials and http://www.efluentials.com/.
52 Wells, M. (2004) 'Kid nabbing', *Forbes*, 2 February, archived at http://tinyurl.com/cuh3a.
53 MarketingSherpa (15 June 2003) *Proven Tactics in Viral Marketing.* MarketingSherpa Inc., pp. 79–81; e-book available at http://sherpastore.com/store/page.cfm/2090.

54 Friedman, H.S., Prince, L.M., Riggio, R.E. and DiMatteo, M.R. (1980) 'Understanding and assessing nonverbal expressiveness: the Affective Communication Test', *Journal of Personality and Social Psychology*, 39 (2): 333–351.
55 Informative (2004) 'Profile the community: find the influencers', unpublished chart, company presentation *Introduction to Informative.*
56 Chura, H. (2004) 'How to calculate word of mouth; Ketchum finds 200 build the buzz', *Advertising Age*, 26 July.
57 Rosen, E. (2000) *The Anatomy of Buzz.* New York: Doubleday.
58 Matathia, I., O'Reilly, A. and Salzman, M. (2003) *Buzz: Harness the Power of Influence and Create Demand.* New York: Wiley.
59 Gladwell, *The Tipping Point.*
60 Interview with Ed Keller.
61 Huba, J. and McConnell, B. (2002) *Creating Customer Evangelists.* Dearborn Trade, a Kaplan Professional Company.
62 Interview with Jackie Huba.
63 Chura, 'How to calculate word of mouth'.
64 Toppo, G. (2005) 'Education Dept. paid commentator to promote law', *USA Today*, 7 January.
65 Merriam-Webster (2004) 'Merriam-Webster announces 2004 Words of the Year', press release, 30 November.
66 'The state of viral marketing', interview with Jonathan Carson, avantmarketer.com, November 2004.
67 Mathew, M. (2005) *P&G's Tremor: Reinventing Marketing by Word of Mouth.* ICFAI Business School Case Development Centre, Hyderabad, India.
68 Rodgers, Z. (2004) 'Marketers pay their way to the youth audience', *ClickZ*, 28 May, archived at www.clickz.com/news/article.php/3360711.
69 Wells, 'Kid nabbing'.
70 Silverman, *Secrets of Word of Mouth Marketing.*
71 Interview with Jackie Huba.
72 Interview with Dave Evans, on 11 November 2004.
73 Interview with Ed Keller.
74 Frost, R. (2004) 'Gaining influence through word of mouth', Brandchannel.com, 9 February.
75 Berenstein, R. (2004) *Op-Ed: Youth Marketers.* ImediaConnection, 18 October.
76 Cherkoff, J. (2005) 'What is open source marketing?', archived at http://www.changethis.com/14.OpenSourceMktg.

16

Conclusion: the future of connected marketing

Justin Kirby

Managing Director, Digital Media Communications (DMC)

The essence of connected marketing: a stake in the ground

What's going on now in the field of marketing communications is similar to the early days of the Web: marketers and practitioners are trying new approaches, reviving and evolving old ones, and bandying catchy terms around willy-nilly.

It's like the classic Indian parable about an elephant and six blind men: one blind man touches the trunk and thinks he's bumped into a snake, the other the tail and thinks it's a donkey, the other its leg and thinks he has bumped into a tree trunk, and so on. No one can see the whole picture until the zookeeper comes along and presents the wider perspective, the collective view of an elephant.

With this book we, too, have stepped back to provide the bigger picture and put a stake in the ground. The book presents a wide range of practitioners, academics, approaches, research and case studies, all of which come under the connected marketing umbrella. Through them we have demonstrated not only how diverse connected marketing activities are carried out, but also – most importantly – what connected marketing can and should achieve. The whole is greater than the sum of the parts.

Connected marketing – regardless of which specific term or approach you use – is simply any kind of marketing communication (including traditional advertising) that creates within the target market conversations that add measurable value to a brand. The sub-terminology (viral, buzz, word of mouth, etc.) is not only confusing but perhaps even irrelevant; successful connected marketing is not about what you call what you do, but about how you do it and what it achieves.

We hope to have shown in this book that managing successful connected marketing activity is possible through an organized series of decisions and approaches; it's not a hit-or-miss quest for that one groundbreaking buzz idea.

Integrating connected marketing

It may be an overstatement to say that traditional advertising is no longer working, though it's certainly true that it's generally less effective. However, using alternative and non-traditional techniques alone is no guarantee of success either, particularly if you're trying to build and sustain global brands.

Granted, some brands have proved that they can be built from scratch almost entirely by using only non-traditional marketing techniques. Take smoothie-maker Innocent in the UK (though they have since launched a TV ad campaign in summer 2005) and their Australian counterpart Nudie, for example.[1] Or the Brewtopia case study in Chapter 4. These brands have managed to break into the mass market at least locally via mainly word of mouth marketing activities - but will that kind of marketing alone enable them to expand globally as well?

Successful, long-term connected marketing activity not only connects a brand to its clients, customers and consumers and vice versa, but also integrates the different ways in which marketers approach people - from viral, buzz and word of mouth marketing to advertising, direct marketing, sales promotion, PR and more - across the whole marketing mix.

A good start to developing an integrated connected marketing strategy is to listen to what customers are saying about your brand, and to identify and involve the most influential customers in your business.

The people you need to reach with connected marketing

Marketing is no longer simply about trying to identify the 20% of your customers who represent 80% of your revenue, then keeping this group loyal and trying to acquire more who fit their profile. Simply having happy, enthusiastic brand advocates is not enough. Plenty of people would sing the praises of Ferrari, for instance, but how many would actually buy one or really influence the purchasing intent of someone who would?

With connected marketing, the aim is to profile and recruit customers who represent the 10% of society that helps influence the majority of all purchasing decisions (see Chapter 7). These influencers are not necessarily the customers who spend the most money with you, but they are the people who can help amplify and accelerate positive word of mouth about your brand. It is recommendations from this group that have been tracked to business growth.[2]

It doesn't matter if you have an existing product or are about to launch a new one; the point is to make sure you develop two-way relationships with influencers - from giving them a trial before the product is available to the mass market, to going back a step and getting them involved in your research and development. Even if your product or service is not particularly innovative, a buzzworthy connected marketing campaign seeded with influencers first can lead to success, as illustrated by the Virgin.net and Dove promotions mentioned in the Introduction and the FedEx case study mentioned in Chapter 15.

Successful connected marketing is also about spotting and appeasing customers who are not so happy with your brand - or even better reducing the instances of negative experiences in the first place, in order to reduce the instances of negative word of mouth. As Brad Ferguson has shown in Chapter 12, negative word of mouth can reduce your bottom line to a greater extent than positive word of mouth can increase it.

Setting standards in connected marketing

From the wide range of connected marketing experiences and approaches described in this book, we can see that we are not comparing apples with apples. So we can't apply the same set of standards - both in terms of measurement and ethics - to, for example, an overtly branded marketing message that spreads virally and to the use of a product referral agent whose motivation is ambiguous (i.e. it could be spontaneous or incentivized).

Let's look at measurement standards first.

Measurement

Connected marketing is not about control; it's about management. And you can't manage what you can't measure.

While the outcome of any connected marketing campaign should ultimately be the measurement of recommendation rates and thereafter sales, most marketers still work within the brand awareness creation paradigm. So certain connected marketing techniques - particularly those that include buzzworthy advertising - are still being measured by cost per thousand (CPM)-type metrics until the industry learns better. Obviously this kind of measurement is more easily and precisely monitored online. Although the resulting statistics won't tell the whole story - particularly with regard to the extent of third-party brand endorsement - this is still how many connected marketing campaigns will be judged for the time being. At least this kind of measurement can prove that connected marketing is a cost-effective way of creating brand awareness, and it has the added dimension of providing a glimpse of peer-to-peer brand endorsement.

What's needed in the connected marketing field is a standard set of metrics, so marketers can assess the merits of the wide array of connected marketing techniques on offer. However, it's more than likely that there will never be one set of measurements that can be applied to all the diverse approaches being used. Instead, at the very least, a way must be found for any connected marketing campaign to be measured in terms of its impact on customer recommendation rates and the correlation between the increasing instances of these and sales.

Ethics

As with measurement, no 'one size fits all' set of standards can be applied to ethics in connected marketing - or indeed across different marketing disciplines such as advertising, promotions, PR, etc.

You may know that involving customers in a market research study can have a positive impact on their advocacy of your brand, so is that activity cynical and manipulative and, if so, to what degree? What about running a stealth campaign that pays agents to pose as independent brand advocates in online chatrooms, effectively duping consumers?

If you need to incentivize people to talk about a brand, product, or campaign, you need to decide which type of motivation to use. If you choose some kind of 'payment in kind' reward or financial motivation, you must be aware of the risks that this approach can bring in the form of a backlash from customers who feel duped. Word of mouth can work just as quickly and extensively against a brand as it can for it.

One way to avoid any negative outcome for brands and consumers is to make all marketing activity completely transparent, for example being upfront about who you are incentivizing and how.

However, as Greg Nyilasy points out in Chapter 11, the more transparent a connected marketing campaign becomes, the less effective the word of mouth is – simply because the credibility of the recommendation is diminished when it is known that recommender has been incentivized. In fact, incentivized connected marketing could just become another form of clutter and therefore part of the problem for marketers trying to connect with consumers.

One answer is to make sure your product, service, or campaign is buzzworthy enough to inspire voluntary, independent word of mouth – the onus being on the innovation of the product, and/or the idea behind the campaign and its connection with influencers, rather than on the approach you use.

As Justin Foxton describes in live buzz marketing (Chapter 2), and as case studies such as the *Blair Witch Project* (see Introduction and Chapter 15) and the ARG (alternate reality gaming) case studies in Steve Curran's contribution (see Chapter 9) demonstrate, non-disclosure or delayed disclosure of the fact that people are being involved in connected marketing activity can be very successful and form part of the fun of the interaction for customers.

Clearly, pretending you're a client, custoumer or consumer when you're really a marketer is not a good idea. However full disclosure is not always a prerequisite for every type of connected marketing. It may well be vital for buzz marketing projects that want to generate word of mouth recommendations. But they're very different from viral marketing campaigns such as Blair Witch and ARG initiatives where people enjoy entering into the spirit of the entertainment and suspending their disbelief. Those kinds of connected marketing campaign are not looking for advocacy that results in product recommendations; they're looking for advocacy that creates conversations.

There are as yet no agreed ethical standards that fit the wide range of connected marketing (or indeed wider marketing) approaches which practitioners and brands are using. And there may never be one set of standards that are appropriate for all practices. So it's up to businesses to adopt a more commonsense approach, a 'code of practice' that most companies apply to every other aspect of their business: risk management. The point here is to look at your connected marketing activity in terms of consequences, risks and issues, just as you would with any business initiative.

Simply ask yourself: 'Will this activity (whether disclosed or non-disclosed, transparent or not) have a positive or negative outcome for the brand?' If the risk of a negative outcome is high, don't do it.

A specific area of ethics that's being widely discussed is marketing that involves minors (both marketing to minors and marketing with them).

The Tremor initiative (see 'Teen trials - Tremor style' in Chapter 1), illustrates that the way connected marketers (and other marketers) interact and deal with children and teens has to be addressed by the marketing industry.

As Martin Oetting asserts in Chapter 15, protecting the rights of children in relation to marketing is the most important ethics issue - it needs to be looked at and dealt with sooner rather than later. 'Otherwise, and rightly so, the public is likely to turn against certain connected marketing practices to the detriment of the entire field'.

Embracing connected marketing

Academics have a large role to play in providing further credible research into the connected marketing field. They can help legitimize the diverse techniques used and, most critically, demonstrate how those approaches work - we need to know more about the source variables, message variables and receiver variables influencing word of mouth.

In terms of concluding from this book how marketers can create conversations that have a positive influence on a brand, put simply, the message is this: the product or brand should have a stand-out quality in its class to create self-propelling word of mouth and customer recommendations that a connected marketing campaign will amplify and accelerate. If the product is not innovative in itself, you must focus on developing creative executions that generate buzz. You also have to find, connect and collaborate with the people who influence your brand, lead opinions and spread word of mouth. In fact, the most appropriate ideas for your product innovation or connected marketing activity could well come from external stakeholders, not from your marketing agencies and partners.

Clients, customers and consumers already have more power and influence over brand success than most marketers realize. The same goes for employees and businesses.

It's time for business to embrace the new order by developing internal and external connected marketing strategies that integrate new and traditional marketing techniques and media in innovative ways. This approach enables businesses, clients, consumers, employees - in fact, all stakeholders - to connect and collaborate with each other as respected partners in order to achieve mutually beneficial outcomes.

The growing role of mobile connectivity

One field that deserves a special mention in this conclusion for the growing involvement it will undoubtedly have within connected marketing activities is mobile connectivity.

The ongoing development and adoption of mobile connectivity technologies are already resulting in greater user demand for entertainment-focused content via cell phones, turning handsets into a more viable connected marketing channel.

One example of a genuinely viral mobile marketing campaign was The Gadget Shop mobile coupon campaign managed by m-bar-go. It involved sending simple text messages to the cell phones of shoppers on London's Oxford Street in the lead up to Christmas 2003. The text message included an embedded discount barcode that could be read at Gadget Shops and forwarded to friends. The campaign produced a 50% increase in average transaction value.[3]

That case study reveals one of the potential developments of viral marketing. The campaign used mobile handsets to spread the viral agent, and it was tied to generating business as opposed to simply generating awareness from branded entertainment content - in fact, the brand only paid for viral spread that resulted in sales.

In future, we could see online viral marketing being used for e-commerce and mobile viral marketing being used for traditional commerce.

10 predictions for the future of connected marketing

1. Connected marketing will become more strategic, with the focus shifting from promotion (creating remarkable campaigns) to innovation (creating remarkable products).
2. ROI metrics will be mandatory for viral, buzz and word of mouth campaigns. 'Advocacy rates' and 'sales uplift' will become important parts of ROI metrics, displacing traditional measures such as campaign reach.
3. Word of mouth tracking will become a key metric in brand tracking market research.
4. Buzz, viral and word of mouth marketing will be merged into the wider marketing mix, with online viral marketing adopted and

integrated within advertising, word of mouth within promotions and buzz within PR.

5. Managing and avoiding *negative* word of mouth, online and offline, will be an increasingly important area in connected marketing.
6. Online branded entertainment (advertainment, advergaming, alternate reality games) will be used more as key brand touch-points for entertainment brands.
7. Techniques developed in connected marketing initiatives will be adopted for change management and internal communication.
8. Techniques developed in viral, buzz and word of mouth will be increasingly adopted in CRM programmes as both retention and acquisition (turning buyers into advocates) tools.
9. Cell phones will develop rapidly as an important medium for spreading connected marketing promotions, such as mobile invitations, SMS barcode discounts, etc.
10. Marketers will eventually be able to locate influencers by zip/post code, by which point they will be all chasing the same chosen few . . . Prepare for another paradigm shift in marketing?

Notes and references

1 Kirby, Justin (2005) 'Global alternative marketing themes', *DM Weekly*, 21 March, archived at http://www.mad.co.uk/dmweekly/Story.aspx?uid=d788db4f-5755-4cc5-b2e7-4b938d37f20a.

2 Reichheld, F.F. (2003) 'The one number you need to grow', *Harvard Business Review*, 81 (12): 46–54.

3 Buckley, Russell (2004) *Location-Based Advertising: Theory and Practice.* Unstatic White Paper, 27 April, p. 31, archived at http://www.planningaboveandbeyond.com/Downloads/LocationBasedAdvertising.pdf.

Index